# Progress in the Chemistry of Organic Natural Products

More information about this series at http://www.springer.com/series/10169

A. Douglas Kinghorn • Heinz Falk •
Simon Gibbons • Jun'ichi Kobayashi
Editors

# Progress in the Chemistry of Organic Natural Products

Volume 102

With contributions by

G. Appendino
M. Harizani · E. Ioannou · V. Roussis

Springer

*Editors*
A. Douglas Kinghorn
Div. Medicinal Chemistry & Pharmacognosy
The Ohio State University
College of Pharmacy
Columbus, Ohio
USA

Heinz Falk
Institute of Organic Chemistry
Johannes Kepler University Linz
Linz, Austria

Simon Gibbons
Research Department of Pharmaceutical and Biological Chemistry
UCL School of Pharmacy
London, United Kingdom

Jun'ichi Kobayashi
Graduate School of Pharmaceutical Science
Hokkaido University
Sapporo, Japan

ISSN 2191-7043 ISSN 2192-4309 (electronic)
Progress in the Chemistry of Organic Natural Products
ISBN 978-3-319-33170-6 ISBN 978-3-319-33172-0 (eBook)
DOI 10.1007/978-3-319-33172-0

Library of Congress Control Number: 2016943631



Printed on acid-free paper

This Springer imprint is published by Springer Nature
The registered company is Springer International Publishing AG Switzerland

# Contents

# Ingenane Diterpenoids

**Giovanni Appendino**

## Contents



G. Appendino (✉)
Dipartimento di Scienze del Farmaco, Largo Donegani 2, 28100 Novara, Italy
e-mail: giovanni.appendino@uniupo.it



A.D. Kinghorn, H. Falk, S. Gibbons, J. Kobayashi (eds.), *Progress in the Chemistry of Organic Natural Products*, *Vol. 102*, DOI 10.1007/978-3-319-33172-0_1

# 1 Introduction

Ingenane diterpenoids are a biogenetically advanced group of phorboids, a family of polycyclic diterpenoids structurally related to the tigliane skeleton of phorbol (**1**) that also includes the tiglianes and daphnanes. A combination of limited distribution, very low concentration in plant tissues, chemical instability, and difficulty of detection delayed the discovery of ingenane derivatives (ingenoids) compared to the tiglianes, the major class of phorboids. Thus, while phorbol (**1**), the archetypical tigliane, was first obtained in pure form in 1934 [1], the first ingenol derivative (ingenol-3-hexadecanoate = ingenol 3-palmitate, **2**) was only reported in 1968 [2, 3]. On the other hand, while phorbol went into a long series of constitutional and configurational revisions before its structure was clarified unambiguously by X-ray crystallography in 1967 [4], the basic ingenane polyol ingenol (**3**), just like the archetypical daphnane daphnetoxin (**4**) [5], benefited from an early (two years after the isolation) crystallographic determination of its relative and absolute configuration [6]. In retrospect, the early structure elucidation of ingenol, along with the lack of an abundant plant source, is responsible for the limited information still available on its chemistry compared to phorbol, for which the reactivity was thoroughly investigated in attempts to elucidate its structure by classic studies of chemical degradation [7].

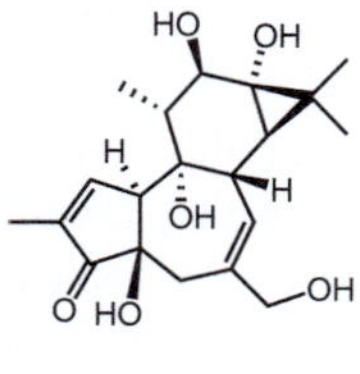

**1** (phorbol)

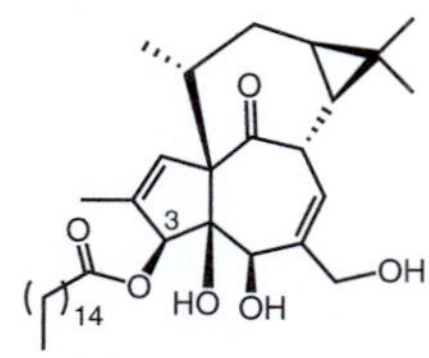

**2** (ingenol-3-palmitate)

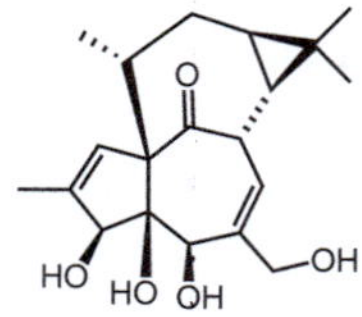

**3** (ingenol)

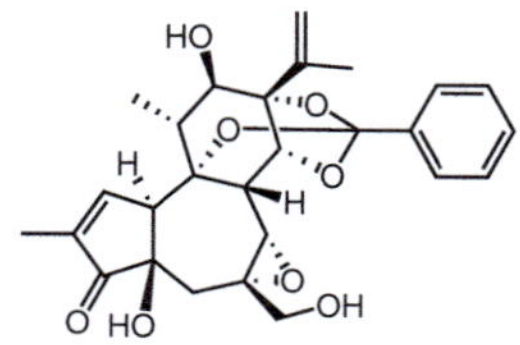

**4** (daphnetoxin)

To detect the presence, often in trace amounts, of ingenol derivatives in crude extracts and chromatographic fractions, the early studies relied extensively on non-selective bioassays common also to other types of phorboids (e.g. cytotoxicity, fish toxicity, skin irritancy). The improvements in analytical techniques and the introduction of hyphenated chromatographic techniques like HPLC-MS have now simplified the detection and the isolation of ingenol esters. Despite these advancements and the growth of the natural products chemistry community, the diversity of ingenol derivatives has not expanded significantly in terms of chemotypes or euphorbiaceous sources since the pioneering work of the groups of Erich Hecker (Plate 1) in Heidelberg and Fred J. Evans (Plate 2) in London during the seventies and the eighties of the past century [8], suggesting that

**Plate 1** Erich Hecker (taken from the brochure "Cancer Research and Cooperation, German-Israeli Cooperation in Cancer Research – The first 20 years". Deutsches Krebsforschungszentrum (1999)

**Plate 2** Fred J. Evans (1943–2007) (photograph provided courtesy of Prof. Elizabeth M. Williamson)

this class of compounds has a very narrow distribution in Nature and a limited structural diversity.

After an initial interest for the oxymoronic tumor-promoting and anticancer properties of ingenol derivatives, activities remained for a long time limited to the chemical synthesis community and the total synthesis of the parent alcohol. On the other hand, the successful and paradoxical development of the (weak) rodent tumor promoter ingenol mebutate (= ingenol-3-angelate, I3A, **5**) into a topical human chemopreventive drug (Picato®) (see Sect. 4.3), has rekindled a multidisciplinary interest for this compound and its biological potential.

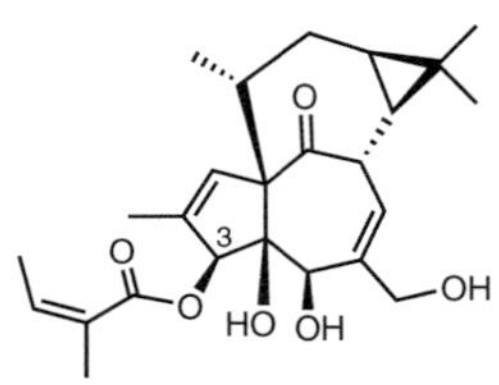
**5** (ingenol-3-angelate, I3A)

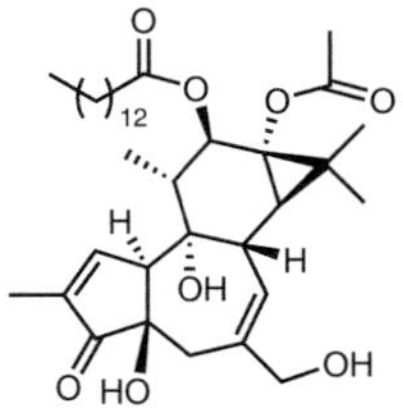
**6** (phorbol-12-myristate-13-acetae, PMA)

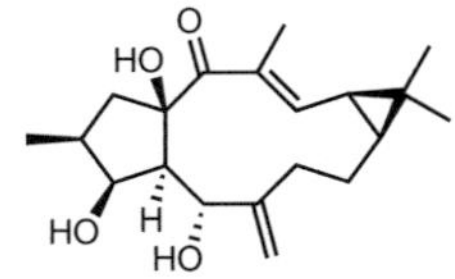
**7** (lathyrol)

Just like phorbol, ingenol does not occur in Nature as a free polyol, but only in esterified form, and its direct isolation from plant extracts is associated with the facile hydrolysis of the 3- and 20-monoesters. The first member of the class was reported in 1968 as part of a screening for co-carcinogenic diterpenoids in euphorbiaceous plants and the products obtained thereof [2, 3]. This systematic search for the distribution of co-carcinogenic compounds was undertaken by the German Cancer Research Centre in Heidelberg in the wake of the identification of phorbol 12-myristate-13-acetate (PMA, **6**) (= tetradecanoylphorbol acetate, TPA) as the major tumor-promoting principle of croton oil (*Croton tiglium* L.) by Hecker [7]. The purification of the compound later structurally elucidated as ingenol 3-hexadecanoate (**2**) was reported in 1968 from the seeds of the caper spurge (*Euphorbia lathyris* L.) [2] and from the latex of the African candelabra tree (*E. ingens* E. Mey.) [3]. Just like croton oil, a major source of phorbol esters, also the sap of the candelabra tree and the oil from caper spurge seeds have long been known for their high irritancy and skin-blistering properties, and the sap of the candelabra tree was also known to display co-carcinogenic properties [9]. Unexpectedly, both sources lacked phorbol esters, containing rather a complex mixture of ingenol esters and macrocyclic diterpenoids. Hecker, who pioneered studies in this field, assigned codes to diterpenoid constituents obtained in his laboratory, possibly to cope with the delay before isolation and structure elucidation that used to plague phytochemistry in the decades before pre-high-field NMR spectroscopic techniques became available. These codes were related to the plant source and to the chromatographic elution order, and the major irritant and co-cocarcinogenic constituents of *E. ingens* and *E. lathyris* respectively, dubbed Factors $I_5$ and $L_1$, turned out to be the same compound, namely, ingenol-3-hexadecanoate (**2**), with the structure of the parent polyol being eventually clarified by X-ray analysis of its crystalline triacetate [6]. Ingenol was named after *E. ingens*, possibly because laboratory work on this plant actually predated that on *E. lathyris*, or, alternatively, because the name lathyrol (**7**) was already being used, as epoxylathyrol [10], for the major macrocyclic diterpenoid polyol of caper spurge oil, a compound previously known as *Euphorbia*-steroid and first reported in 1937 [11].

The total synthesis of ingenol is well covered in the secondary literature (see Sect. 3.2), while the other aspects of its properties have not been comprehensively reviewed since the chapter on ingenoids by Evans and Taylor in Volume of 42 of this Series [8] and the monographic book on phorboids edited by Evans [12]. This work aims at filling this gap, providing a platform for further investigation and the full exploitation of the pharmacological potential of this remarkable and fascinating parent diterpene natural product.

# 2 Phytochemistry

## 2.1 Structure and Spectroscopic Properties

The ingenane skeleton is based on a 5/7/7/3 tetracyclic system featuring a *trans*-bridgeheaded bicyclo[4.4.1] undecane core. The two-dimensional structural representation of ingenol is not straightforward, and several formulas have been used in the literature. The formula A (Fig. 1) is based on the azulenoid A,B ring system, and avoids the ambiguities associated with the two-dimensional representation of the *trans*-bridge, even though it depicts with a wedge/broken line endocyclic bonds

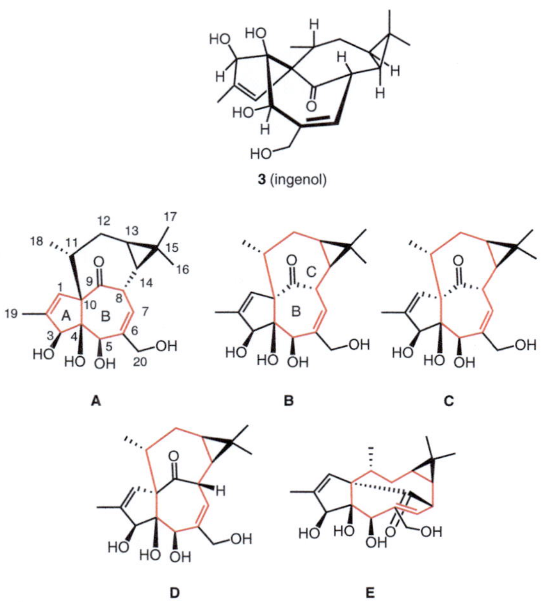


**Fig. 1** Three-dimensional and two-dimensional representations of ingenol (**3**)

centered on stereogenic carbons. Thus, C-11 and C-14 of the cycloheptane C-ring bear an endocyclic wedge/broken line descriptor in addition to the one of their exocyclic substituent. The cyclodecane-based formulas B-D are also documented in the literature, but are ambiguous in terms of *(R),(S)* descriptors for the stereogenic centers at the bridge. As a consequence of this, *Chemical Abstracts Service* (Columbus, Ohio, USA) uses a different cyclodecane-based representation (*E*) for ingenol, unambiguous from the standpoint of translation in *R,S* configurational descriptors, but rather unfamiliar in the chemical literature, ambiguous in terms of configuration of the ring B double bond, and reminiscent of an Escher painting in the way the bridge is drawn. In conformational terms, the cyclodecane rim of the B, C-ring system approaches a sofa conformation, with C-10 upwardly oriented outside the mean plane through the other atoms, while the C-9 bridge and ring A are oriented toward the opposite ($\alpha$-) face of the molecule.

The *trans*-bridge between C-8 and C-10 is an example of a conformationally locked *in,out*-bridge junction [13], a very rare structural element within the pool of natural products of which the major examples among the isoprenoids belong to the *seco*-trinervitane class of insect diterpenoids [14, 15]. The configuration is *in* at C-8 (angle C-10/C-8/H-8 $<$ 90°, Fig. 1) and *out* at C-10 (angle C-8/C-10/C-11 $>$ 90°, Fig. 1), and the bridge results, when referred to the de-apicalized cyclodecane system, from the merging on the C-9 methylene of a $\beta$-bond at C-8 and of an $\alpha$-bond at C-10 (Fig. 2). The homeomorphic interconversion [13] of the *in*/*out* and the *out*/*in* forms is impossible in ingenol because of the small size of the ring system and because of the presence of the junction with the cyclopentane A-ring and the cyclopropane D-ring. Models show angular strain at several carbon atoms, with flattening at C-5, C-8 and C-11. Although the original X-ray data on ingenol triacetate [6] were not made available, the X-ray structural analysis of 20-deoxyingenol (**3**) found indeed that the bond angles of these carbons are significantly larger than the tetrahedral value (125.0°, 126.4° and 119.5°, respectively) [16].

In general, the relative stabilities of isomeric bicyclic [x.1.y] ring systems depend on their size. Isomers of the *out,out*-type are only possible in rings up to cyclononane, and they are also the most stable form up to cycloundecane, with the *in,out* becoming the preferred form for cyclotridecane and larger rings [13]. Unsurprisingly, the C-8 epimer of ingenol, the *out,out*- stereoisomer **8**, is significantly (24.7 kJ/mol) more stable than the natural polyol, and is characterized by a reduced angular strain [17]. The esters of this isomer, at least in the tris-*nor* version lacking the D ring (**9**), are devoid of tumor-promoting properties, and have been assumed to

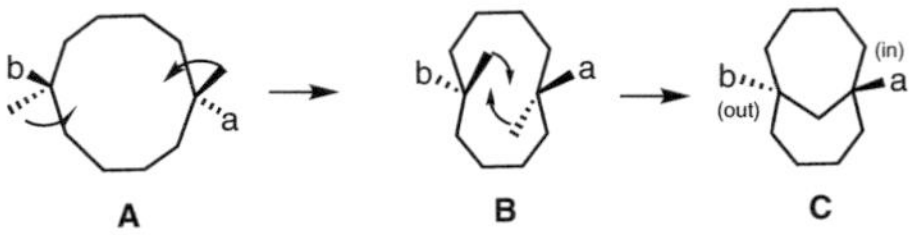


**Fig. 2** Effect of de-apicalization on the configurational descriptors of the bridgehead carbons in a cyclodecane bearing a 1,6-*trans* bridge (bold and hashed wedges are used to indicate the orientation of the substituents a and b compared to the mean plane through the cyclodecane ring)

be biologically inactive [18]. The parent tris-*nor* polyol was synthesized by Paquette in 1988 [18], 20 years after the isolation of ingenol and 14 years before the first total synthesis of the natural product by Winkler [19]. In the light of the articulated and multifaceted nature of the biological profile of ingenol esters and of the lack of ring D, this assumption should be better evaluated before generalization.

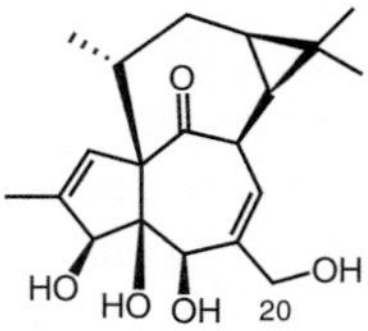


**8** (8-epi-ingenol)

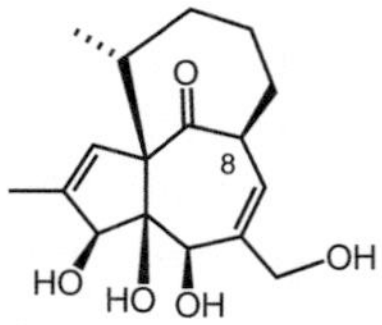


**9** (8-epi-tris-*nor*ingenol)

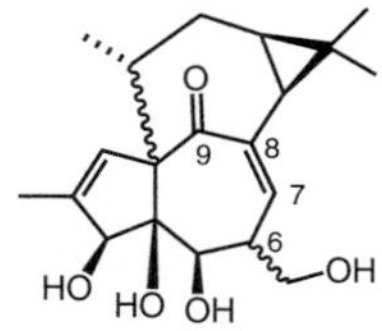


**10** (isoingenol)

The isomerization of ingenol does not lead to 8-epi-ingenol (**8**), but to iso-ingenol (**10**), a compound where the double bond has shifted to the bridgehead carbon and is formally conjugated with the C-9 carbonyl, although the two π-systems cannot achieve planarity [20]. There is, unfortunately, a certain confusion in the literature between epi-ingenol (**8**) and isoingenol (**10**), and the situation is further compounded by confusion between epi-ingenol (**8**), a compound so far unknown, and its synthetic tris-*nor* derivative **9**. Isomerization of ingenol to isoingenol (**10**) leads to total loss of tumor-promoting properties of its 3-esters, and iso-ingenol 3-esters have been used as negative controls in biological assays just like 4-epi-phorbol (4$\alpha$-phorbol) esters have been used for the activity of tumor-promoting phorbol esters [20].

The $^{1}$H- and $^{13}$C-NMR features of ingenol are summarized in Fig. 3 [21]. The most remarkable features of the NMR spectra are the downfield resonance of the non-oxygenated carbon C-10 ($\delta$ 72.6) and of the H-8 methine ($\delta$ 4.13), and the upfield resonance ($\delta$ 3.80) of the 5-proton, presumably because of its location in the anisotropy cone of the C-9 carbonyl. Although the $^{1}$H NMR spectrum of the *out,out* isomer **9** was not assigned, all oxymethines resonated at $\delta > 4$ ppm [18], as expected from a different orientation of the C-9 carbonyl and H-5. Esterification causes predictable α,γ-downfield and β-upfield shifts in the proton and carbon resonances of the parent polyol. Remarkable differences were noticed between

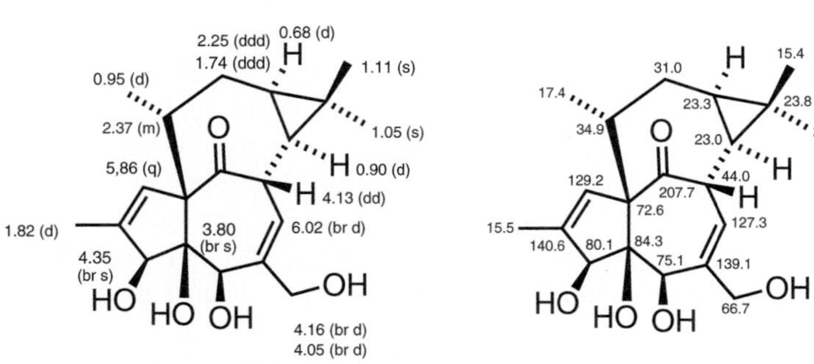


**Fig. 3** $^{1}$H- and $^{13}$C NMR data for ingenol (**3**) ($CDCl_3$, 400 and 100 MHz, respectively). *J* values: 1,19 = 1.5 Hz; 7,8 = 4.5 Hz; 8,14 = 12.5 Hz; 11,12 = 3.0 Hz; 11,12′ = 8.5 Hz, 12,12′ = 15; 12,13 = 6.0 Hz; 12′,13 = 8.5 Hz; 13,14 = 8.5 Hz, 20,20′ = ca. 7 Hz [21]

the IR carbonyl frequencies of ingenol and its *out,out* model analog **9** ($\nu = 1705$ and 1675 $cm^{-1}$), presumably because of different degrees of deviation from planarity, of ground-state strain energy, or of a combination of both [18]. An extensive compilation of $^{13}C$ NMR data for ingenol esters has been published, facilitating the characterization of the acyl decoration of its esters [22].

The MS fragmentation of ingenol is significantly different from those of phorbol and deoxyphorbol, and this makes it possible to selectively detect its derivatives in complex extracts using MS-MS techniques [23]. The mass spectra of ingenol esters are characterized by the stepwise loss of the ester groups, and their fragmentation profile has been investigated in the positive-ion mode, which readily ionizes them [23]. While the nature of the acyl moiety can be inferred from the pattern of mass fragmentation loss, the elimination order is of limited general value for the establishment of the esterification profile, since it was found to be dependent on the experimental conditions and the nature of the ester group [23].

Ingenol triacetate and ingenol-3-angelate (I3A, **5**) [24] have been investigated by X-ray diffraction, and such data on several semi-synthetic analogs obtained during various total synthesis campaigns are also available [16]. Crystalline forms of I3A were obtained from ethanol by addition of water, and from acetonitrile by cooling. Apparently, the crystalline forms are easier to formulate than the amorphous form, and their preparation has been patented [24]. In accordance with observations made on 20-deoxyingenol, angular strain was particularly evident at the angles C4–C5–C6 (123.0°) and C10–C13–C14 (119.3°). Angular strain was also present at the carbonyl bridge, with the angle C8–C9–C10 being only 111.2°. Linear strain was especially marked at the bond C4–C10, which is considerably longer than normal [1.602 (5) Å] [24].

Ingenol was originally reported as dextrorotatory ($[\alpha]_D^{25} = +41° cm^2 g^{-1}$, methanol) [21], and further studies confirmed a positive optical rotation also for the synthetic product. Nevertheless, a high-purity sample prepared in the course of a large-scale isolation of the product at LEO Pharma, the company that developed Picato®, showed a negative optical rotation ($[\alpha]_D^{25} = -18° cm^2 g^{-1}$, methanol), confirmed with an ad hoc prepared synthetic sample ($[\alpha]_D^{25} = -13° cm^2 g^{-1}$, methanol), with the absolute configuration of the levorotatory sample supported by X-ray analysis of a heavy-atom containing derivative (3-(4-bromobenzoate) of the 5,20-acetonide) [16]. More than the presence of impurities, for which the profiles should be different between samples of natural and of synthetic origin, it is possible that the reversal of the optical rotation might be related to degradation, since ingenol is not stable (see Sect. 2.3). Ingenol esters are mostly dextrorotatory, while those of 13-hydroxyingenol show an opposite optical rotation [8]. Some exceptions are known, but, since the purification of ingenol esters is difficult and the observed optical rotations are sometimes small, this could be related to the presence of impurities. Similar considerations hold also for the CD spectra. Ingenol esters show a positive Cotton effect at 300 nm, while 13-hydroxyingenol esters show a negative Cotton effect at 290 nm [8]. No systematic study has been done so far on the chiroptical properties of ingenoids.

## 2.2 Biogenesis

Euphorbiaceous plants are prolific producers of diterpenoid skeletal types, many of which, like the ingenanes, are unique to this large plant family [25]. According to their derivation from a linear tetraprenyl phosphate, *Euphorbia* diterpenoids can be broadly divided into compounds derived from the cascade cyclization and those derived from the head-to-tail cyclization of a tetraprenyl phosphate precursor through the effects of type II and type I cyclases, respectively (Fig. 4).

The first cyclization mode is at the origin of the so-called polycyclic diterpenoids, represented in euphorbiaceous plants mainly by abietanes, pimaranes, kauranes, atisanes, and trachylobanes, while the second one generates, in a step-wise fashion and with the cation processes typical of the isoprenoid cyclizations, macrocyclic diterpenoids, namely, the monocyclic cembranes, the bicyclic casbanes, and the tricyclic lathyrane derivatives (Fig. 5). Alternatively, cyclopropane ring opening generates jatrophanes from lathyranes (Fig. 5).

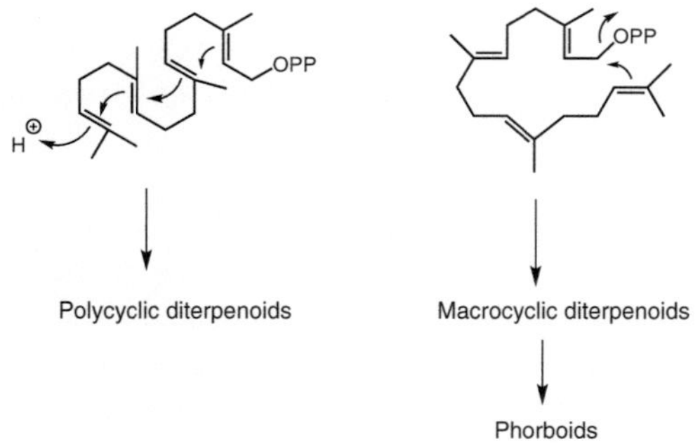


**Fig. 4** Cascade and head-to-tail cyclization mode of a tetraprenyl phosphate

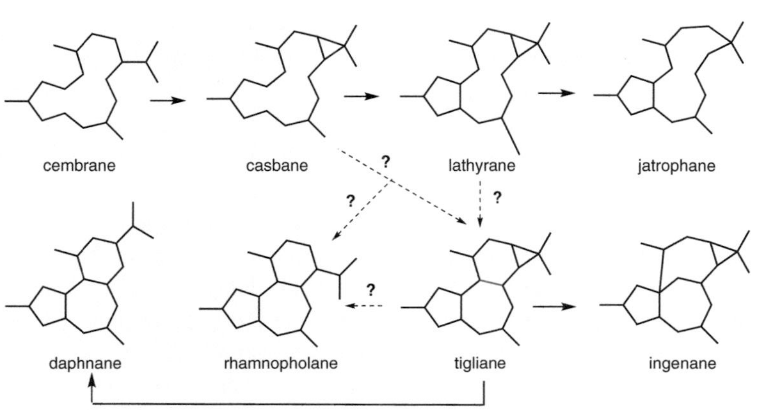


**Fig. 5** Topological relationship between the macrocylic diterpenoid skeleta (upper line) and those of phorboids (lower line), and their possible biogenetic relationship via lathyrane derivatives (topological proposal) or via casbane derivatives (Jakupovic proposal)

From a structural standpoint, phorboids are polycyclic diterpenoids, but are biogenetically derived by macrocyclic diterpene precursors. Therefore, they are considered products of a "head-to-tail" cyclization mode of a tetraprenyl phosphate, in one of the many oxymoronic relationships between biogenesis and structure that plague natural products chemistry, as exemplified by the isoquinoline alkaloid quinine, an indole alkaloid from a biogenetic standpoint. The relationship between macrocyclic diterpenoids and phorboids has long remained a mechanistic black box, despite obvious structural similarities between the lathyrane and tigliane skeleta. Tiglianes can next, at least formally, be converted to the other structural types of phorboids by cationic rearrangements typical of isoprenoid chemistry, and the present mechanistic uncertainties on the biogenesis of phorboids center, therefore, on the relationship between lathyranes and tiglianes and, in particular, on the formation of the single bond that differentiates compounds representative of these two classes.

Despite the structural similarity between the carbon-carbon connectivity of the lathyrane and tigliane skeleta, several issues make questionable a direct relationship between compounds of the two classes:

(a) While tiglianes have a broad distribution, occurring in several species of Euphorbiaceous plants, lathyranes have a much more limited distribution, exclusive to only a few species of the genus *Euphorbia*. If compounds from the two types were biogenetically related, one would expect a better overlapping occurrence, and not a "traceless" biogenetic relationship.

(b) As exemplified by phorbol (**1**) and lathyrol (**7**), the cyclopropyl carbon C-13 (tigliane numbering, C-11 in terms of lathyrane numbering) is hydroxylated in all natural tiglianes, as well as in many ingenanes, while no lathyrane discovered so far in Nature is hydroxylated at the cyclopropane carbons. In biological systems, C-H functionalization is basically an iron-mediated radical process [26], and the direct hydroxylation of a cyclopropane ring would therefore be biased toward the formation of a cyclopropylalkyl radical rather than a cyclopropyl radical [26]. In turn, the cyclopropylalkyl radical would rapidly generate the 3-butenyl radical, with eventual opening of the cyclopropane ring [27]. These considerations suggest distinct mechanisms of formation for the cyclopropane ring in lathyranes and tiglianes.

(c) In topological terms, the conversion of lathyranes to tiglianes involves the formation of a bond between the carbons 8 and 14 (corresponding to the carbons 8 and 9 of the tiglianes, cf. **1** and **7**), but no lathyrane derivative isolated so far bears the correct functionalization for this cyclization, which would require complementarity in terms of nucleophilic and electrophilic properties between these two carbons. Thus, C-14 of the lathyranes is generally a carbonyl, while C-8 is either hydroxylated, and therefore electrophilic, or is part of an unfunctionalized sequence of methylenes, as exemplified by lathyrol (**7**). Furthermore, most lathyranes contain at least one endocyclic (*E*)-double bond [25], which, apart from preventing a spatial interaction between the functionalities at C-8 and C-14, makes the cyclization to tiglianes impossible, since (*E*)-double bonds can only exist in eight-membered or higher ring systems. Unsurprisingly, the cationic cyclization of lathyranes has provided a wealth of rearranged diterpenoid skeleta, with none of these, however, bearing any resemblance with the tigliane skeleton [28].

A high-yield lathyrane-to-tigliane conversion was reported via an anionic process that involved the intramolecular aldol reaction of an ad hoc prepared semi-synthetic lathyrane 7,14-diketone devoid of unsaturations on the eleven-membered ring (**11**) [29]. The resulting tigliane (**12**) was assumed tentatively to have the same configuration as natural phorbol esters at the stereogenic carbons at the B,C ring-junction, but the functionalization of these two rings was completely different from that of the naturally occurring compounds of this class. Furthermore, aldol-type anionic cyclizations, although well documented in the chemistry of isoprenoids [30], are rarely involved in their biogenesis.

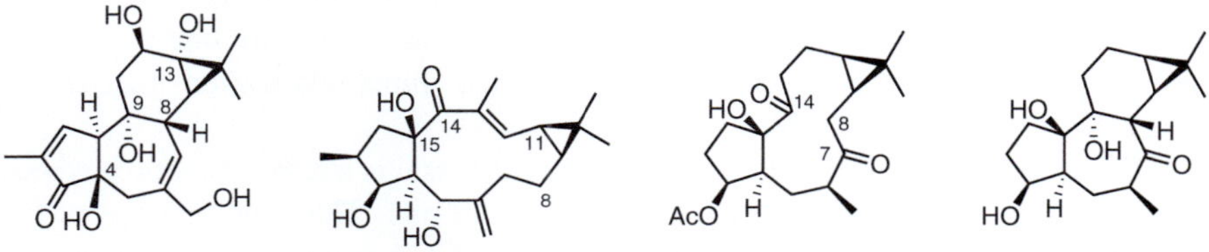


(d) Naturally occurring lathyranes and tiglianes show an opposite hydroxylation at the A,B ring junction, as exemplified by comparison of phorbol (**1**) and lathyrol (**7**).

Owing to these differences, and the implausibility of the hydroxylation of a cyclopropane carbon under biomimetic conditions, the biogenetic relationship between lathyranes and tiglianes has been questioned, without, however, proposing a possible biogenetic alternative [31]. As a result, tiglianes and ingenanes may be added to the thapsane family of sesquiterpenoids, the ryanodane family of diterpenoids, and the *Iboga* alkaloids as veritable biogenetic enigmas.

The late Jasmin Jakupovic (Plate 3) (1949–2005) of the Technische Universität Berlin and later Analyticon elaborated an elegant biogenetic proposal to solve this

**Plate 3** Jasmin Jakupovic (1949–2005) (photograph provided courtesy of Fotini Tsichrtzis)

riddle, providing a unifying mechanistic setting for the generation of lathyranes and tiglianes from casbanes that explains the differences in the functionalization pattern of these two class of compounds. This proposal has important implications for the biogenesis of ingenol, and is also capable of shedding light on the biogenesis of rhamnofolanes, another biogenetically enigmatic class of phorboids. The premature death of Jasmin Jakupovic has precluded the dissemination of his ideas within the natural products community, and his views remained confined to laboratory meetings, private discussions, and a Ph.D. thesis [32]. Thanks to the help of the Jakupovic family and his collaborator Frank Jeske, as well as to ideas and comments exchanged during personal discussions before Jakupovic's death, sufficient material has been retrieved to present in an organic fashion what will be referred to as the "Jakupovic biogenetic proposal" for phorboids. Despite radically deviating from some current ideas in the area, this proposal is the only one capable of rationalizing the biogenetic derivation and the functionalization pattern of phorboids in light of the isoprenoid rule.

The starting point for the elaboration of this proposal is a retrosynthetic analysis of the tigliane skeleton according to the isoprenoid rule. Thus, the reversal of the cyclopropane ring closure by protonation of the C-13 α-carbon unambiguously connects the tigliane skeleton (**13**) to the cembranyl cation **14** (Fig. 6). The alternative opening of the cyclopropane by protonation at C-14 is, in fact, incompatible with the isoprenoid rule, which sets the relative location of the alkyl substituents on isoprenoid chains. Conversely, the topologically symmetric cembranyl cation **14** could be generated by two distinct foldings of the acyclic tetraprenyl precursor. The folding **15** of pathway A seems more obvious, since it leads to tiglianes with the correct location of the cyclopropane on ring C. Conversely, the folding **17** of pathway B is apparently counterintuitive, since, by (formal) insertion

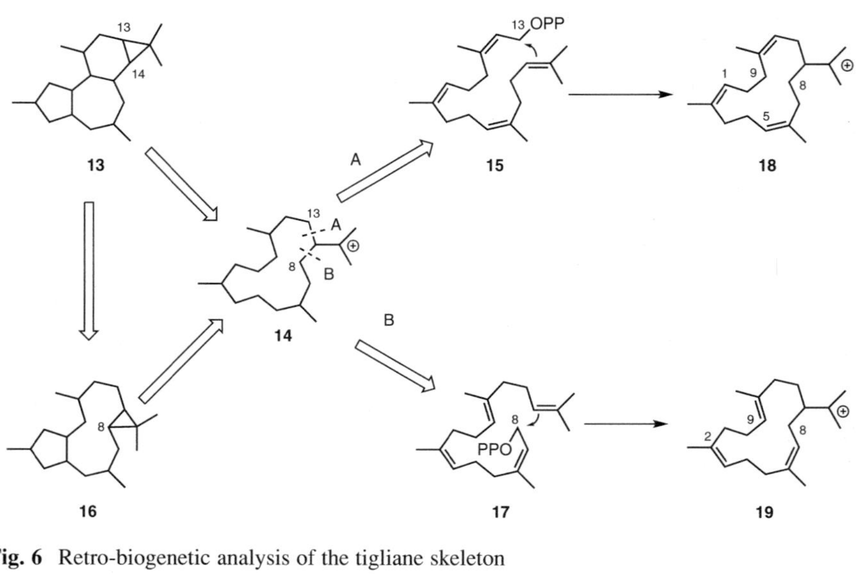


**Fig. 6** Retro-biogenetic analysis of the tigliane skeleton

of the side-chain cation center into a C(8)-H sigma bond, it would generate compounds with the unnatural location of the other cyclopropane bond at C-8 (represented in **16** in its lathyrane version), and a rearrangement would be needed to bring it to its natural location at C-13. This rearrangement is, indeed, at the core of the Jakupovic biogenetic proposal, and provides a mechanistic explanation for all the "unknowns" of the biogenesis of phorboids. Thus, pathway A requires the triply-isomerized all-(*Z*) tetraprenyl precursor **18**, and will lead to the "functionalization" "cul-de-sac" discussed before, since neither C-8 nor C-9 are properly functionalized to form the critical B, C ring junction. Furthermore, it leaves unresolved the issue of the hydroxylation of the cyclopropane carbon.

The configuration issue for the double bond located in what will become the cyclopentane ring A ($\Delta^1$ in **18** and $\Delta^2$ in **19**) is irrelevant, since the isomerization can be associated with functionalization and the generation of the five-membered A-ring, with a similar argument holding also for the allylic C-9 methylene of **18**. However, this issue is difficult to resolve by simply advocating functionalization for the $\Delta^5$ double bond of **18**, since this occurs as an unfunctionalized methylene in most tiglianes. Conversely, in **19** the configuration of the endocyclic double bond is not an issue, since the $\Delta^9$ double bond can be accommodated in an (*E*)-configuration, and the $\Delta^6$ double bond is allylic to the phosphate group, and therefore can isomerize via the classic rearrangement of primary γ,γ-dimethylallyl phosphates to tertiary α-dimethylvinyl phosphates [33]. In general, the isomerization of the pyrophosphate starter from a terpenyl-geranyl to the terpenyl-linalyl type (Fig. 7,

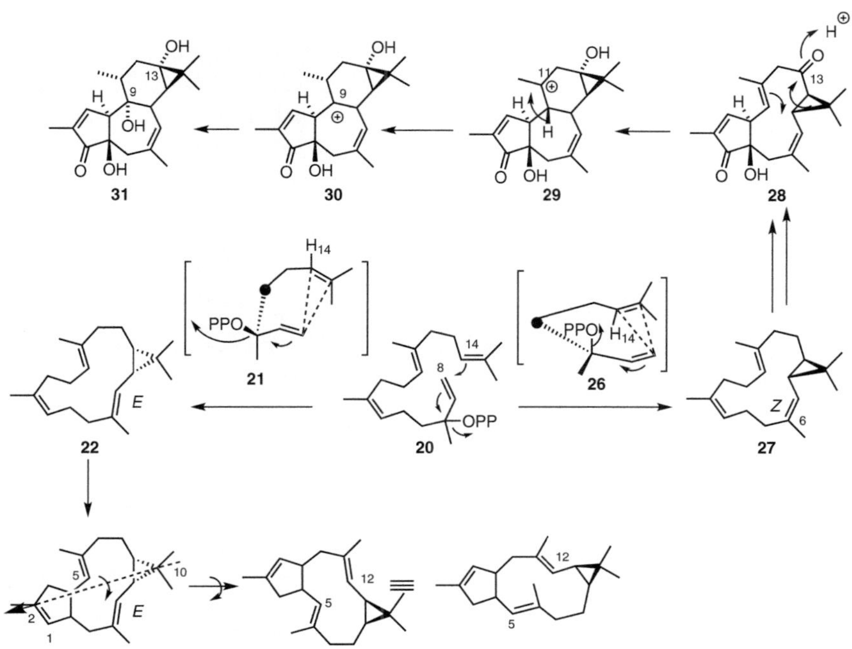


**Fig. 7** The Jakupovic proposal for the biogenesis of phorboids and lathyranes

**20**) is associated with an increase of the possibility of cyclization modes, and this conformational diversity (Fig. 7, **21** and **26**) might ultimately be responsible for the generation of both lathyranes and tigliane from the same precursor. Thus, electrophilic attack on the *re*-face (H-14 "up") of the terminal double bond from the vinyl antiperiplanar to the phosphate leaving group, would translate into an *E*-configuration of the newly generated trisubstituted double bond, and, when the resulting casbene derivative is viewed in a tigliane orientation, an α-orientation of the cyclopropane ring (Fig. 7, **22**). Conversely, attack on the *si*-face of the terminal double bond (H-14 "down") from the vinyl group *synclinal* to the phosphate leaving group will generate an opposite configuration of the double bond and the cyclopropane ring [27]. After formation of the cyclopentane ring A, rotation around the C-2–C-10 axis (lathyrane numbering) of the casbene **23** generates a lathyrane with the double bond locations ($\Delta^5$ and $\Delta^{12}$) and configuration typical of this class of products. Conversely, the casbene **27** has the correct double bond configuration and location to generate, after closure of the A-ring, the phorboid skeletal of tiglianes, ingenanes and rhamnofolanes. Thus, introduction of a carbonyl function at C-13 (tigliane numbering) triggers the formation of a stable non-classic cyclopropylmethyl cation, which is then quenched by the C-9 (tigliane numbering) endocyclic double bond and closure of the cyclopropane ring at C-13 (see **28**), with formation of the critical C-8—C-9 bond and of the hydroxylated version of the cyclopropane D ring. Hydride migration and quenching by water will eventually provide a tigliane (**31**) with hydroxy groups at C-9 and C-13, the hallmark of all natural compounds of this class (Fig. 7).

For ingenanes that lack a 13-hydroxy group, like the parent polyol ingenol, the electrophilic trigger for the migration of the cyclopropane could be provided by the generation of an oxonium ion at C-13 (Fig. 8), while their rearranged skeleton might, in principle, derive from an alternative quenching of the C-9 cation **34**.

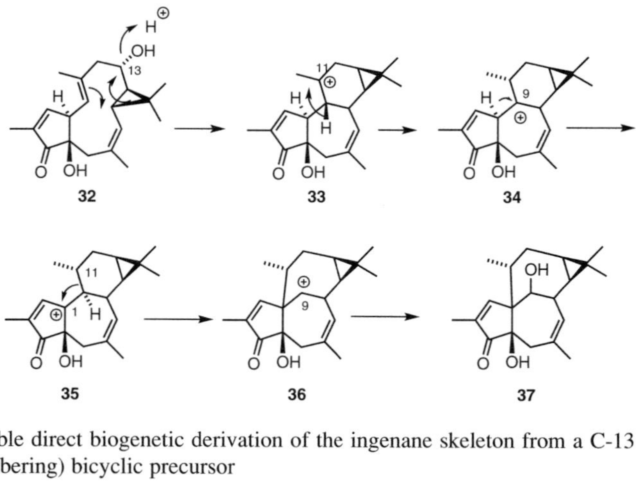

**Fig. 8** Possible direct biogenetic derivation of the ingenane skeleton from a C-13 hydroxylated (tigliane numbering) bicyclic precursor

Rather than directly undergoing water trapping like the cation **30**, quenching could occur after a 1,2 hydride shift to generate a C-1 cation that next undergoes a 1,2-alkyl shift, eventually rearranging to the ingenyl cation **36**. A detailed study on the tigliane-to-ingenane rearrangement during the Baran synthesis of ingenol [16] has highlighted the role of the presence of a hydroxy substituent at C-9 to overcome the thermodynamic barrier involved in the reaction, suggesting that a direct derivation of the ingenane skeleton from a lathyrane-type bicyclic precursor seems unlikely.

Finally, oxidation of the casbene **27** at C-16 (tigliane numbering) and the generation of lathyranes like **38** (Fig. 9) may trigger a different mode of transannular opening of the cyclopropane ring, generating the typical ring C functionalization of rhamnofolanes (**25**) via the backside attack of the double bond to the cyclopropane C-8 carbon.

The two different retro-biogenetic disconnections of the tigliane skeleton according to the isoprenoid rule (Pathways A and B in Fig. 6) could be distinguished easily by labeling experiments. Thus, labeling C-1 of geranyl-geraniol will lead, via the phosphate **40**, to a tigliane labeled at C-13 according to pathway A (**41**), but, via the phosphate **42**, to a tigliane labeled at C-8 (**43**), if pathway B is indeed the one occurring in Nature (Fig. 10).

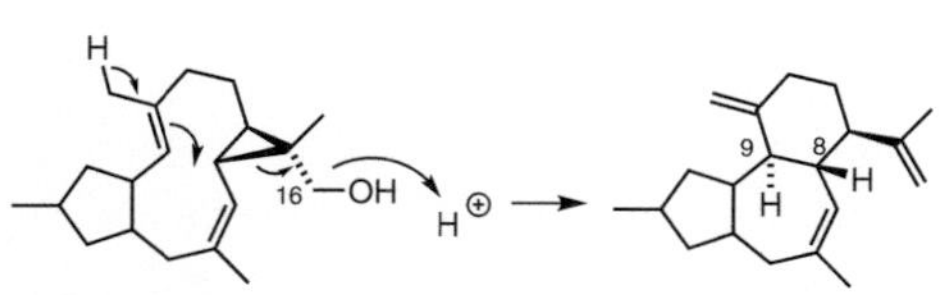


**38** **39**

**Fig. 9** Generation of the rhamnofolane skeleton by rearrangement of a C-16 hydroxylated lathyrane precursor (tigliane numbering throughout)

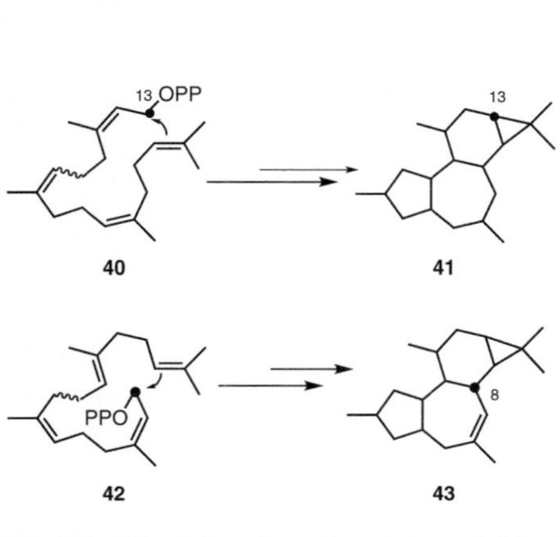


**40** **41** **42** **43**

**Fig. 10** Location of the labeling of the C-1 carbon of a tetraprenyl phosphate precursor according the two possible retro-biogenetic disconnections of the tigliane skeleton

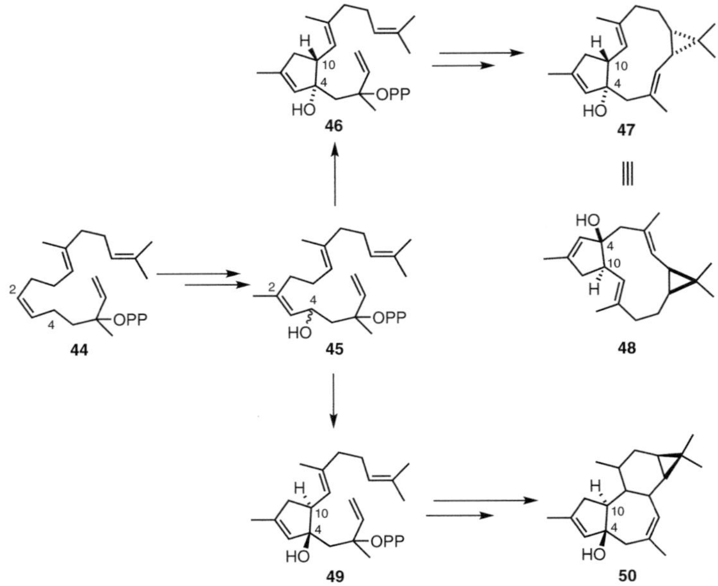


**Fig. 11** Possible rationalization of the opposite hydroxylation of the A,B-ring junction in lathyranes and tiglianes (tigliane numbering throughout)

According to the Jakupovic biogenetic proposal, lathyranes and tiglianes have, in biogenetic terms, a different location of the cyclopropane ring. The conventional representation of this class of compounds hides this, and generates inconsistencies between the functionalization pattern of these compounds and phorboids. Regarding the opposite functionalization of the cyclopentane A-ring in lathyranes and tiglianes, it is tempting to imagine that the allylic oxidation of the $\Delta^2$ double bond (tigliane folding) of a tetraprenyl precursor like **44** could trigger formation of the cyclopentane A ring, and that the different configuration of this hydroxy group, which generates diastereoisomers in a tetraprenyl starter of the geranyl-linalyl type, might be involved in the pre-organization of the cyclization step involving the pyrophosphate group (Fig. 11), thus explaining the apparently different hydroxylation of the A, B ring junction in lathyranes and tiglianes.

The Jakupovic proposal provides a mechanistic framework to relate biogenetically the functionalization pattern of lathyranes and phorboids, elegantly by-passing the "cul-de-sac" represented by the apparently unrelated configuration and location of the double bonds within these compounds and providing a mechanistic rationale for the oxygenation of the cyclopropane ring in tiglianes and ingenanes. Since it rationalizes all the many "unknowns" of the biogenesis of phorboids, its wider appreciation will, hopefully, trigger studies aimed at its confirmation or invalidation.

## 2.3 *Isolation*

Many ingenol esters are potent irritant and vesicant agents as well as tumor promoters, and during their extraction and purification great care has to be taken to avoid contact with the skin, mucous membranes, and the eyes. Accidental eye contact is characterized by massive lacrimation and acute pain, and may require hospital treatment to be managed correctly. In turn, skin contact induces within 2–3 h the formation of an erythema that can persist for well over 24 h, but which if the exposed area is limited, can be soothed by topical treatment with local anesthetics and ice blocks [34], or with a solution of a polyamphoteric detergent [35]. Hospitalization might, however, be necessary in case of the accidental contamination of a large skin area [34]. Based on observations done with I3A, the inflammatory response to ingenoids is transient and resolves spontaneously without scarring within 2–4 weeks, during which topical treatment with corticosteroids may be recommended to control pain and inflammation [34]. Severe allergic reactions, including throat tightness, difficulty in breathing, feeling faint, or swelling of the lips or tongue, have been reported with Picato® [36], and they might well result also from contact with other ingenoids. Facial, especially the eyes, and skin protection is therefore mandatory during the laboratory manipulation of ingenol esters, as well as of plant material, extracts, and chromatographic fractions containing compounds of this type. Glassware should be treated with bases before washing, and contaminated laboratory spaces should be treated in the same way. It is known that 5% KOH in methanol quickly hydrolyzes ingenol esters, and is preferred to washing with aqueous solutions that can not quickly penetrate fatty matrices [37].

The literature is rife with exaggerated statements on the difficulties involved in the isolation of ingenol esters, and, in general, of phorboids. Thus, the claim that "*the isolation (of phorboids) must be conducted in oxygen-free conditions; solvents must be degassed and extraction should be conducted under continuous flow of nitrogen or argon*" [38] is exaggerated, even though ingenol derivatives are unstable toward hydrolysis and transesterification, and are easily oxidized by atmospheric oxygen. Undoubtedly, their isolation in the native form requires care and the need to avoid acidic and basic conditions, which even if mild, can promote degradation and/or isomerization reactions, but these features are not exclusive to this class of natural products, and do not preclude working under normal laboratory conditions. Although no systematic study has yet been done, ingenoids are probably also easily degraded by heating, explaining the edible uses of the seeds of the caper spurge (*E. lathyris*), which after roasting, were used as a coffee replacement during the shortages associated with World War I [39]. Heating is also use to detoxify this oil from this species before its technical uses may be applied [39].

Ingenane polyols do not occur as free alcohols, but only in the esterified form. Ingenol esters are lipophilic and insoluble in water. They are extracted from plant material with organic solvents, and, depending upon their degree of acylation and the nature of the ester groups, they show different polarities and therefore partition differently between hydrocarbon and more polar solvent phases. Extraction from a plant biomass typically involves the use of solvents like ethanol, acetone or chloroform, evaporation, and next partition between hexane and water-methanol, with ingenol esters ending up in the more polar and hydrophilic phase. Conversely, when the primary extract is partitioned between water and chloroform, or between water and ethyl acetate, ingenol esters end up in the organic phase. Partition is especially critical to recover long-chain ingenol esters from fatty matrices like seed oils. In this case, the system acetonitrile-cyclohexane was found to work much better than other partition systems [21]. Crude ingenoid mixtures can then be partially resolved by various chromatography techniques, namely, gravity chromatography on silica gel (both in the normal and in the RP mode) or Sephadex. For final purification, HPLC is generally necessary, since ingenol esters often represent minor constituents of the diterpenoid bouquet of the plant under investigation. During solid-liquid chromatographic separation, acyl rearrangement can take place, and ingenoids having a free hydroxy group at C-3 and ester groups at C-5 and C-20 are most likely to be isolation artifacts (see Sect. 3.1). The early phytochemical work on ingenoids made extensive use of droplet counter-current chromatography for their purification [8], a technique that minimizes acyl rearrangements. While now obsolete in its original form, this method could benefit from many recent rapid developments in the field [40], and holds potential especially for large scale purifications. However, this possibility is still largely unexploited for the phorboids.

The isolation of ingenol has been detailed from the seed oil of *E. lathyris*, an agricultural commodity (Plate 4), and the methods developed could be applied also to other oily sources. The original method by Hecker [41] involved a series of liquid-liquid and solid-liquid partitions. Thus, a methanolic extract from the seeds

**Plate 4** *Euphorbia lathyris* L. at the time of fruit maturation (photograph provided courtesy of Dr. Gianpaolo Grassi, Cra, Rovigo, Italy)

(9.7% yield) was first partitioned between a half saturated NaCl solution and ethyl acetate to remove the polar fractions containing mono- and dimeric coumarins and other polar constituents. The organic phase was then evaporated, and the residue (7.7%) was emulsified with methanol. An oily phase made up by glycerides separated from the emulsion, and, after evaporation, the methanol phase (3.2%) was adsorbed onto silica gel. The cake was then washed sequentially with petroleum ether, to further remove glycerides, and with ethyl acetate, to recover the diterpenoid fraction (0.58%). Zemplén methanolysis of the latter and partitioning between water and ethyl acetate of the residue of the hydrolysis eventually gave a polyol fraction (0.5%) that was adsorbed onto silica gel and then washed with chloroform to remove the ester fraction. Washing with ethyl acetate afforded a purified polyol fraction (0.13%) that was subjected to a sequence of gravity column chromatographic purifications, eventually affording ingenol in an overall 0.025% yield (250 mg/kg of seeds).

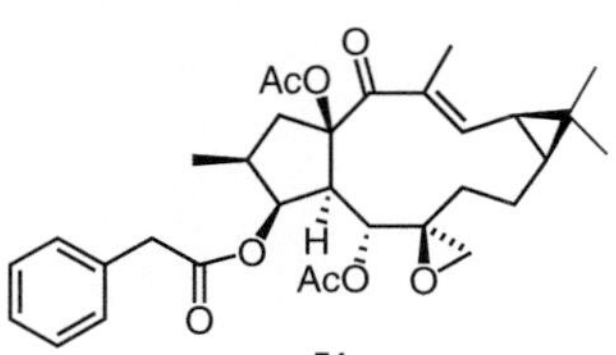


**51**
(*Euphorbia* Factor L1))

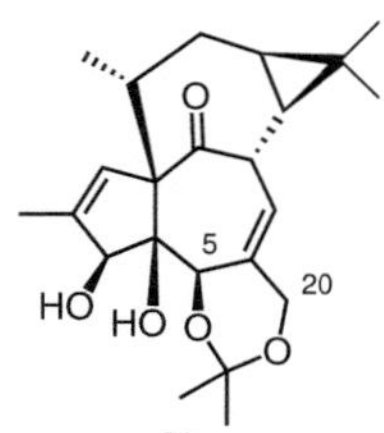


**52**
(ingenol 5.20-acetonide)

A simplified method was later developed [21], based on the possibility of directly separating the diterpene fraction of the oil by partition of a methanolic extract between acetonitrile and cyclohexane, a system originally devised to remove lipophilic tin derivatives in the work-up of $Bu_3SnH$ reductions [42]. The diterpenoid fraction was then treated with 8% KCN in methanol to hydrolyze selectively the macrocyclic diterpenoid esters, of which the major member, the epoxylathyrol ester $L_1$ (**51**), directly precipitated from the reaction mixture. Ingenol, a more polar compound, could then be easily separated from the macrocyclic esters by gravity column chromatography on silica gel, with an overall isolation yield of ca. 250–600 mg/kg of seeds [21]. The rationale for the chemoselectivity of the hydrolysis is that the presence of nearby free hydroxy groups catalyzes the deacylation of ester groups [43], a condition holding for ingenol esters but not for lathyrol and epoxylathyrol esters. Zemplén methanolysis with freshly prepared sodium methylate in methanol gave lower yields and was less selective for ingenol vs. lathyrol and epoxylathyrol esters, but would, undoubtedly, be more practical for a large-scale process. When obtaining macrocyclic polyesters is not important, a further simplification is possible, directly extracting the crushed seeds with methanolic sodium methylate. In this step, a significant precipitation of the highly crystalline macrocyclic polyols lathyrol and epoxylathyrol takes place, and, after filtration, neutralization, and evaporation, a fraction containing a mixture of ingenol and the partially hydrolyzed macrocyclic diterpene moiety is obtained. This can then be purified by gravity colum chromatography on silica gel [44]. This method minimizes the exposure to ingenol esters, since the crude crushed seeds are extracted with an active solvent that "detoxifies" them, and, comparative experiments with other procedures have highlighted the superiority of this protocol, with yields consistently in the range of 500 mg/kg of seeds [44].

Other sources of ingenol, like the latex of *E. ingens*, turned out not not to be reliable due to large variations in their concentrations [41]. Since all succulent African *Euphorbia* species are now included in the CITES list [45], the trade of *E. ingens* would, in any case, be subjected to severe regulatory limitations. The latex of a South-American chemotype of *E. tirucalli* L., a plant known as "Aveloz" in Brazilian folk medicine, has also been claimed to be an excellent source of ingenol, and used for the development of semi-synthetic antiviral agents (see Sect. 4.2), without, however, disclosing details of the yield and the isolation protocol [46]. "Aveloz" is a popular medicinal plant in Brazil, but in 2011 the Brazilian Agency for Sanitary Vigilance (ANVISA) prohibited its use in herbal medical products because of the presence of co-carcinogenic diterpenoids [47].

Ingenol obtained by hydrolysis of its esters occurs as an amorphous glass, which degrades easily even in the cold, and is better stored as its crystalline 5,20-acetonide, **52**. Alternatively, it has been stored as such in frozen DMSO or in benzene.

## *2.4 Distribution and Diversity*

The distribution of ingenoids is limited to three closely related genera within the Euphorbiaceace family (*Euphorbia*, *Elaeophorbia* and *Mabea*) [48]. The tropical genera *Elaeophorbia* and *Mabea* are small (less than 100 species each), with the genus *Elaeophorbia* often being included in the genus *Euphorbia* [48]. The latter, with 1600–2000 species, depending on the classification system, is the second largest among higher plants after the genus *Astragalus* [49]. Most ingenoids have therefore been reported from plants belonging to the genus *Euphorbia*, to the point that this genus is often quoted as the only source of ingenoids in Nature. For comparison, tigliane derivatives have been reported from twenty-three genera belonging to two families (Euphorbiaceae and Thymelaeaceae) [50]. Ingenol esters occur both in spurges and in cactus-like members of the genus, and no clear trend associating their presence to certain morphological features is evident from the phytochemical literature. Just like phorboids in general, ingenol esters are absent in the New World species of the genus *Euphorbia*. The concentration of ingenoids in plants is generally rather low, often in ppm (mg/kg of) of fresh biomass. The best source of ingenol remains the caper spurge (*E. lathyris*), one of its two original sources, where it occurs in the seeds at a concentration up to ca. 0.5–0.8 g/kg [44]. For comparison, the concentration of phorbol esters in croton oil is an order of magnitude higher [7]. A systematic study on the concentration of ingenol in the latex of various *Euphorbia* species was carried out after hydrolysis of the diterpenoid mixture and GC analysis of the triacetate [51]. Typical concentration levels in the latices analyzed ranged between 1% and 3% [51].

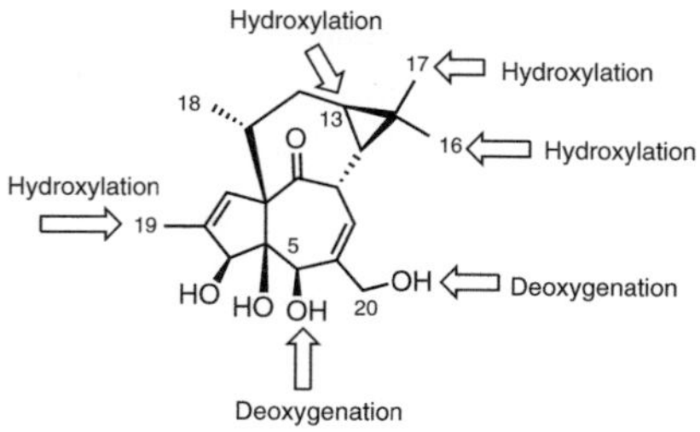


**Fig. 12** Functionalization diversity within ingenol esters

Chemical diversity within ingenol derivatives is associated with modification of the hydroxylation pattern of the parent polyol and/or to the nature of the acyl decoration. The latter can include, apart from fatty acids and their derivatives, also reasonably complex anthranilate-derived oligopeptides. Deoxygenation can occur at C-5 and C-20, and hydroxylation at C-13, C-16, and C-19, with the possibility also of combining deoxygenation and hydroxylation, as in 20-deoxy-16-hydroxyingenol derivatives (Fig. 12). It is, however possible that the structure of all 16-hydroxyingenoids should actually be revised to their corresponding 17-hydroxy derivatives (see Sect. 2.4.8).

In some species of *Euphorbia*, ingenol esters occur in a mutually exclusive manner with phorbol esters, but the co-occurrence of ingenane, tigliane and daphnanes has been reported in some species, in particular in *E. resinifera* O. Berg. [52], with tiglianes and ingenanes co-occurring in few other species. Ingenol derivatives are sometimes isolated as very complex mixtures of homologous polyunsaturated esters that could not be individually characterized. The compounds listed in Tables 2–19 might therefore only represent the tip of a much larger "iceberg" of compounds of this type that still await complete structure elucidation.

Some acyl groups are unique to ingenol esters. The most interesting case is that of the anthranilate-based tripeptides occurring in milliamines (sic) and related compounds isolated from *E. milii* Des Moul., an ornamental plant native to Madagascar and popularly known as "crown of thorns" (Table 1). 2,6-Dimethyl substitution as well as polyunsaturation (up to five conjugated double bonds) are not uncommon in the acyl decoration of ingenol esters (Table 1). The configuration of these acyl chains has not generally been assigned, but in some cases isomeric forms of polyunsaturated acids have been isolated.

Tables 2–19 report all the known ingenol esters, along with their plant source. Authority abbreviations for the binomial were assigned using the International Plant Name Index (http://www.ipni.org/ipni/plantnamesearchpage.do). Authority abbreviations are not included in the Tables for the species mentioned in the text. Compounds are sorted out into six basic chemotypes:

(a) ingenol esters
(b) ingenol mono (13-hydroxy-, 16-hydroxy-, 17-hydroxy-, 19-hydroxyingenol) and dihydroxy (13,16-dihydroxy-, 13,17-dihydroxyingenol) esters
(c) tetrahydroingenol esters

**Table 1** Ester groups found in ingenoids

| Abbreviation | Name | Formula |
|---|---|---|
| Ac | Acetic acid | 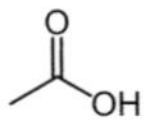 |
| Prop | Propionic acid | 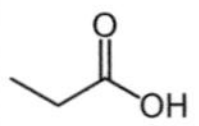 |
| *i*Bu | Isobutyric acid | 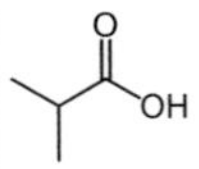 |
| 2Me-$C_4$ | 2-Methylbutyric acid | 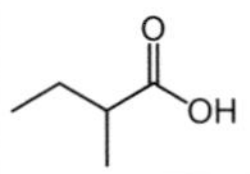 |
| 2,3-Me-$C_4$ | 2,3-Dimethylbutyric acid | 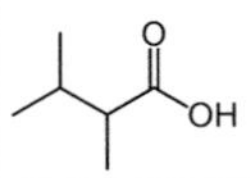 |
| $C_5$ | Pentanoic acid | 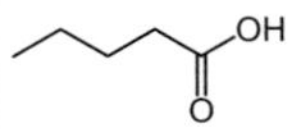 |
| Ang | Angelic acid | 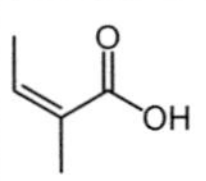 |
| Tigl | Tiglic acid | 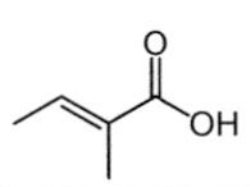 |
| $C_6$ | Hexanoic acid | 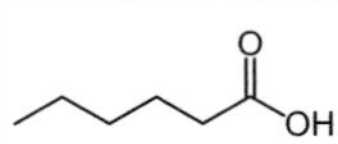 |
| $\Delta^{2,4}$-$C_6$ | 2,4-Hexadienoic acid | 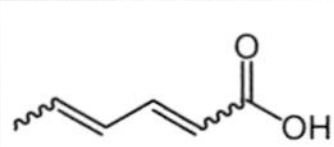 |
| $C_8$ | Octanoic acid | 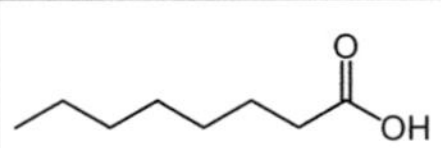 |
| $\Delta^{2}$-$C_8$ | 2-Octenoic acid | 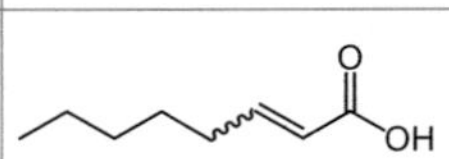 |
| 2,6-Me-$C_8$ | 2,6-Dimethyloctanoic acid | 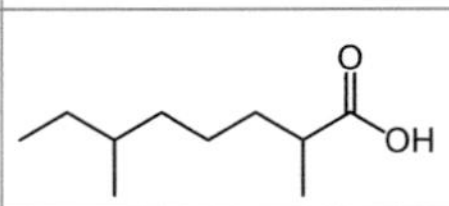 |
| 2-Me-$C_9$ | 2-Methylnonanoic acid | 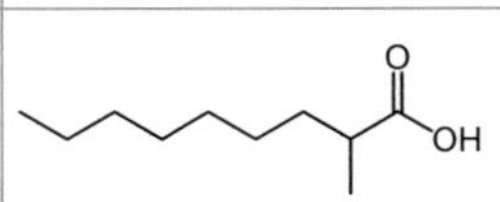 |

(continued)

**Table 1** (continued)

| Abbreviation | Name | Formula |
|---|---|---|
| 2,6-Me-$C_9$ | 2,6-Dimethylnonanoic acid | 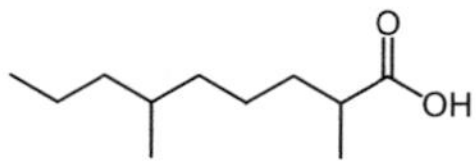 |
| $C_{10}$ | Decanoic acid | 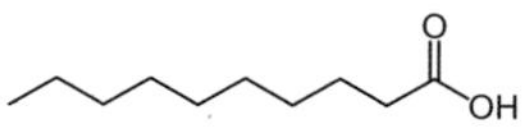 |
| D2,4-$C_{10}$ | 2,4-Decadienoic acid | 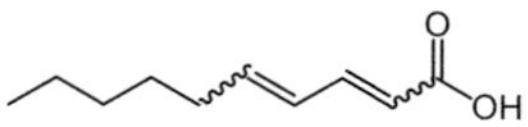 |
| (*E*,*Z*)-D2,4-$C_{10}$ | (2*E*,4*Z*)-Decadienoic acid | 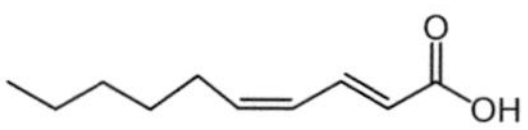 |
| (*E*,*Z*)-D2,4-$C_{10}$ | (2*E*,4*E*)-Decadienoic acid | 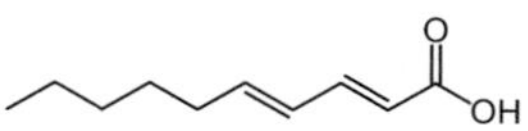 |
| D2,4,6-$C_{10}$ | (2*E*,4*E*,6*E*)-Decatrienoic acid | 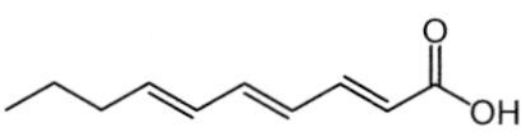 |
| 2-Me-$C_{10}$ | 2-Methyldecanoic acid | 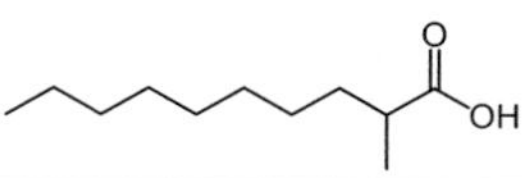 |
| 2,6-Me-$C_{10}$ | 2,6-Dimethyldecanoic acid | 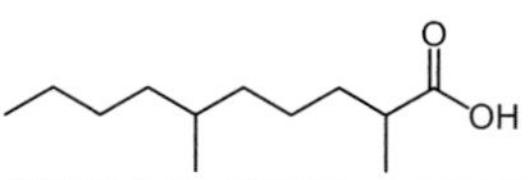 |
| 2-Me-$C_{11}$ | 2-Methylundecanoic acid | 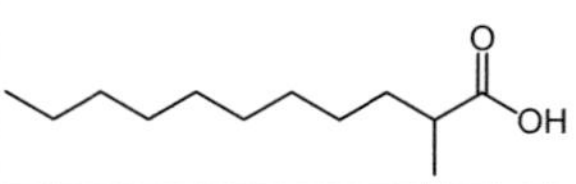 |
| 2,6-Me-$C_{11}$ | 2,6-Dimethylundecanoic acid | 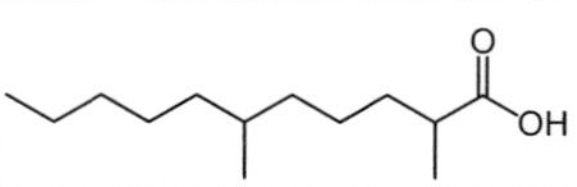 |
| $C_{12}$ | Dodecanoic (lauric) acid | 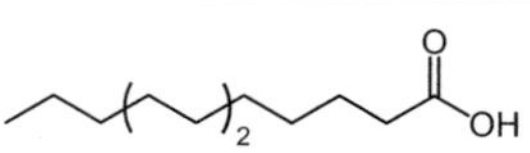 |
| $\Delta^{2,4}$-$C_{12}$ | 2,4-Dodecadienoic acid | 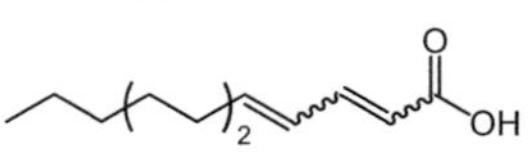 |
| $\Delta^{2,4,6}$-$C_{12}$ | 2,4,6-Dodecatrienoic acid |  |
| $\Delta^{2,4,6,8}$-$C_{12}$ | 2,4,6,8-Dodecatetraenoic acid | 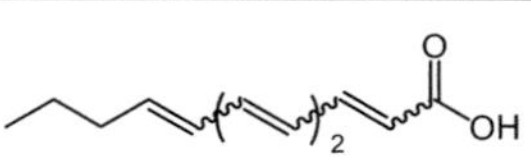 |

(continued)

**Table 1** (continued)

| Abbreviation | Name | Formula |
|---|---|---|
| $C_{14}$ | Tetradecanoic (myristic) acid | 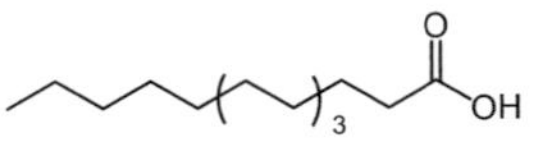 |
| $(E,Z)$-$\Delta^{2,4}$-$C_{14}$ | (*E*,*Z*)-2,4-Tetradecadienoic acid | 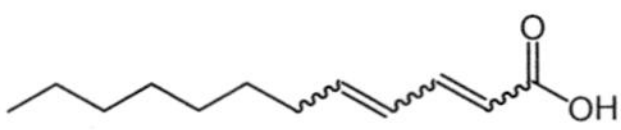 |
| $\Delta^{2,4,6,8}$-$C_{14}$ | 2,4,6,8-Tetradecatetrienoic acid | 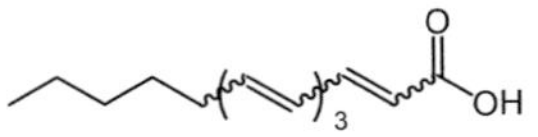 |
| $\Delta^{2,4,6,8,10}$-$C_{14}$ | 2,4,6,8,10-Tetradecapentaenoic acid | 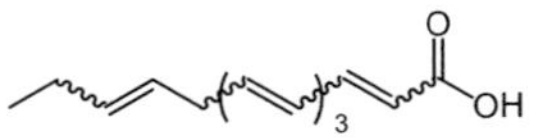 |
| $C_{15}$ | Pentadecanoic acid | 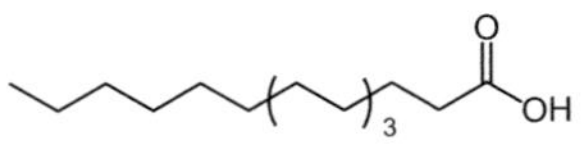 |
| $\Delta^{2,4,6,8}$-$C_{15}$ | 2,4,6,8-Pentadecatetraenoic acid |  |
| $C_{16}$ | Hexadecanoic (palmitic) acid | 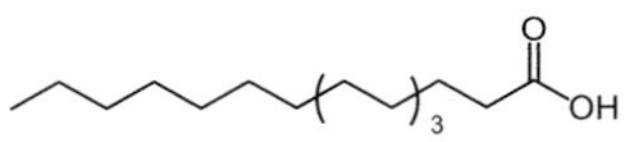 |
| $(Z)$-$\Delta^{9}$-$C_{18}$ | Oleic acid | 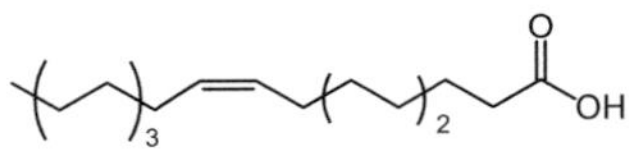 |
| 12-OAc-$(Z)$ $-\Delta^{9}$-$C_{18}$ | 12-*O*-Acetylricinoleic acid | 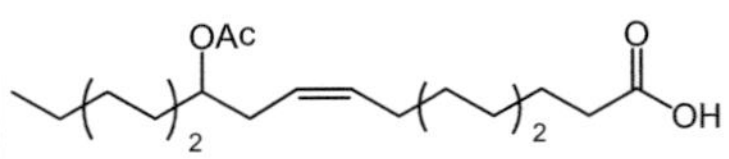 |
| $C_{20}$ | Eicosanoic acid | 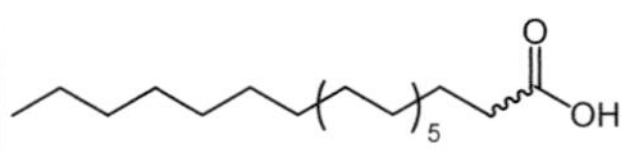 |
| $\Delta^{11}$-$C_{20}$ | 11-Eicosenoic acid | 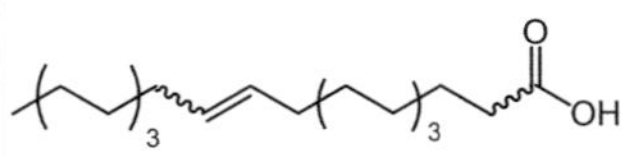 |
| $\Delta^{11,14}$-$C_{20}$ | 11,14-Eicosadienoic acid | 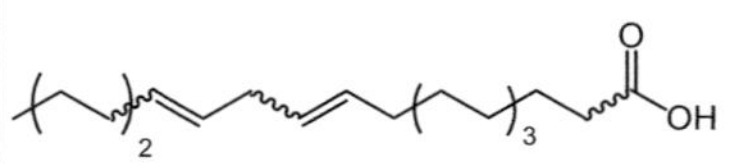 |
| Bz | Benzoic acid | 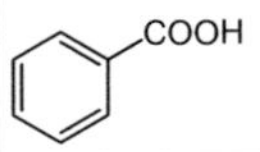 |
| PhAc | Phenylacetic acid | 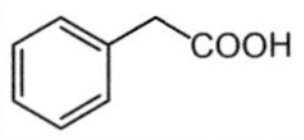 |

(continued)

**Table 1** (continued)

| Abbreviation | Name | Formula |
|---|---|---|
| Nic | Nicotinic acid | 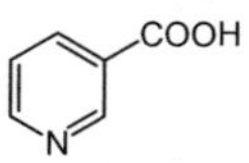 |
| P1 | 2-(2-Aminobenzamido)benzoic acid | 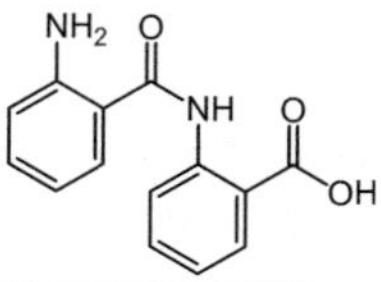 |
| P2 | 2-(2-(Benzamido)-3-hydroxybenzamido)benzoic acid | 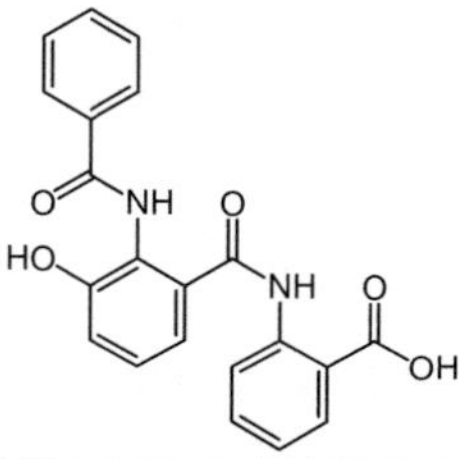 |
| P3 | 2-(2-(2-(Dimethylamino)benz-amido)-3-hydroxybenz-amido)benzoic acid | 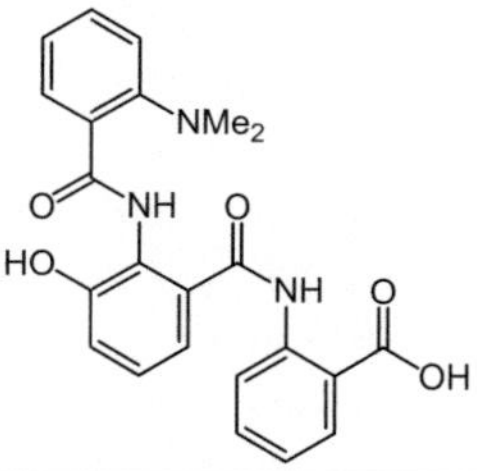 |
| P4 | 2-(3-Hydroxy-2-(2-methyl-amino)benzamido)benz-amido)benzoic acid | 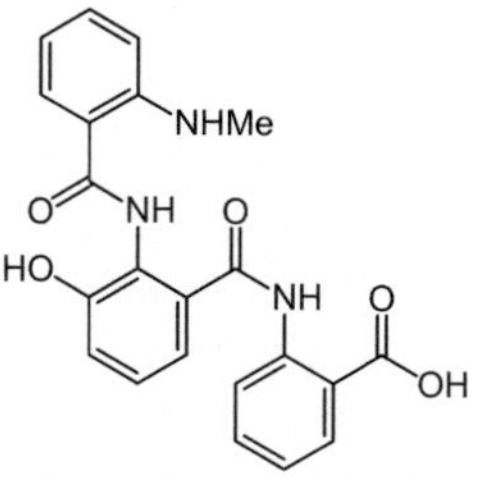 |
| P5 | (*E*)-2-(3-Hydroxy-2-(2-methylbut-2-enamido)benzamido)benzoic acid | 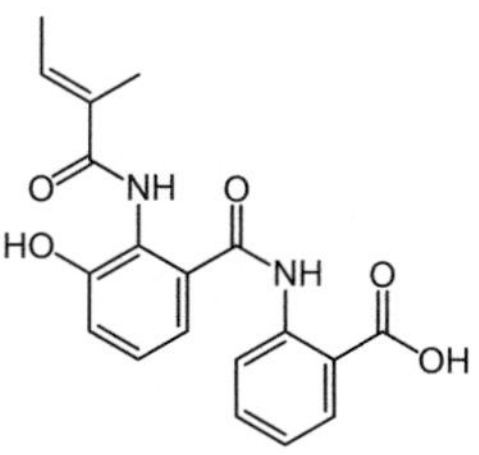 |

(d) 20-deoxyingenol esters
(e) mono (13-hydroxy-, 16-hydroxy-, 17-hydroxy)- and dihydroxy (13,17-dihydroxy-) 20-deoxyingenol esters
(f) 5-deoxyingenol esters.

Within each class, compounds are ordered according to increasing acylation, from monoesters to tetraesters, and with an increasing number of carbons of the longest linear sequence of their side chains. Abbreviations are used for the acyl group (Table 1). Thus, Δ is used to indicate the presence of an unsaturation, and Me for methyl branching, while the chain length is indicated using a "Cn" code, where n is the number of the longest linear carbon sequence of the acyl group. Irritancy data, mostly from the mouse ear erythema assay, are the only data available for most of these compounds, and will be discussed briefly, in an anticipation of the general coverage in Sect. 4.4.1.

### 2.4.1 Ingenol 3-Monoesters

Compounds of this type are by far the largest group of ingenoids. The acyl group is generally aliphatic, sometimes extensively unsaturated, and with methyl branching at the α(2)- and the ε(6)-carbons. The configuration at the stereogenic double bonds and carbons of these side chain has, in general, not been assigned, and it is not clear if diastereomeric forms occur, as shown for unsaturated acids like 2,4-decadienoic acid, which has been found to occur both in the (2*E*,4*Z*) and (2*E*,4*E*) diastereomers. Many individual esters could not be obtained in pure form, but were only isolated as mixtures, and the nature of the acyl groups was identified by GC analysis of the methyl esters obtained by transesterification of the mixture of the corresponding 5,20-acetonides. Of special taxonomic relevance is the occurrence of the structurally unique anthranilate-base tripeptides named milliamines. The biogenesis of these tripeptides, especially the derivation of the 2-amino-3-hydroxybenzoyl moiety, is unknown. While esterification with acyl long chains is associated with both irritancy and tumor-promoting activity, these properties can be dissected, since unsaturation decreases tumor promotion, which is completely absent in the highly irritant tripeptide esters [52]. Interestingly, unsaturation as well as shortening of the acyl moiety accelerate the attainment of the peak of irritancy in the mouse ear erythema assay. For instance, this is ca. 24 h for ingenol 3-tetradecanoate, but only ca. 5 h for ingenol 3-angelate (**5**) [52].

**Table 2** Naturally occurring ingenol 3-esters

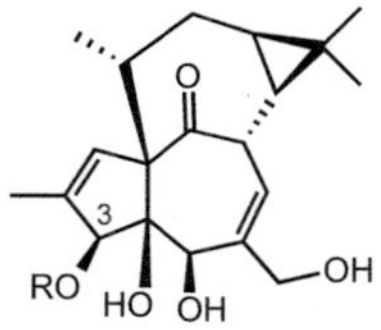

**53a–53w**

| Compound | R | Source | Ref |
|---|---|---|---|
| **53a** (= **5**) | Ang | *E. antiquorum* L. | [53] |
| | | *E. canariensis* | [54] |
| | | *E. drummondii* Boiss. | [55] |
| | | *E. hirta* L. | [55] |
| | | *E. helioscopia* | [56] |
| | | *E. paralias* | [57] |
| | | *E. peplus* | [58, 59] |
| | | *E. virgata* Waldst. & Kit | [60] |
| **53b** | $C_8$ | *E. peplus* | [61] |
| **53c** | 2,6-Me-$C_8$ | *E. resinifera* | [52] |
| **53d** | 2-Me-$C_9$ | *E. resinifera* | [52] |
| **53e** | 2,6-Me-$C_9$ | *E. resinifera* | [52] |
| **53f** | $C_{10}$ | *E. kamerunica* Pax | [62] |
| **53g** | 2-Me-$C_{10}$ | *E. resinifera* | [62] |
| | | *E. virgata* | [60] |
| **53h** | $\Delta^{2,4}$-$C_{10}$ | *E. esula* | [56] |
| | | *E. helioscopia* | [56] |
| | | *E. lathyris* | [60] |
| | | *E. tirucalli* | [63] |
| **53i** | (*E*,*E*)-$\Delta^{2,4}$-$C_{10}$ | *E. kansui* | [64, 65] |
| **53j** | (*E*,*Z*)-$\Delta^{2,4}$-$C_{10}$ | *E. helioscopia* | [56] |
| | | *E. kansui* | [65, 66] |
| **53k** | $\Delta^{2,4,6}$-$C_{10}$ | *E. ingens* | [67, 68] |
| | | *E. esula* | [56] |
| | | *E. lathyris* | [60] |
| | | *E. tirucalli* | [63] |
| **53l** | 2,6-Me-$C_{10}$ | *E. resinifera* | [52] |
| **53m** | 2-Me-$C_{11}$ | *E. resinifera* | [52] |
| **53n** | 2,6-Me-$C_{11}$ | *E. resinifera* | [52] |
| **53o** | $C_{12}$ | *E. kamerunica* | [62] |
| | | *E. esula* | [56] |
| **53p** | $\Delta^{2,4}$-$C_{12}$ | *E. lathyris* | [60] |
| **53q** | $\Delta^{2,4,6}$-$C_{12}$ | *E. esula* | [56] |
| | | *E. lathyris* | [69, 70] |
| | | *E. tirucalli* | [63] |

(continued)

**Table 2** (continued)

| Compound | R | Source | Ref |
|---|---|---|---|
| **53r** | $\Delta^{2,4,6,8}$-$C_{12}$ | *E. esula* | [56] |
| | | *E. tirucalli* | [63] |
| **53t** | $C_{14}$ | *E. broteri* Daveau | [71] |
| | | *E. wallichii* Hook.f. | [72] |
| **53u** | (*E*,*Z*)-$\Delta^{2,4}$-$C_{14}$ | *E. petiolata* | [73, 74] |
| **53v** | $\Delta^{2,4,6,8}$-$C_{14}$ | *E. nematocypha* Hand.-Mazz | [75] |
| | | *E. tirucalli* | [63] |
| **53w** | $\Delta^{2,4,6,8,10}$-$C_{14}$ | *E. lathyris* | [69, 70] |
| | | *E. esula* | [56] |
| | | *E. jolkini* Boiss. | [76] |
| | | *E. tirucalli* | [63] |
| **53x** | $\Delta^{2,3,6,8}$-$C_{15}$ | *E. nematocypha* | [75] |
| **53y** (=**2**) | $C_{16}$ | *E. ingens* | [3] |
| | | *E. lathyris* | [2] |
| | | *E. serrata* L. | [77] |
| | | *E. wallichii* | [72] |
| **53z** | P2 | *E. milii* | [78, 79] |
| **53aa** | P3 | *E. milii* | [78, 79] |

### 2.4.2 Ingenol 5-Monoesters

Ingenol 3-monoesters rearrange to their corresponding 20-esters with the intermediate formation of 5-monoesters (for a mechanistic discussion of the reaction, see Sect. 3.1.6), and represent therefore essentially extraction and/or fractionation artefacts. Ingenol 5-monoesters are not stable, and rapidly generate ingenol 20-esters under acidic or basic conditions. Mild treatment of the ingenol 20-esters

**Table 3** Naturally occurring ingenol 5-monoesters

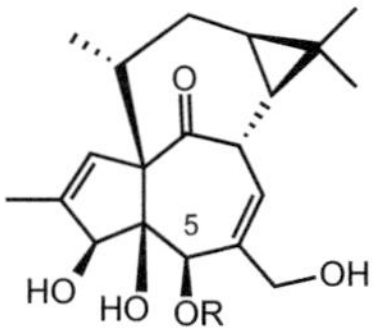


**54a–54d**

| Compound | R | Source | Ref |
|---|---|---|---|
| **54a** | $\Delta^{2,4}$-$C_6$ | *E. kamerunica* | [62] |
| **54b** | $\Delta^{2}$-$C_8$ | *E. kamerunica* | [62] |
| **54c** | (*E*,*Z*)–$\Delta^{2,4}$-$C_{10}$ | *E. kansui* | [64] |
| **54d** | $C_{14}$ | *E. broteri* | [71] |

can, however, lead to the generation of 5-esters, at least with saturated acyl groups [53]. Interestingly, ingenol-5-decanoate shows irritant activity, possibly related to partial retro-equilibration to its corresponding 3-decanoate [52], but the general instability of these compounds, and their difficult preparation, have precluded a systematic investigation of their biological profile.

### 2.4.3 Ingenol 20-Monoesters

Compounds of this type are the stable final products of the acyl rearrangement of ingenol 3-monoesters, and are, most probably, isolation artifacts, devoid of the significant irritancy associated with their corresponding 3-esters. Since they are

**Table 4** Naturally occurring ingenol 20-monoesters

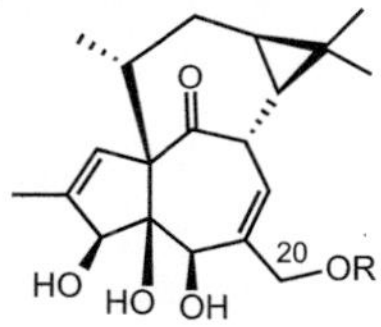


**55a–55l**

| Compound | R | Source | Ref. |
|---|---|---|---|
| **55a** | Ac | *E. segetalis* L. | [199] |
| | | *E. kansui* | [64] |
| **55b** | *i*Bu | *E. cotinifolia* L. | [200] |
| **55c** | $C_6$ | *E. serrata* | [77] |
| **55d** | $C_8$ | *E. peplus* | [61] |
| **55e** | (*E*,*E*)-$\Delta^{2,4}$-$C_{10}$ | *E. kansui* | [64] |
| **55f** | (*E*,*Z*)-$\Delta^{2,4}$-$C_{10}$ | *E. kansui* | [64] |
| **55g** | $C_{14}$ | *E. broteri* | [71] |
| | | *E. wallichii* | [72] |
| **55h** | $C_{16}$ | *E. ingens* | [67, 68] |
| | | *E. lathyris* | [69] |
| | | *E. nematocypha* | [201] |
| | | *E. quinquecostata* Volkens | [202] |
| | | *E. serrata* | [77] |
| | | *E. sieboldiana* Morren & Decne. | [203] |
| | | *E. wallichii* | [72] |
| **55i** | $C_{20}$ | *E. iberica* Boiss. | [204] |
| | | *E. kansui* | [64] |
| **55j** | Bz | *E. virgata* | [60] |
| **55k** | PhAc | *E. virgata* | [60] |
| **55l** | P3 | *E. milii* | [77] |

generated under physiological conditions, it would, nevertheless, be interesting to investigate their biological profile and evaluate if, and to what extent, they contribute to the complex biological profile of ingenol 3-esters.

### 2.4.4 Ingenol 3,5-Diesters

Compounds of this type are unstable towards acyl rearrangement and quickly generate their corresponding 5,20- and 3,20-diesters.

**Table 5** Naturally occurring ingenol 3,5-diesters

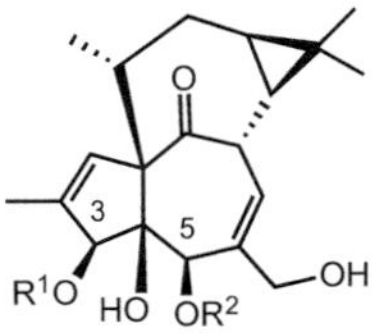


**56a–56g**

| Compound | $R^1$ | $R^2$ | Source | Ref |
|---|---|---|---|---|
| **56a** | P2 | Ac | *E. milii* | [79] |
| **56b** | P3 | Ac | *E. milii* | [78, 79] |
| **56c** | Prop | *i*Bu | *E. cotinifolia* | [205] |
| **56d** | Prop | 2-Me-$C_4$ | *E. cotinifolia* | [205] |
| **56e** | *i*Bu | *i*Bu | *E. cotinifolia* | [205] |
| **56f** | (*E,E*)-$\Delta^{2,4}$-$C_{10}$ | Ac | *E. kansui* | [64] |
| **56g** | (*E,Z*)$\Delta^{2,4}$-$C_{10}$ | Ac | *E. kansui* | [64] |

### 2.4.5 Ingenol 3,20-Diesters and 5,20-Diesters

It is not clear if compounds of this type are genuine natural products, or if, conversely, they are the result of acyl rearrangement from the corresponding and, sometimes co-occurring, 3,5-diesters. The 5,20-diesters are stable to transesterification, and acyl rearrangement is slow in 3,20-diesters, highlighting the relevance of a free primary 20-hydroxy group as an anchor for the migration of secondary acyl groups from the 3- and 5-hydroxy functionalities. The most representative member of the 3,20-diesters is the antitumor agent ingenol dibenzoate, which was subjected to a structure-activity study for this type of effect (see Sect. 4.1.4).

**Table 6** Naturally occurring ingenol 3,20-diesters and 5,20-diesters

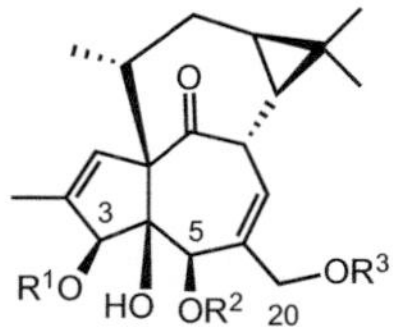


**57a–57s**

| Compound | $R^1$ | $R^2$ | $R^3$ | Source | Ref. |
|---|---|---|---|---|---|
| **57a** | Ang | H | Ac | *E. canariensis* | [54] |
| | | | | *E. helioscopia* | [56] |
| | | | | *E. hermentiana* Lem. | [105] |
| | | | | *E. kamerunica* | [62] |
| | | | | *E. peplus* | [59] |
| | | | | *E. trigona* Mill. | [206] |
| **57b** | Tigl | H | Prop | *E. cotinifolia* | [200] |
| **57c** | 2-Me-$C_4$ | H | Prop | *E. cotinifolia* | [200] |
| **57d** | Prop | H | *i*Bu | *E. cotinifolia* | [200] |
| **57e** | *i*Bu | H | *i*Bu | *E. cotinifolia* | [200] |
| **57f** | $\Delta^2$-$C_8$ | H | Ac | *E. kamerunica* | [62] |
| **57g** | (*E*,*E*)-$\Delta^{2,4}$-$C_{10}$ | H | Ac | *E. broteri* | [71] |
| | | | | *E. kansui* | [83, 207] |
| **57h** | (*E*,*Z*)-$\Delta^{2,4}$-$C_{10}$ | H | Ac | *E. kansui* | [64] |
| | | | | *E. petiolata* | [74] |
| **57i** | $\Delta^{2,4}$-$C_{10}$ | H | Ac | *E. biglandulosa* Desf. | [208] |
| **57j** | $\Delta^{2,4,6}$-$C_{10}$ | H | Ac | *E. biglandulosa* | [208] |
| **57k** | Bz | H | Bz | *E. esula* | [145] |
| **57l** | P1 | H | Ac | *E. milii* | [79] |
| **57m** | H | P1 | Ac | *E. milii* | [79] |
| **57n** | P4 | H | Ac | *E. milii* | [79] |
| **57o** | P3 | H | Ac | *E. leuconeura* L. | [209] |
| | | | | *E. milii* | [103] |
| **57p** | P3 | Ac | H | *E. milii* | [103] |
| **56q** | P2 | H | Prop | *E. leuconeura* | [209] |
| **57r** | H | P2 | Ac | *E. milii* | [79] |
| **57s** | H | P3 | Ac | *E. milii* | [79] |

### 2.4.6 Ingenol 3,5,20-Triesters

Many compounds of this class contain anthranilate-based short peptides and come from *E. milii* and a few closely related species. Remarkably, all the statistical possibilities of the combination of three acyl groups may be present in the same plant, suggesting a randomized biogenetic strategy of acylation, in sharp contrast to glycerides, where a preference for a specific position exists depending upon the nature of the acyl group.

**Table 7** Naturally occurring ingenol 3,5,20-triesters

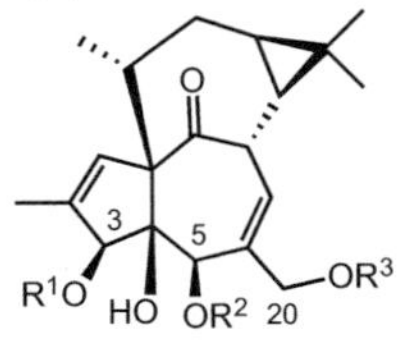

**58a–58o**

| Compound | $R^1$ | $R^2$ | $R^3$ | Source | Ref. |
|---|---|---|---|---|---|
| **58a** | Ac | Ac | $C_{14}$ | *E. broteri* | [71] |
| **58b** | Ang | Ac | Ac | *E. acrurensis* N.E. Br. | [210] |
| | | | | *E. canariensis* | [54] |
| | | | | *E. esula* | [73, 74] |
| | | | | *E. petiolata* | [73, 74] |
| **58c** | $C_{14}$ | Ac | Ac | *E. broteri* | [71] |
| **58d** | Ac | $C_{14}$ | Ac | *E. broteri* | [71] |
| **58e** | Ac | (*E,Z*)-$\Delta^{2,4}C_{14}$ | Ac | *E. petiolata* | [74] |
| **58f** | (*E,Z*)-$\Delta^{2,4}$-$C_{14}$ | Ac | Ac | *E. petiolata* | [74] |
| **58g** | P1 | Ac | Ang | *E. cornigera* | [160] |
| **58h** | P1 | Ang | Ac | *E. cornigera* | [160] |
| **58i** | Ac | P1 | Ang | *E. cornigera* | [160] |
| **58j** | Ac | P1 | Ac | *E. milii* | [79] |
| **58k** | Ac | Ang | P1 | *E. cornigera* | [160] |
| **58l** | Ang | Ac | P1 | *E. cornigera* | [160] |
| **58m** | Ang | P1 | Ac | *E. cornigera* | [160] |
| **58n** | Ac | P1 | Ac | *E. cornigera* | [160] |
| **58o** | P1 | Ac | Ac | *E. cornigera* | [160] |

### 2.4.7 13-Hydroxyingenol Esters

In this relatively small group of compounds, for which the first member was reported in 1974 from *E. kansui* Liou ex S.B. Ho the tertiary 13-hydroxy group is generally acylated, and the acylation profile shows the variation expected from the possibility of intramolecular migration of acyl groups from the 3- to the 5- and 20-hydroxy groups. It is therefore not clear what compounds are genuine natural products, and so they have consequently been grouped in one single class.

**Table 8** Naturally occurring 13-hydroxyingenol esters

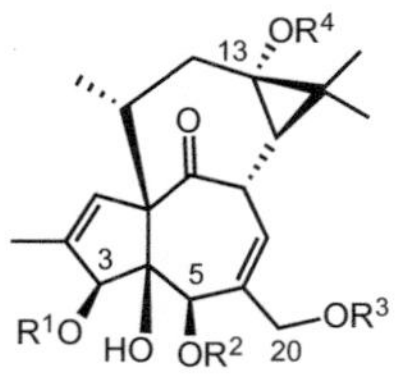

**59a–59o**

| Compound | $R^1$ | $R^2$ | $R^3$ | $R^4$ | Source | Ref. |
|---|---|---|---|---|---|---|
| **59a** | H | H | H | $C_{12}$ | *E. kansui* | [211] |
| **59b** | 2,3-Me-$C_4$ | H | H | $C_8$ | *E. esula* | [212] |
| **59c** | 2,3-Me-$C_4$ | H | H | $C_{10}$ | *E. cyparissias* | [213] |
| | | | | | *E. kansui* | [207, 214] |
| **59d** | 2,3-Me-$C_4$ | H | H | $C_{12}$ | *E. cyparissias* | [213] |
| | | | | | *E. kansui* | [207] |
| **59e** | 2,3-Me-$C_4$ | H | Ac | $C_{12}$ | *E. kansui* | [214] |
| **59f** | H | 2,3-Me-$C_4$ | H | $C_{12}$ | *E. cyparissias* | [213] |
| **59g** | H | H | $C_6$ | $C_{12}$ | *E. kansui* | [104] |
| **59h** | H | H | Ang | Ac | *E. cornigera* | [82] |
| **59i** | H | Ang | H | Ac | *E. cornigera* | [82] |
| **59j** | H | H | 2,3-Me-$C_4$ | $C_{12}$ | *E. cyparissias* | [213] |
| | | | | | *E. kansui* | [214] |
| **59k** | C10 | H | C6 | H | *E. esula* | [212] |
| **59l** | Bz | H | H | Bz | *E. esula* | [212] |
| **59m** | 2,3-Me-$C_4$ | Ac | $C_6$ | $C_{12}$ | *E. cornigera* | [82] |
| **59n** | Bz | H | H | $C_8$ | *E. esula* | [212] |
| **59o** | 2,3-Me-$C_4$ | Ac | Ac | C12 | *E. kansui* | [211] |

### 2.4.8 16-Hydroxyingenol Esters and 17-Hydroxyingenol Esters

In the early studies on ingenoids, hydroxylation at C-16 was established based exclusively on considerations of chemical shift differences with the parent polyol ingenol. Whenever rigorous 2D-NMR measurements were done, the hydroxylation site on the geminal methyls was, however, always located at C-17. Since the chemical shift considerations underlying the location of the ester group at C-16 also hold for their 17-isomers, the structure of the C-16 esters should be probably revised to their corresponding C-17 isomers. Indeed, the $^1$H NMR spectroscopic data of compounds from the two series are very similar, although the spectra of the 16-esters have not been fully assigned and a comparison of the whole NMR spectrum is therefore not possible. Interestingly, phorbol esters are indeed generally hydroxylated at the 16-methyl [3], and this might have biased the original structure assignment of the ingenol esters. Hydroxylation at C-16 of ingenol esters was found detrimental for both irritancy and tumor promotion [67].

**Table 9** Naturally occurring 16-hydroxyingenol esters

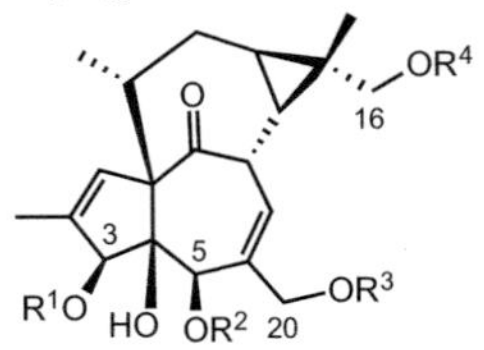


**60a–60g**

| Compound | $R^1$ | $R^2$ | $R^3$ | $R^4$ | Source | Ref. |
|---|---|---|---|---|---|---|
| **60a** | Ang | H | H | Bz | *E. canariensis* | [215] |
| **60b** | $\Delta^{2,4,6}$-$C_{10}$ | H | H | Ang | *E. hermentiana* | [105] |
| | | | | | *E. ingens* | [67] |
| **60c** | Ang | H | Ac | H | *E. hermentiana* | [105] |
| **60d** | Ang | H | Ac | Ac | *E. hermentiana* | [105] |
| **60e** | Ang | H | Ac | Bz | *E. cornigera* | [82] |
| **60f** | Ang | Ac | Ac | Ac | *E. hermentiana* | [105] |
| **60g** | Ac | H | Ang | Bz | *E. canariensis* | [215] |

**Table 10** Naturally occurring 17-hydroxyingenol esters

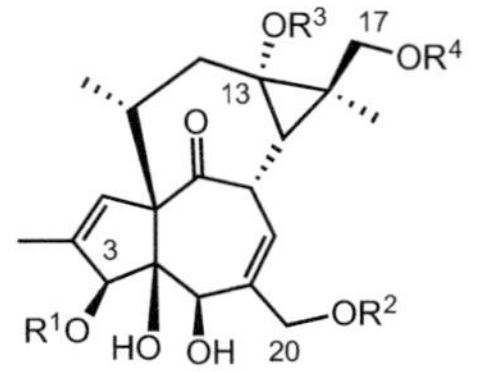


**62a–62l**

| Compound | $R^1$ | $R^2$ | $R^3$ | $R^4$ | Source | Ref. |
|---|---|---|---|---|---|---|
| **61a** | H | H | C16 | H | *E. quinquecostata* | [202] |
| **61b** | H | H | H | $\Delta^{2,4,6}$-C10 | *E. caducifolia* Haines | [216] |
| **61c** | Ang | H | H | $\Delta^{2,4,6}$-$C_{10}$ | *E. caducifolia* | [216] |
| **61d** | Ac | H | Ang | H | *E. caducifolia* | [216] |
| **61e** | Ang | H | H | Ac | *E. caducifolia* | [216] |
| **61f** | Ang | H | Ac | H | *E. caducifolia* | [216] |
| | | | | | *E. canariensis* | [54] |
| **61g** | Ang | H | H | Bz | *E. caducifolia* | [216] |
| | | | | | *E. canariensis* | [54] |
| **61h** | H | Ang | H | Bz | *E. canariensis* | [54] |
| **61i** | Ang | H | Ac | Bz | *E. canariensis* | [54] |
| | | | | | *E. caducifolia* | [216] |
| **61j** | Ang | Ac | Ac | Ac | *E. trigona* | [206] |
| | | | | | *E. kamerunica* | [217] |
| **61k** | Ac | Ang | H | Bz | *E. caducifolia* | [216] |
| **61l** | Ac | Ac | Ac | Ac | *E. kamerunica* | [217] |

Some data on the chemical behavior of these compounds have been reported. Thus, in basic conditions, hydrolytic removal of the secondary ester at C-3 is often easier than that of the neopentylic primary ester at C-16, while the opposite holds when the reaction is carried out under acidic conditions, so that selective hydrolyses are possible. Also worth mentioning is the hydroxylation of I3A to its 16-hydroxy derivative by cell cultures of barley (*Hordeum vulgare* L.) [81]. Interestingly, activity against melanoma cells was retained by hydroxylation of the 16-methyl group [82]. Furthermore in this case, however, the location of the hydroxylation needs confirmation.

### 2.4.9 13,17-Dihydroxyingenol Esters

Most compounds from this type were isolated from a single Chinese collection of *E. esula* L., but are, otherwise, very rare in spurges. What is also remarkable is the replacement of the angelate residue with its reduced version. Interestingly, other collections of *E. esula* gave instead less hydroxylated ingenol esters (see Sect. 4.4).

**Table 11** Naturally occurring 13,17-dihydroxyingenol esters

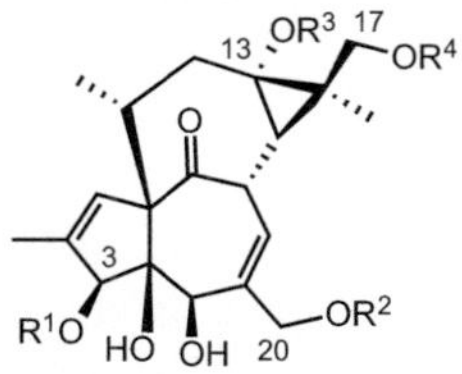


**62a–62m**

| Compound | $R^1$ | $R^2$ | $R^3$ | $R^4$ | Source | Ref. |
|---|---|---|---|---|---|---|
| **62a** | Bz | H | Ac | H | *E. segetalis* L. | [199] |
| **62b** | Ang | H | Ac | H | *E. segetalis* | [199] |
| **62c** | 2,3-Me-$C_4$ | H | 2,3-Me-$C_4$ | Bz | *E. esula* | [212] |
| **62d** | H | 2,3-Me-$C_4$ | 2,3-Me-$C_4$ | Bz | *E. esula* | [212] |
| **62e** | H | H | $C_8$ | Bz | *E. esula* | [212] |
| **62f** | Bz | H | $C_8$ | Bz | *E. esula* | [212] |
| **62g** | H | Bz | $C_8$ | Bz | *E. esula* | [212] |
| **62h** | 2,3-Me-$C_4$ | H | $C_8$ | Bz | *E. esula* | [212] |
| **62i** | H | 2,3-Me-$C_4$ | Bz | Bz | *E. esula* | [212] |
| **62j** | H | 2,3-Me-$C_4$ | C8 | Bz | *E. esula* | [212] |
| **62k** | Bz | H | Bz | Bz | *E. esula* | [212] |
| **62l** | 2,3-Me-$C_4$ | H | Bz | Bz | *E. esula* | [212] |
| **62m** | Bz | H | 2,3-Me-$C_4$ | Bz | *E. esula* | [212] |

### 2.4.10 13,19-Dihydroxyingenol Esters

Hydroxylation at the allylic C-19 methyl is very rare, and has so far been observed only in a single species (*E. cyparissias* L.). All compounds bear a 2,3-dimethylbutanoyl residue of unknown configuration at C-19.

**Table 12** Naturally occurring 13,19-dihydroxyingenol esters

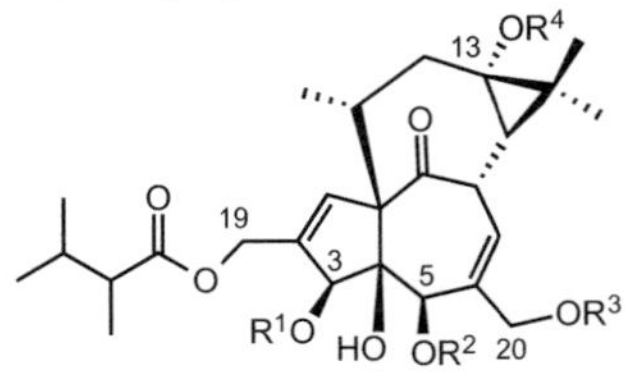


**63a–63c**

| Compound | R[1] | R[2] | R[3] | R[4] | Source | Ref. |
|---|---|---|---|---|---|---|
| **63a** | 2,3-Me-$C_4$ | H | H | 2,3-Me-$C_4$ | *E. cyparissias* | [213] |
| **63b** | H | 2,3-Me-$C_4$ | H | 2,3-Me-$C_4$ | *E. cyparissias* | [213] |
| **63c** | H | H | 2,3-Me-$C_4$ | 2,3-Me-$C_4$ | *E. cyparissias* | [213] |

### 2.4.11 Tetrahydroingenol Esters

Ingenoids of this type have so far only be isolated from *E. erythradenia* Boiss., a spurge endemic to Iran. These compounds are characterized by hydroxylation at C-7, C-11, and C-13, and represent a unique class of phorboids. Also remarkable is the esterification pattern, which includes nicotinic acid, a residue common in the *Euphorbia* macrocyclic diterpenoids, but uncommon in phorboids. Some spectroscopic features of these compounds are also unusual for ingenoids,

**Table 13** Naturally occurring tetrahydroingenol esters

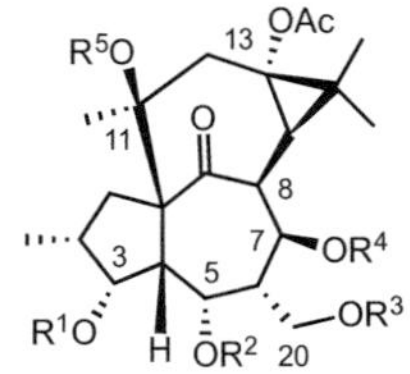


**64a–64d**

| Compound | R[1] | R[2] | R[3] | R[4] | R[5] | Source | Ref. |
|---|---|---|---|---|---|---|---|
| **64a** | But | Ang | Nic | H | H | *E. erythradenia* | [218] |
| **64b** | Prop | Ang | Nic | Ac | H | *E. erythradenia* | [218] |
| **64c** | Prop | Ac | Nic | Bz | H | *E. erythradenia* | [218] |
| **64d** | But | Ang | H | H | Nic | *E. erythradenia* | [218] |

especially the relatively upfield shift of C-10 ($\delta$ 48.6, vs. ca. $\delta$ 70 ppm in ingenol esters). The nature of the bridge was not clearly evaluated, and was suggested to be of the *out,out*-type. Further investigation on these compounds is surely warranted, owing to the marked deviation from the basic structural chemotype of the class also in terms of configuration of the C-3 and C-5 hydroxy groups, and in terms of deoxygenation at C-4.

### 2.4.12 20-Deoxyingenol Esters

This type of ingenoids was first reported in 1974 from *E. kansui* [83], and is of considerable interest because of the potent irritancy still exerted by its 3-esters, a surprising observation because the primary 20-hydroxy group was previously considered essential for the irritancy of ingenane derivatives. Thus, **65a**, the 20-deoxy derivative of I3A (**5**), was reported to be one order of magnitude more potent as a skin-irritant than **5** in the mouse ear erythema assay [57, 84], and even showed a more rapid onset of activity. The molecular targets involved in the potent irritancy induced by these compounds are unknown, and the activity might, indeed,

**Table 14** Naturally occurring 20-deoxyingenol esters

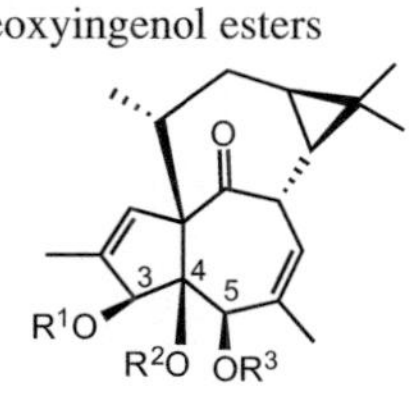


**65a–65l**

| Compound | $R^1$ | $R^2$ | $R^3$ | Source | Ref. |
|---|---|---|---|---|---|
| **65a** | Ang | H | H | *E. helioscopia* | [56] |
| | | | | *E. paralias* | [84] |
| | | | | *E. peplus* | [61] |
| **65b** | Ac | Ang | | *E. kansui* | [219] |
| **65c** | $C_6$ | H | H | *E. paralias* | [84] |
| **65d** | (*E,Z*)-$\Delta^{2,4}$-$C_{10}$ | H | H | *E. kansui* | [214] |
| **65e** | (*E,E*)-$\Delta^{2,4}$-$C_{10}$ | H | H | *E. kansui* | [214] |
| **65f** | $\Delta^{2,4,6}$-$C_{10}$ | H | H | *E. biglandulosa* | [208] |
| **65g** | H | H | $\Delta^{2,4,6}$-$C_{10}$ | *E. biglandulosa* | [208] |
| | | | | *E. palustris* | [220] |
| **65h** | Bz | H | H | *E. kansui* | [83] |
| **65i** | H | H | Ang | *E. peplus* | [61] |
| **65j** | H | H | Bz | *E. kansui* | [83] |
| | | | | *E. palustris* | [220] |
| **65k** | Bz | H | Bz | *E. palustris* | [220] |
| **65l** | H | Ac | Bz | *E. palustris* | [220] |

be independent from PKC. The Baran synthesis of ingenol (see Sect. 3.2) can provide an expeditious access to this type of compounds, and could make possible the clarification of this important issue.

### 2.4.13 16- and 17-Hydroxy-20-deoxyingenol Esters

The same considerations referred to above for the 16- and 17-esters of ingenols also hold for the corresponding esters of 20-deoxyingenol, namely, that the structure of the 16-hydroxy derivatives probably need revision, and location of the neopentyl

**Table 15** Naturally occurring 16-hydroxy-20-deoxyingenol esters

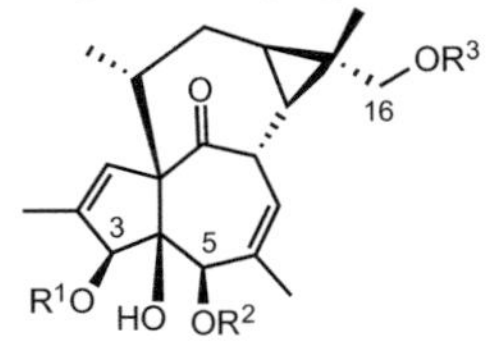


**66a–66c**

| Compound | $R^1$ | $R^2$ | $R^3$ | Source | Ref. |
|---|---|---|---|---|---|
| **66a** | Ang | H | Ac | *E. quadrialata* Pax | [220] |
| | | | | *E. hermentiana* | [105] |
| **66b** | Bz | H | Bz | *E. quadrialata* | [220] |
| **66c** | H | Bz | Bz | *E. quadrialata* | [220] |

**Table 16** Naturally occurring 17-hydroxy-20-deoxyingenol esters

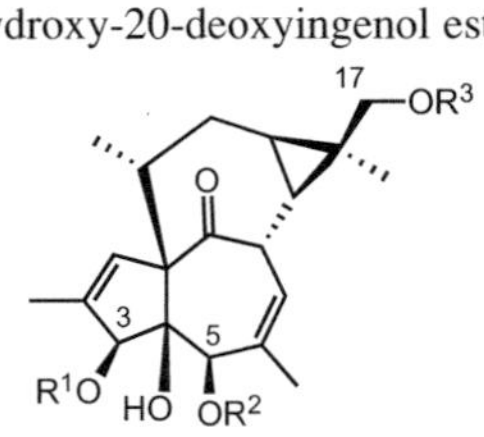


**67a–67i**

| Compound | $R^1$ | $R^2$ | $R^3$ | Source | Ref. |
|---|---|---|---|---|---|
| **67a** | Ang | H | Ac | *E. trigona* | [206] |
| **67b** | Ang | H | Ang | *E. esula* | [212] |
| **67c** | Bz | H | Bz | *E. esula* | [212] |
| **67d** | Ang | H | Bz | *E. esula* | [212] |
| **67e** | 2,3-Me-$C_4$ | H | Bz | *E. esula* | [212] |
| **67f** | H | Ang | Tigl | *E. neriifolia* L. | [221] |
| **67g** | Ac | Ang | H | *E. neriifolia* | L. [221] |
| **67h** | Bz | H | Bz | *E. quadrialata* | [220] |
| **67i** | H | Bz | Bz | *E. quadrialata* | [220] |

hydroxylated residue should be moved to C-17. In fact, the spectroscopic data of compounds **66a** and **66b** are very similar to those of the 17-hydroxyingenol esters **67a** and **67c**.

### 2.4.14 16- and 17-Hydroxy-20-deoxy-13-hydroxyingenol Esters

Esters of 13,16-dihydroxy-20-deoxyingenol have so far been isolated from a single plant source (*E. palustris* L.), and probably need revision to the corresponding derivatives of 13,17-dihydroxy-20-deoxyingenol derivatives.

**Table 17** Naturally occurring 13,16-dihydroxy-20-deoxyingenol esters

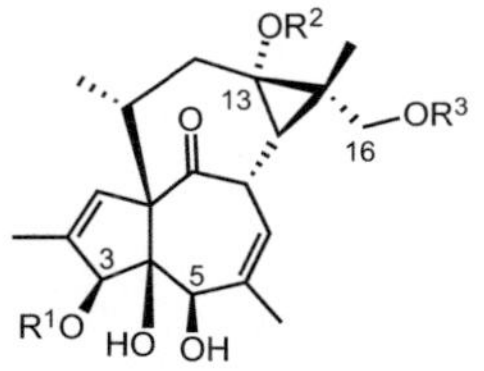


**68a–68d**

| Compound | $R^1$ | $R^2$ | $R^3$ | Source | Ref. |
|---|---|---|---|---|---|
| **68a** | Bz | H | Bz | *E. palustris* | [220] |
| **68b** | Bz | Bz | Bz | *E. palustris* | [220] |
| **68c** | Bz | PhAc | Bz | *E. palustris* | [220] |
| **68d** | Bz | Ac | Bz | *E. palustris* | [220] |

**Table 18** Naturally occurring 13,17-dihydroxy-20-deoxyingenol esters

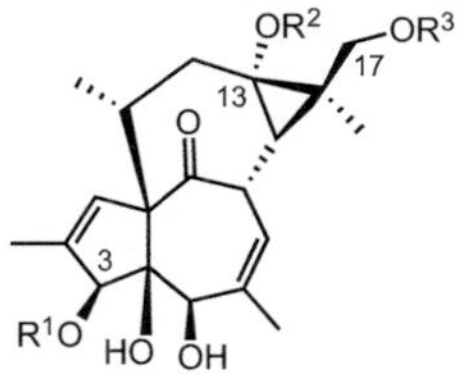


**69a–69c**

| Compound | $R^1$ | $R^2$ | $R^3$ | Source | Ref. |
|---|---|---|---|---|---|
| **69a** | 2,3-Me-$C_4$ | 2,3-Me-$C_4$ | Bz | *E. esula* | [212] |
| **69b** | Bz | $C_8$ | Bz | *E. esula* | [212] |
| **69c** | 2,3-Me-$C_4$ | Bz | Bz | *E. esula* | [212] |

### 2.4.15 5-Deoxyingenol Esters and Miscellaneous Ingenoids

The major source of these compounds is Chechum (*Mabea excelsa* Standl. & Steyerm.), a Central American highly irritant and toxic plant that was related to the development of a debilitating inflammatory condition experienced by British soldiers in Belize [85]. The toxic principle of the plant was named "Chechum toxin", and turned out to be a complex mixture of 5-deoxy-16-hydroxyingenol esters that can potently activate PKC [85]. There is also a report on the occurrence of a 5-deoxyingenol derivative in the latex of *E. canariensis* L. [54]. The presence of 5-deoxyingenol esters has been suggested as being typical of *Euphorbia* species from the section *Tithymalus* [48].

The 5,20-acetonides of some 3-esters of ingenol (tetradecanoate, 3-*O*-(2*E*,4*Z*)-tetradecadienoate, palmitate) were reported as natural products from *E. petiolata* Banks & Sol. [74] and *E. eubracteolata* Hayata [86], but are most probably isolation artifacts since acetone was used in their extraction and/or purification.

**Table 19** Naturally occurring 5-deoxyingenol esters

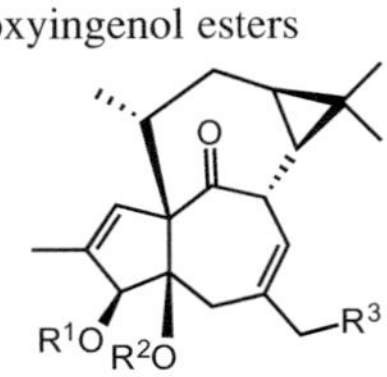


**70a–70g**

| Compound | $R^1$ | $R^2$ | $R^3$ | Source | Ref. |
|---|---|---|---|---|---|
| **70a** | H | H | O-$C_5$ | *Mabea excelsa* | [80] |
| **70b** | H | H | O-$C_{14}$ | *Mabea excelsa* | [80] |
| **70c** | H | H | O-(Z)-$\Delta^9C_{18}$ | *Mabea excelsa* | [80] |
| **70d** | H | H | O-12OAc-$\Delta^9$-$C_{18}$ | *Mabea excelsa* | [80] |
| **70e** | H | H | O-$\Delta^{11}$-$C_{20}$ | *Mabea excelsa* | [80] |
| **70f** | H | H | O-$\Delta^{11,14}$-$C_{20}$ | *Mabea excelsa* | [80] |
| **70g** | Ang | Ac | H | *E. canariensis* | [54] |

## 2.5 *Ecology*

The skin-irritant properties of most ingenoids makes it reasonable that they play an ecological role as feeding deterrents, complementing the physical protection of spines in the succulent species of the genus *Euphorbia*. This strategy is successful with cattle, but goat and sheep can feed on spurges, while camels, dromedaries and rhinos can also consume the thorny succulent species with impunity [87]. The physiological basis for the insensitivity to oral irritation by phorboids observed in these animals is unknown, but excretion in milk represents a major detoxification route in lactating animals, making the milk and the dairy products derived therefrom unsuitable for human consumption [87]. Ingenoids often co-occur with large amounts of macrocyclic diterpenoid esters [88, 89]. These, by inhibiting transport proteins like P-gp that are overexpressed in lactating mammary tissues [88, 89] could interfere with the transfer of diterpenoids from plasma to milk, potentially exacerbating the toxicity of spurges (see Sect. 4.4).

Also the larvae of some moths can feed with impunity on *Euphorbia* species, accumulating toxic phorboids in their body and carrying them into adult life as a protection against predators [90]. The hawk moth (*Hyles euphorbiae*) is capable of accumulating ingenol esters [91], and was even considered for the biological control of the leafy spurge (*Euphorbia esula*), a highly invasive species in the North American plains (see Sects. 4.2.4 and 4.4.2).

# 3 Chemistry

## *3.1 Reactivity*

Ingenol, the parent polyol of ingenoids, is unstable to basic conditions as well as to atmospheric oxidation, but is relatively stable in mildly acidic media (pH 3–5), undergoing, however, skeletal rearrangements at more acidic values of pH [92]. No specific degradation product of treatment of ingenol with bases has been characterized, but the instability in these conditions might be related to the presence of the aldol motif (a carbonyl at C-9 and a hydroxy group at C-4), which makes possible retro-aldol equilibration at low pH values. This, rather than eventually leading to epimerization at C-4 as observed for phorbol [7], triggers a massive degradation process, presumably due to the instability of the bridged polycyclic system of ingenol. Quite remarkably, the 4,5-carbonate of synthetic 20-deoxyingenol (**71**) can, however, be deprotected to the parent polyol (**72**) under strong basic conditions (1 h stirring in 1:1 THF-10% NaOH) [16] (Fig. 13).

Reversible acyl migration to the primary 20-hydroxy group takes place in 3-acyl ingenol derivatives, scrambling the acylation pattern and potentially dramatically affecting the biological profile. These changes might underlie the practice of

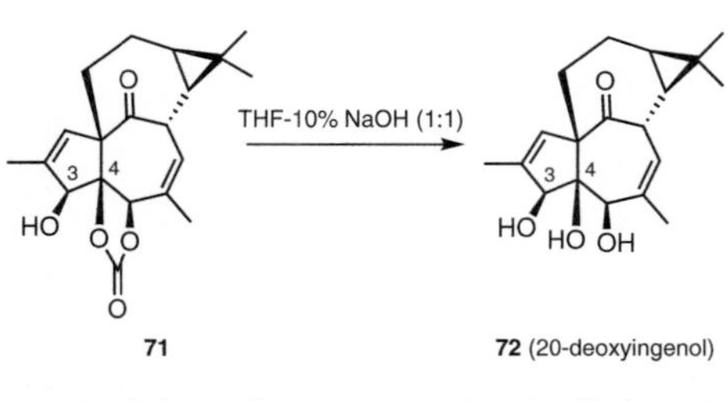


**71** **72** (20-deoxyingenol)

**Fig. 13** Successful removal of the carbonate protection in 20-deoxyingenol carbonate under strongly basic conditions. (The reaction was part of a telescoped three-step process that gave an overall yield of 43%) [17]

vinegar baking associated with the medicinal use of certain plants like Kansui (*E. kansui*), a celebrated east Asian medicine for edema and asthma but also a rich source of ingenol esters [93].

Much data on the reactivity of ingenol has been obtained in the course of studies on its total synthesis rather than by modification of the natural polyol, but the development of ingenol mebutate (I3A, **5**) as a drug has rekindled systematic studies on the reactivity of the natural product.

### 3.1.1 Epimerization and Isomerization

Interest in the epimerization α- to the C-9 carbonyl was spurred by the availability of the totally synthetic C-8 epimerized analog **9**, potentially amenable of conversion to the natural product [18]. However, the epimerization to **73** never materialized, presumably because of its counter-thermodynamic trait, while ingenol is rapidly degraded in basic media, in accordance with the presence of a sugar-like string of adjacent hydroxy groups susceptible to retro-aldolization. On the other hand, the 5,20-acetonide **52** was converted in good yield into iso-ingenol 5,20-acetonide (**75**) by refluxing with NaH in anhydrous THF (Fig. 14) [20]. Surprisingly,

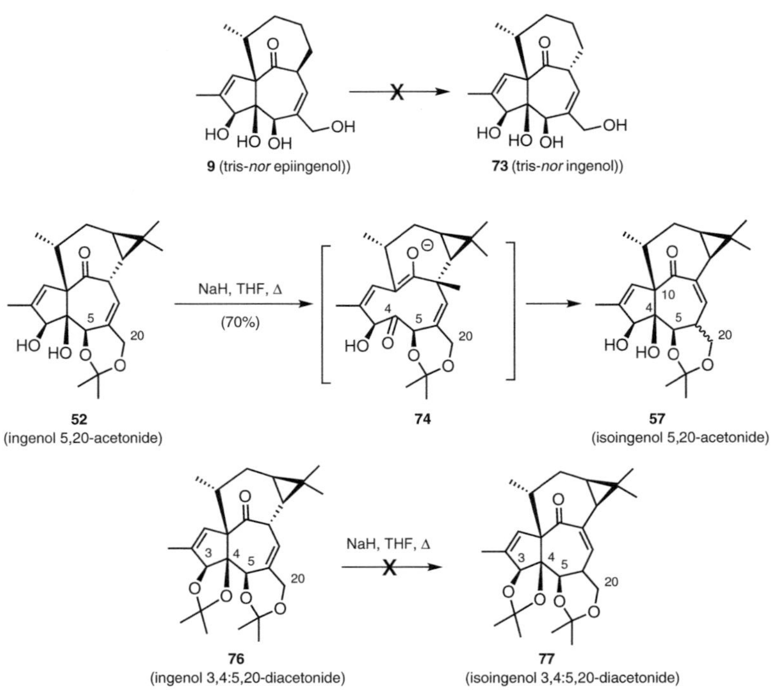


**Fig. 14** Reaction of ingenoids with bases

the 3,4:5,20-bis acetonide **76** was stable under these conditions, as well as under a variety of other basic treatments used to epimerize carbonyl derivatives [41]. This observation suggests that epimerization takes place after retro-aldolization of **52** to **74**, and is next followed by aldolization. If so, the configuration of isoingenol at C-4 and C-10 should be confirmed. Esterification of isoingenol (**10**) led to 3-esters substantially devoid of irritancy [20], an observation reminiscent of the effect of epimerization at C-4 in phorbol esters [7]. The isomerization of ingenol to iso-ingenol is surprising, owing to the generation of an anti-Bredt, yet "isolable" olefin [94]. Apparently, the partial pyramidalization is compensated by a reduction of angular strain in the undecane cyclic core, but the generation of an anti-Bredt olefin by isomerization of a non-bridgehead double bond is nevertheless unusual, although not unprecedented [95].

### 3.1.2 Skeletal Rearrangements

On account of its considerable angular strain, the bridged tetracyclic skeleton of ingenol can rearrange to more stable polycyclic skeletons devoid of the transannular bridge. The first rearrangement of this type was observed with the (9*R*) reduction product of ingenol diacetonide (**78**), which, upon treatment with mesyl chloride in pyridine, afforded the trienic tigliane **82** [41]. The mechanism of the reaction involves activation of the 9-hydroxy group by mesylation, followed by its diplacement triggered by the anionotropic migration of C-11 and, eventually, by proton loss. The concerted nature of the rearrangement, which involves the back-side displacement of the mesylate, was confirmed by the results observed with the (9*S*) epimer **83**, where migration of C-4 and formation of the isomeric naphthalenyl diene **86** were observed instead [41] (Fig. 15).

A more direct rearrangement was discovered serendipitously while investigating the chemoselective deacylation of ingenol polyesters. Thus, overnight treatment of ingenol and its esters in $10^{-2}$ *N* $HClO_4$ in methanol triggered a retro-pinacol rearrangement initiated by protonation of the 9-carbonyl and terminated in a vinylogous fashion by stereoselective methanol trapping of a C-2 cation [92] (Fig. 16). The reaction is interesting since it provides access to uniquely functionalized ingenane derivatives, and, apart from ingenol esters, it could also be carried out on the native polyol itself. Its facile nature means that the acidic deprotection of ingenol acetonides, important synthetic intermediates, should be carefully controlled in terms of conditions and duration.

Remarkably, the opposite rearrangement, the formation of a C-9 oxo-ingenane from a C-10 hydroxylated tigliane, was critical in the Baran synthesis of ingenol (Fig. 17) [16]. This fundamentally counter-thermodynamic transformation could be steered into kinetic control and the generation of the ingenane derivative **90** by judicious choice of the reaction conditions (temperature, Lewis acid activation, quenching) and by exploiting the pre-organizing effect of carbonate protection of

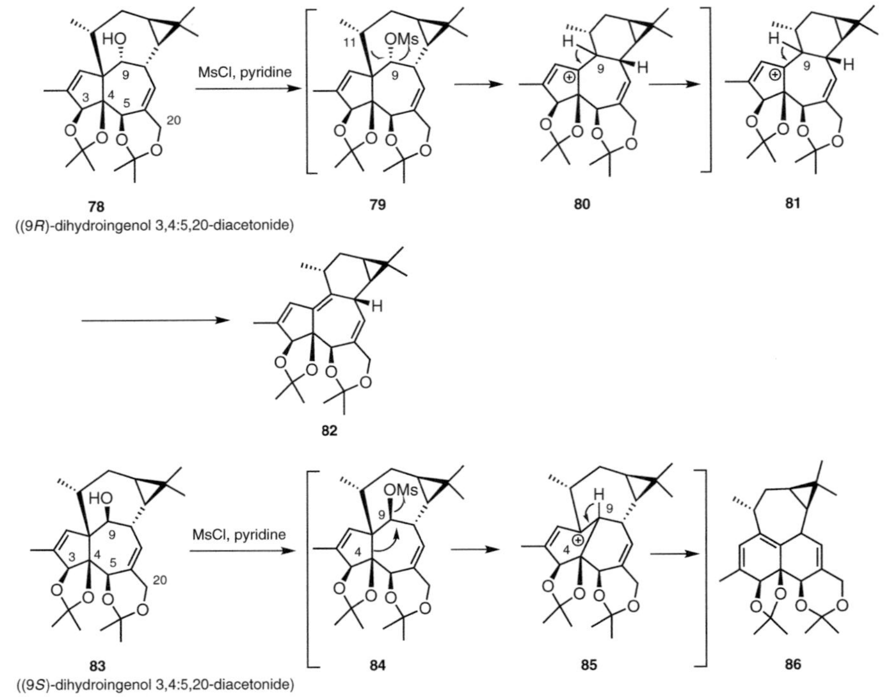


**Fig. 15** Rearrangement of C-9 epimeric dihydroingenol esters

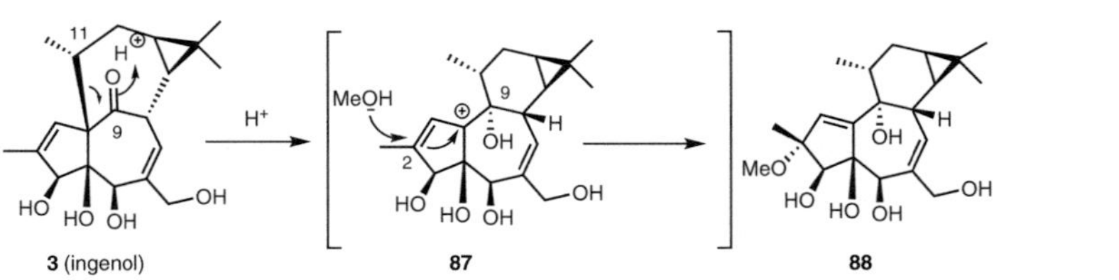


**Fig. 16** Acidic rearrangement of ingenol to a tigliane derivative

the 4,5-diol group, with silylation of the C-10 hydroxy group being also critical to avoid oxygen trapping of the C-10 cation **88**. DFT calculations were used to rationalize some of these observations. Thus, the anionotropic migration of C-11 to C-10 in the vinylogous pinacol rearrangement is favored by silylation of the C-9 hydroxy group because of the possibility of hyperconjugative silyl β-stabilization of the resulting oxonium ion [96]. A plausible mimic of this process could be associated to Brønsted base activation of the 10-hydroxy group, for instance, by the ω-amino group of a lysine or by the basic imidazole nitrogen of a terpenoid cyclase. In this way, the positive charge on the oxonium ion generated by the anionotropic

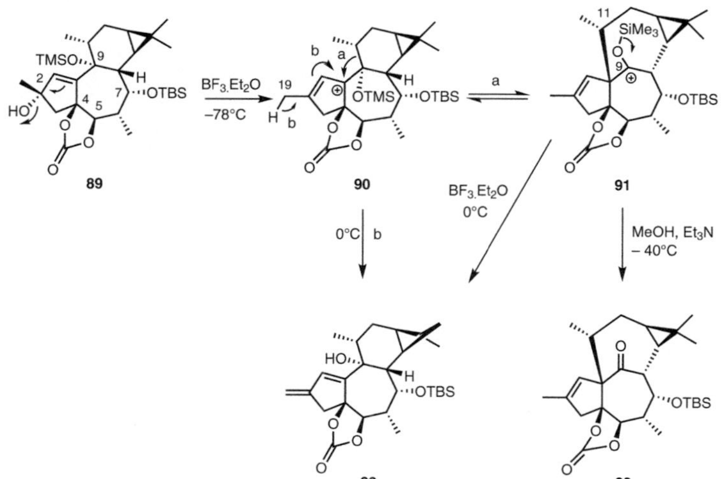


**Fig. 17** The tigliane-to-ingenane rearrangement involved in the Baran synthesis of ingenol [16]

migration of the carbon-carbon bond could be stabilized by proton transfer to the Brønsted base, *overall turning an oxonium ion in a more stable ammonium ion,* an interesting and potentially general concept for the mechanism of activity of isoprenoid cyclases. Calculations also highlighted the stabilizing effect of oxygenated functions on ring A to favor the generation of an allyl cation by Lewis acid-promoted activation of the C-2 hydroxy group, presumably because of the associated inductive effect that tends to move the positive charge on the adjacent ring. The calculations showed that in the rearrangement of **89** to **92**, the effects of β-silyl stabilization, the presence of oxygenated functions on ring A, and the pre-organization effect of the carbonate, all combine to make the vinylogous pinacol rearrangement of the allylic cation on ring A favored in terms of both thermodynamics and in kinetics. On the other hand, the rearrangement is fundamentally reversible, and quenching at 0°C produced the thermodynamically favored exomethylene tigliane olefin **93**, while treatment of the ingenane **92** with $BF_3{\cdot}Et_2O$ at 0°C afforded the same tigliane olefin. Low-temperature quenching with methanol-triethylamine at −40°C eventually proved optimal to isolate the ingenane product of vinylogous pinacol rearrangement [16].

A different type of rearrangement of the bridged system was discovered in 13-hydroxyingenol esters by Hirata [97]. Since the 13-ester group could not be removed hydrolytically, a reductive removal with an excess $LiAlH_4$ was attempted. This triggered, via the cyclopropylogous version of a 1,3-sigmatropic rearrangement, the formation of a 12,13-enol ester (**95**), which then, after reductive cleavage, collapsed via an intramolecular aldol reaction, eventually affording the cyclobutane derivative **96**, characterized as a triacetate. This 9,12-cycloingenoid could be further rearranged after reduction of the C-13 carbonyl. Activation of

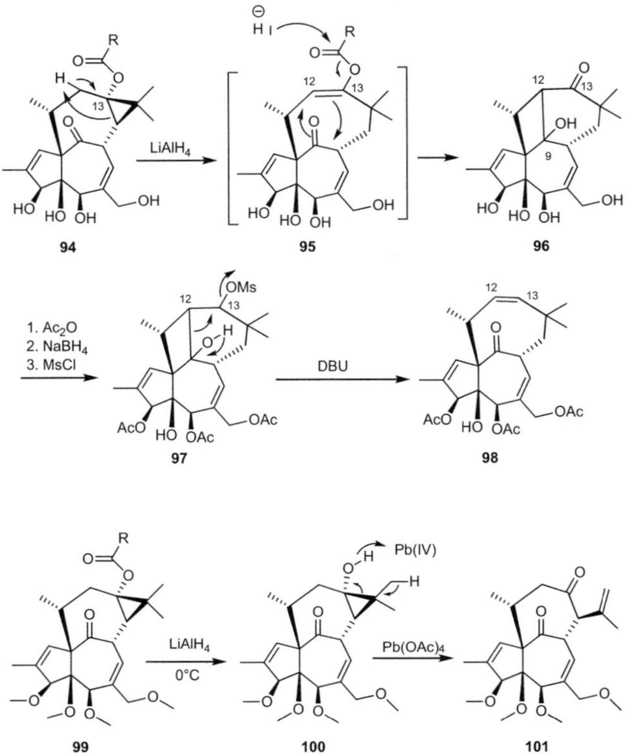


**Fig. 18** Rearrangement of 13-hydroxyingenol

the resulting alcohol by mesylation, and treatment with diazabicycloundecane (DBU) triggered a Grob fragmentation that regenerated the 9-keto group and expansion of ring C to an eight-membered ring, eventually affording the D-*seco*-derivative **96**. With only a moderate excess of $LiALH_4$ and at low temperature, the permethylated derivative **99** afforded the expected C-13 alcohol, which, when treated with lead tetraacetate, underwent oxidative opening of the cyclopropane ring to generate the isopropenyldiketone **101** (Fig. 18).

### 3.1.3 Reduction

Reduction of the C-9 ketone carbonyl was investigated under several reaction conditions using the diacetonide **76** as a substrate since the natural polyol proved unstable to basic conditions. In general, this hindered keto group proved resistant to mild reducing agents, like $NaBH_4$, and stronger reducing agents were necessary [41]. The configuration of the resulting C-9 hydroxy group was found to be

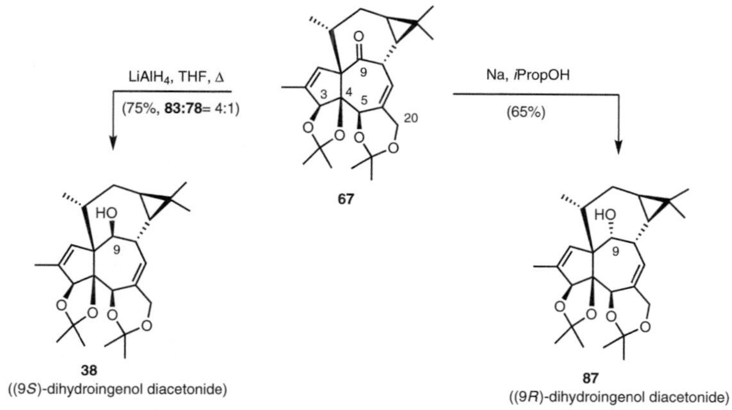


**Fig. 19** Reduction of the C-9 carbonyl of ingenol diacetonide (**76**)

dependent on the reducing agent, with stereo-complementary results being obtained on hydride ($LiAlH_4$) and dissolving metal (Na/isopropanol, Bouveault-Blanc reduction) reductions (Fig. 19). The C-9 carbonyl was also not reduced by sodium amalgam (Na-Hg), and the chemoselective desulfurization of an ingenol 20-sulfone is featured in the final steps of the Wood total synthesis of ingenol [98]. Oxidation of the allylic hydroxy group at C-3 in ingenol 5,20-acetonide led to a 3-dehydro derivative that could not be reduced back to ingenol with hydrides or boranes [99]. The difficulty of reducing the C-9 carbonyl is predictable from its orientation on the concave and sterically encumbered $\alpha$-face of the bridged system, but the situation with the C-3 carbonyl is unexpected, since this group can be easily reduced in phorbol. On the other hand, reduction of a C-3 carbonyl is featured in advanced ingenane intermediates in the total syntheses of ingenol by Winkler [19] and Wood [98].

Hydrogenation of ingenol triacetate led to the exclusive non- stereoselective saturation of the $\Delta^6$ double bond, and was accompanied by hydrogenolysis of the primary 20-hydroxy group, with the $\Delta^1$ double bond being unaffected [83].

### 3.1.4 Oxidation

The oxidation of the allylic 3-hydroxylated group was investigated in ingenol 5,20-acetonide (**52**). $MnO_2$ afforded a mixture of unidentified compounds, but the reaction succeeded with Cr(VI)-based reagents. The best results (80% yield) [99] were obtained with the Snatzke reagent ($CrO_3$-$H_2SO_4$ in DMF) [100].

The allylic oxidation of a 20-deoxyingenol (**103**) is a critical step in two total syntheses of ingenol, and is indeed the final step of the Baran synthesis [16], where it was efficiently carried out in terms of regiochemistry (exclusive formation of the primary 20-alcohol) and chemoselectivity (avoidance of over-oxidation to a carbonyl derivative) under the Shibuya conditions ($SeO_2$-formic acid) [101].

Epoxidation of I3A (**5**) occurred chemo- and stereoselectively on the $\beta$-face of the $\Delta^6$ double bond even with an excess of *meta*-chloroperbenzoic acid (MCPBA), affording the epoxide **104** as the only reaction product [102]. The oxidation of the allylic 20-hydroxy group of ingenol 3,5-dibenzoate (**57j**) failed with a variety of reagents ($MnO_2$, PCC, PDC). Using $CrO_3$ in the presence of 18-crown-6, the 20-enal **105** could be obtained, but as an inseparable mixture with the exomethylene enone **106** resulting from the [3]-sigmatropic rearrangement of the intermediate chromate ester [92]. A similar rearrangement occurred also during the reaction of an ingenyl 20-chloride with $NaN_3$, affording via the 20-azide **109**, the rearranged azide **110** as the only reaction product [92] (Figs. 20 and 21).

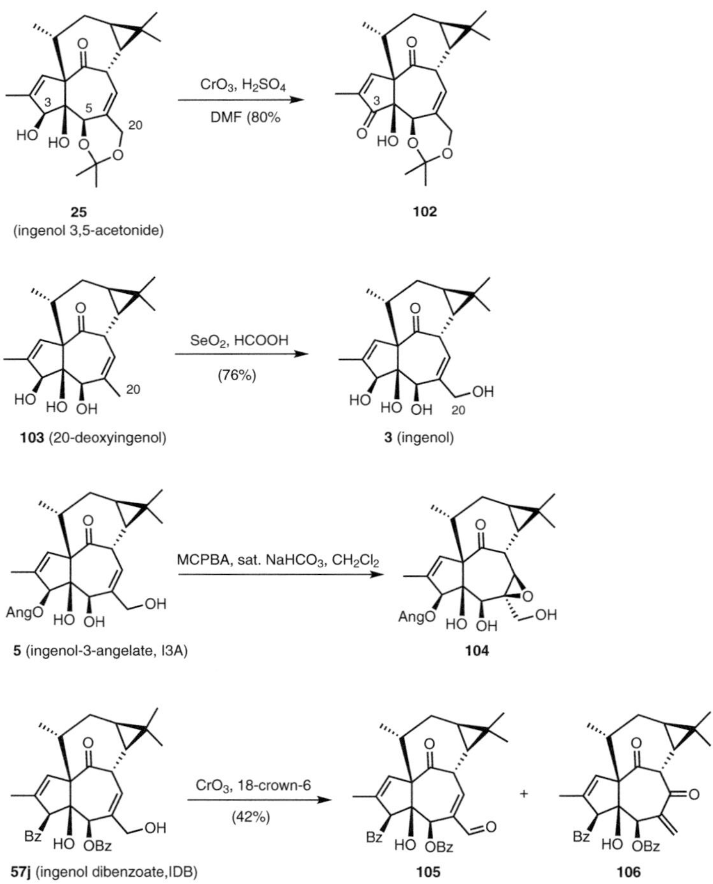


**Fig. 20** Oxidative modifications of ingenoids

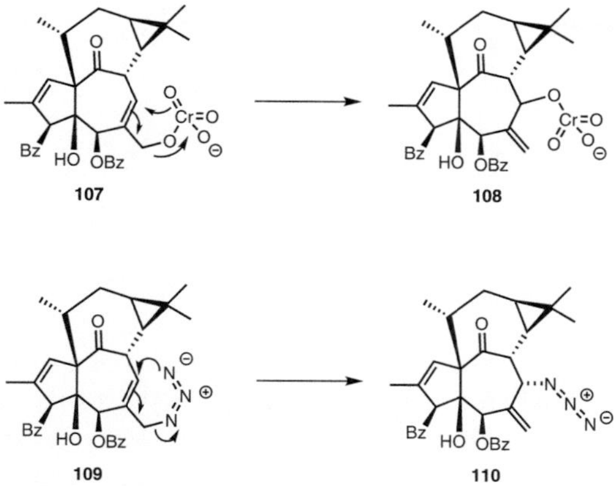


Fig. 21 [3.3] Sigmatropic rearrangement of ingenoids derivatized at C-20

### 3.1.5 Etherification and Acetalization

The primary 20-hydroxy group of ingenol (**3**) can be chemoselectively silylated and tritylated with, respectively, TBDMS-Cl (*tert*-butyldimethylsilyl chloride)-imidazole in DMF or trityl chloride in pyridine [92]. Removal of these protecting groups under controlled acidic conditions did not trigger acyl rearrangement from 3- and 5-esters to the 20-hydroxy group position [92]. Conversely, removal of the 20-silyl protection by fluoridolysis triggered acyl migration [92]. Reaction of ingenol with acetone in the presence of acids affords chemoselectively the seven-membered 5,20-acetonide **52** and only traces of the regio-isomeric 3,4-acetonide (**111**) [99]. The acetonide **52** is a stable and crystalline compound, and is a suitable storage form for the natural polyol, which is otherwise rapidly degraded in the solid state. Reaction of ingenol or its 5,20-acetonide (**52**) with the more reactive acetalizating agent isopropenyl methyl ether affords a 3,4:5,20-bis acetonide (**76**) [99]. The protection of the hydroxy groups with carboinyldimidazole gave similar results, with easier formation of the 5,20-carbonate compared to the 3,4-carbonate [103] (Fig. 22).

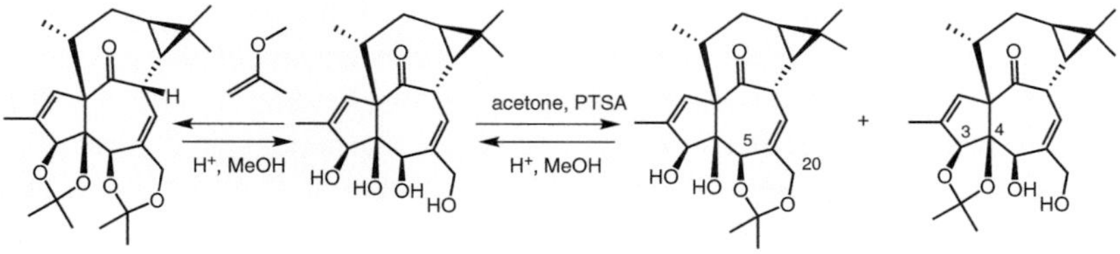


**76** (ingenol diacetonide)
**3** (ingenol)
**52** (ingenol 5,20-acetonide) (73%)
**111** (ingenol 3,4-acetonide) (4.5%)

**Fig. 22** Acetalization-deacetalization of ingenol

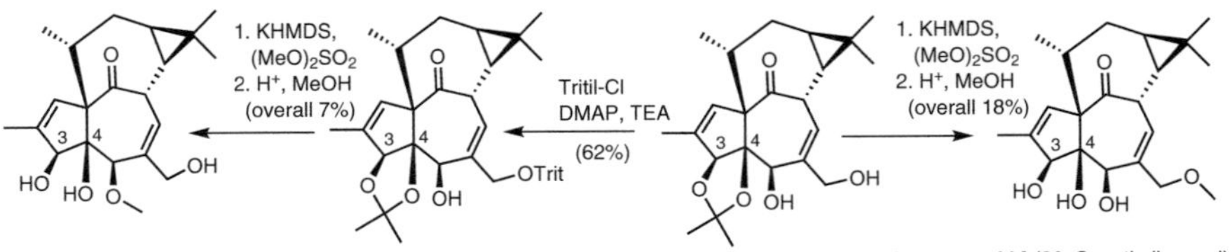


**Fig. 23** Synthesis of ingenol 5- and 20-methyl ethers **113** and **114**

The chemoselectivity of the acetalization (formation of a seven-membered ring rather than a five-membered ring) is probably related to the strain induced by formation of an acetonide involving the tertiary hydroxy group at the A,B-ring junction. The acetonides could be hydrolyzed in acidic methanol [0.34% $HClO_4$ (70% *w*/*w*) in methanol] [92]. Starting from the minor 3,4-acetonide **111** formed as a side-product in the acetalization reaction, the primary hydroxy group could be selectively methylated by treatment with dimethyl sulfate and KHMDS. Methylation of the 5-hydroxy group could be achieved in the same conditions after protection of the 20-hydroxy group as trityl ether [102] (Fig. 23).

### 3.1.6 Hydrolysis, Acylation, and Acyl Rearrangements

Acyl groups can be easily removed from the 3-, 5- ,16-, and 20-hydroxy groups by basic hydrolysis, while hydrolysis of the 13-ester unit is more difficult and much slower [104]. This observation is in sharp contrast to what is observed for phorbol esters, where the tertiary 13-hydroxy group can be easily deacylated (as well as re-esterified) [7]. The difficulties encountered in the hydrolysis of 13-acyl ingenols are not uncommon for hindered tertiary hydroxy groups, while the unusual reactivity of the tertiary 13-hydroxy functionalities of phorbol might be mediated by acyl migration to the adjacent secondary C-12 hydroxy group.

Acyl groups at the primary and allylic 20-hydroxy group can be selectively removed by solvolysis in acidic methanol [105]. The rate of the reaction is critically dependent on the size of the ester group, sharply decreasing with the growing length of the acyl group. Mild basic conditions can remove the ester group at the primary 20-hydroxy group, then promoting next the migration of a 3-acyl group to the secondary 5-hydroxy unit, and next to the primary 20-hydroxy group, affording mixtures dominated by the 20-ester (Fig. 24). This rearrangement plagued the early studies on the structural elucidation of ingenol esters, leading to a wrong assignment of the acylation pattern. Acyl migration is indeed a major stability issue for all ingenol 3-esters having free hydroxylated substituents at C-3 and C-20, and is reversible, as shown by the observation that acetylation has been reported to afford the same 5,20-diacetate from both ingenol 3-ester and ingenol 20-esters [64].

The primary 20-hydroxy group is only slightly more reactive than the two secondary hydroxy groups, but could be chemoselectively esterified with the Steglich protocol at low temperature [92], as exemplified by the synthesis of

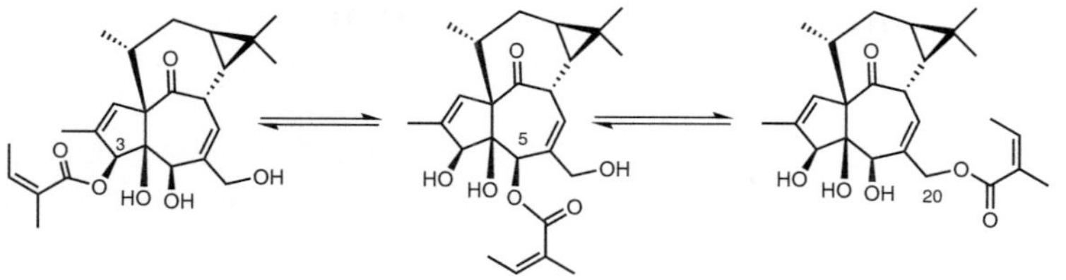


**5** (ingenol-3-angelate, I3A) (5%)
**115** (ingenol-5-angelate) (5%)
**116** (ingenol-20-angelate) (90%)

**Fig. 24** Acyl rearrangement of 3-acyl ingenol derivatives as exemplified by the equilibrium mixture of ingenol monoangelates at pH 7.4 (water-DMSO) and room temperature

ingenol-20-benzoate by treatment of ingenol in the Steglich esterification conditions [benzoic acid, 1-ethyl-3-(3-dimethylaminopropylcarbodiimide (EDC) and 4-dimethylaminopyridine (DMAP), 41% yield] [92]. The modest difference of reactivity between the two secondary and the tertiary hydroxylated units of ingenol is reminiscent of the situation in sugars, and is presumably due to hydrogen-bonding network effects. Esterification of the 3-hydroxy could be accomplished only after protection of the 20-hydroxy, for instance by treatment of ingenol 5,20-acetonide with an acid in the presence of a carbodiimide and a hypernucleophilic species like DMAP (Steglich esterification), or by treatment with an acyl chloride in the presence of pyridine and DMAP. Acidic deprotection can next be controlled, to avoid acyl rearrangement. The esterification of ingenol with angelic acid is the key step in the semisynthesis of Picato®, and has been thoroughly investigated [106]. The reaction is plagued by two problems, one general to ingenol 3-esters, and the other specifically associated with angelic acid. 3-Acylingenol derivatives are unstable toward acyl rearrangement, which eventually affords the corresponding and more stable 20-acyl isomers via transient 5-acyl derivatives (Fig. 24). At equilibrium, a 1:1:18 mixture of 20-, 3-, and 5-esters is generated from I3A. The acyl rearrangement is promoted by acids and bases, with complementary mechanisms, namely, ester carbonyl activation in acidic medium and alcoholate formation under basic conditions. Stability peaks at mildly acidic pH values, around 3–4, and rapidly declines outside this range of values, and is faster in basic media than in acidic ones. Stability is also solvent-dependent, being favored by aprotic media, and this issue been the subject of proprietary research [107]. Picato® is indeed a gel with a limited shelf life, and an analogous problem of acyl migration, for instance, once plagued the development of topical 17-esters of corticosteroids [108]. The stability of ingenol-3-acylates was investigated as a water emulsion at pH 7.4, measuring recovery after 16 h. Recovery for ingenol mebutate (I3A) was 60%, and the rate of the acyl migration was dependent from the acidity and the bulkiness of the acyl group. With acids stronger than angelic acid ($pK_a = 4.3$) the acyl rearrangement was faster, while steric hindrance had a stabilizing effect [106]. Thus, replacement of angelic acid with its *nor*-analog (*E*)-crotonic acid led to complete degradation, while addition of an extra methyl increased recovery to 75%. Aliphatic acids were less stable due to their higher acidity, while the recovery of the benzoate was only 30% presumably on account of the lack of an α-methyl, as suggested by the increased stabilizing effect of *ortho*-substitution [102, 106, 109].

The second issue is the facile isomerization of angelic acid to tiglic acid upon activation of the carboxylate group. Angeloylation is a notorious synthetic problem because of the unreactivity of angelic anhydride (**117**) and its propensity to undergo isomerization to tiglic acid by reversible Michael addition with the nucleophilic agents used to accelerate acyl substitution reactions (pyridine, DMAP). Given the hydrolytic stability of angelic anhydride, a compound that can be purified by column chromatography, it is not unconceivable that nucleophilic alcohol attack only occurs after Michael addition, which removes the transmission of the electron-donating inductive effects of the methyl and the mesomeric electron-donation of the double bond to the carbonyl of the anhydride (Fig. 25). Various literature strategies were evaluated to overcome this difficulty, and, eventually, the best results were obtained with LiHMDS (lithium hexamethyldisilazide). The reaction can be carried out at room temperature in THF, affording in ten minutes a complete conversion of ingenol 5,20-acetonide to its angeloyl ester, with a very rewarding 86% isolation yield. The corresponding sodium and potassium amides gave lower yields, suggesting that lithium, on account of a higher oxyphilicity, can activate the anhydride by behaving as a Lewis acid. Therefore, LiHMDS co-activates in a complementary fashion both the 3-hydroxylated site of ingenol 5,20-acetonide, by acting as a Brønsted base, and the acyl carbonyl of angelic anhydride, by behaving as a Lewis acid. Removal of the acetonide protection was uneventful, and I3A (**5**) was eventually obtained in 62% yield (Fig. 26).

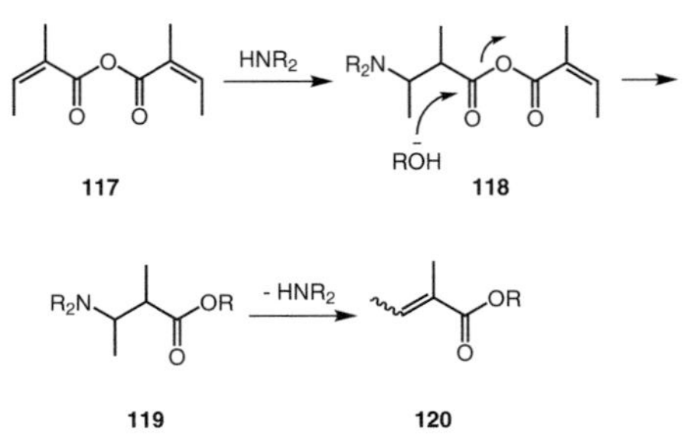


**Fig. 25** Possible mechanism of the angelate to tigliate isomerization during acylation reactions

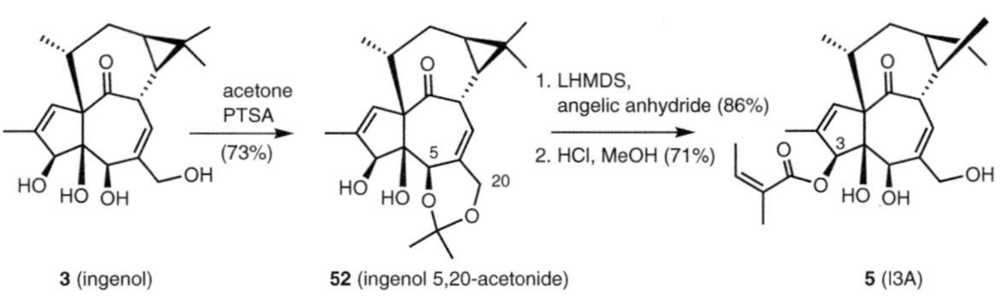


**Fig. 26** Semisynthesis of ingenol mebutate

### 3.1.7 Other Functional Group Modifications

The two double bonds of ingenol show a different type of reactivity, and the ring B $\Delta^6$-double bond of 3-acyl ingenols can be reacted selectively with electrophilic reagents like MCPBA [102]. Substitution reactions at the 5-hydroxy group substituent tend to proceed in a vinylogous fashion, and 3-benzoyl-20 trityl ingenol and 20-acetyl I3A were shown to react with DAST (diethylaminosulfur trifluoride) to give the corresponding $\Delta^5$-7$\alpha$-fluorides, as observed for the 20-azide (Fig. 21 [102]).

The primary 20-hydroxy unit of ingenol 3,5-dibenzoate (**121**) may be replaced by a fluorine by reaction with DAST at low temperature, while, using Mitsunobu-type chemistry, the 20-hydroxy group of ingenol can be substituted with an amino group via the corresponding 20-phthalimide (**123**) (Fig. 27) [92]. 20-Deoxy-20-aminoingenol (**124**) was not found to be stable, but can be trapped as an amide or an urea by acylation or reaction with isocyanates, respectively [92].

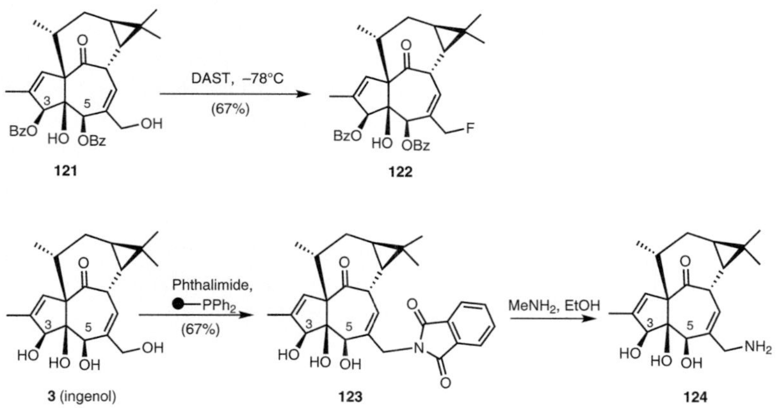


**Fig. 27** Functional group modification of ingenoids

## 3.2 *Total Synthesis*

The challenging structure of ingenol has long fascinated organic chemists, and this compound has been a favored target for total synthesis ever since its isolation. Four decades of efforts have culminated in four total syntheses of ingenol, one synthesis of its 13-hydroxylated derivative, and countless frustrated efforts, which, despite missing the final target, nevertheless represent a monumental treasure-trove of ingenuity to serve as a source of inspiration for synthetic chemists. There is no shortage of reviews on the synthetic approaches to ingenol [110–112], and only the successful total syntheses will be discussed, focusing on the generation of the tricyclic core of ingenol and its functionalization to the natural product.

A comparative analysis of the published ingenol syntheses exemplifies some important concepts of general relevance for natural products synthesis. The first one is the clash between the brilliant generation of a scaffold and the tribulations of its functionalization that characterizes the pre-Baran approaches. The generation of the scaffold did not succeed, or was beyond control in terms of stereochemistry, with functionalized precursors, and its decoration with functional groups had therefore to be carried out in a painfully linear strategy plagued by the unpredictability of functional group manipulation in the strained ingenoid framework. The tribulations of the syntheses of Winkler, Wood, Kuwaijima, and Kigoshi with the sugar-like lower part of ingenol are instructive for the ingenuity of the solutions eventually found to get the synthesis "out of the swamps" of functional group manipulation and to pave the way to the final target. The second concept relates to the logic underlying the choice of a natural product as a synthetic target. Until recently, ingenol was simply a curiosity, a minor relative of phorbol of little practical relevance. Its derivatives never surged to the status of biological tools like the more stable and easily available phorbol esters, and there was little really exciting in the biomedical literature to justify the titanic efforts of the synthetic community to achieve a total synthesis of this compound. Indeed, apart from phytochemical reports of occurrence of specific esters, until recently the literature on ingenol was dominated by studies on its synthesis. One therefore wonders why ingenol has so long fascinated synthetic organic chemists. The answer is probably the challenges associated by its synthesis, in particular the presence of an exotic scaffold with an unconventional *in,out* rather than *out,out* configuration at the bridge, and the clash between its hydrocarbon-like northern sector and the sugar-like nature of the lower rim, with three contiguous and *cis*-oriented hydroxylated functionalities. In this context, ingenol became an iconic compound, a sort of chemical "Mount Everest" waiting to be conquered.

### 3.2.1 The Winkler Synthesis (2002)

This synthesis was the first one described for ingenol [19]. The key step is the intramolecular dioxenone-alkene photoisomerization of the bicyclic allylic chloride **125**. The [2+2] photocycloaddiction (De Mayo reaction) was followed by a base-induced retro-Claisen fragmentation that generated the cyclopentane-fused *trans*-bridged cycloundecane skeleton of the natural product (**126**). The missing carbons from the cyclopropane ring could be installed after chloride elimination to **127** and stereoselective addition of the dibromocarbene followed by halogen-methyl exchange via a mixed cuprate (Corey-Posner reaction), a sequence also featured in the Kuwajima-Tanino synthesis (see Sect. 3.2.2). The following functionalization of the lower rim had to rely only on the C-6 hydroxymethyl of **128** to functionalize a string of seven adjacent carbons, in a real tour-de-force of ingenuity. The final missing carbon (C-14) was introduced after carboxyallylation of the ketone **129**, which made possible α-monoalkylation and the palladium-catalyzed oxidative

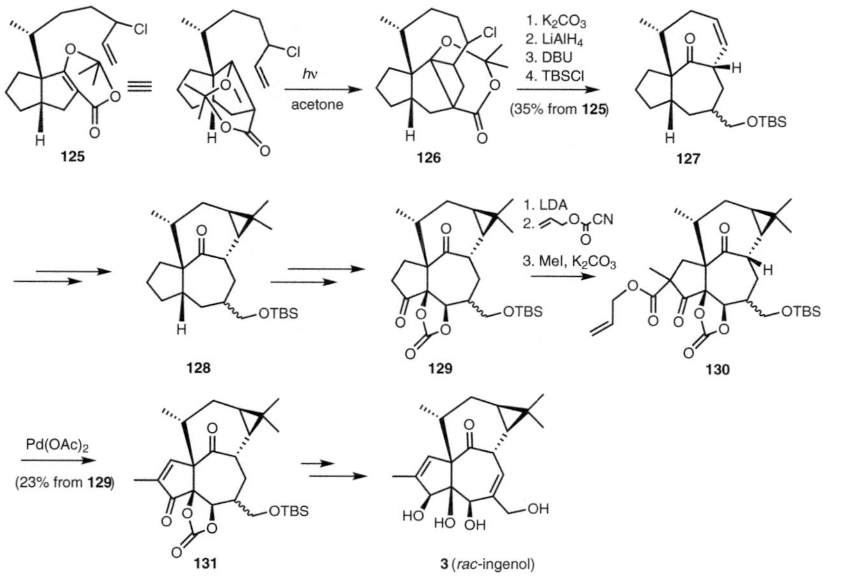


**Fig. 28** The Winkler synthesis of ingenol (DBU: diazabicycloundecene, TBSCl: *tert*-butyldimethylsilyl chloride, TBS: *tert*-butyldimethylsilyl, LDA lithium diisopropylamide)

deallylation to a C-3 enone. This could then be reduced stereoselectively with $NaBH_4$-$CeCl_3$ according to the Luche protocol, delivering, after global deprotection, the racemic natural polyol. Overall, the synthesis required 45 linear steps. Even though the average yield per step was high (ca. 80%), the final yield was exceedingly low <0.1%. The asymmetric synthesis of one intermediate from the Winkler synthesis was reported by Kigoshi based on a ring-closing metathesis approach [113] (Fig. 28).

### 3.2.2 The Kuwajima-Tanino Synthesis (2003)

This racemic synthesis was published only a few months after that by Winkler [114], and relies on a completely different approach to build the *trans*-intrabridgehead tricyclic core, generated via the type-3 semi-pinacol rearrangement of the epoxyalcohol **132** promoted by the Lewis acid $AlMe_3$. Just like in the synthesis of Winkler, the functionalization of the tricyclic compound also proved tricky, especially the introduction of the 20-carbon, which required eight steps from the *nor*-ingenoid **136**, and was eventually achieved by dissolving a metal (Zn, $NH_4Cl$) for the reduction of the bromoepoxide **137**. The number of steps was similar to the synthesis by Winkler, and the overall yield similar (<0.1%). Within the many remarkable reactions involved, the bis-dihydroxylation of the diene **134** to **135** is remarkable for its efficiency (59% yield) and stereoselectivity (Fig. 29).

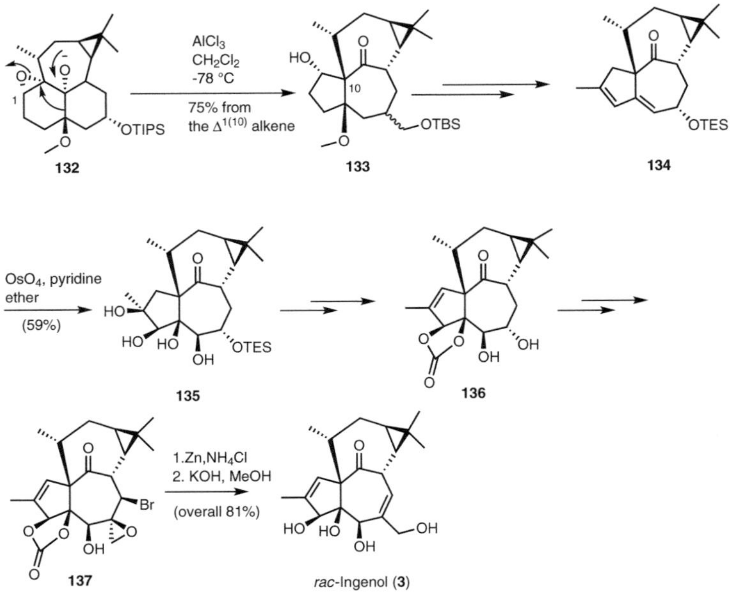


**Fig. 29** The Kuwajima-Tanino synthesis of ingenol (TBS: *tert*-butylddimethylsilyl TES: triethylsilyl)

### 3.2.3 The Wood Synthesis (2004)

This synthesis, the first asymmetric such method [98], is considerably shorter than the two previous ones, and relies on an olefin metathesis reaction to assemble the tricyclic core of ingenol from the optically active diene precursor **140**. The resulting adduct was already better functionalized than those involved in the previous syntheses, and this explains the shortening of the whole sequence, now reduced to 32 linear steps from the so called Funk's keto-ester (**139**), a compound available in five steps from (+)-carene (**138**). In spite of this, the overall yield remained around 0.1%. A critical reaction found was the regio- and stereoselective allylic oxidation of the diene **142**, which afforded in almost 50% yield only one of the four possible regioisomeric products of allylic oxidation (**143**). The functionalization of ring B was achieved in a quite remarkable way. The enone **144** was stereoselectively α-hydroxylated and the 4-hydroxy group next served as a handle to stereoselectively epoxidize, via the VO(acac)$_2$-*t*-BuOOH protocol, the $\Delta^5$ double bond, to afford **145**. The epoxide ring, as observed also in the Kuwaijima synthesis, proved recalcitant to opening, and this could eventually be effected only in a rather roundabout way. After functional group modulation that involved reduction of the enone and acetal protection of the 3,4-diol system, the 20-hydroxy group of **146** was replaced by a phenylsulfonyl function by mesylation, reaction with thiophenol, and oxidation to **147**. The epoxide ring was then opened via an $E_2$ reaction that capitalized on the sulfur-induced acidity of the 20-methylene, and the resulting exomethylene olefin easily isomerized to the endocyclic isomer, and was eventually

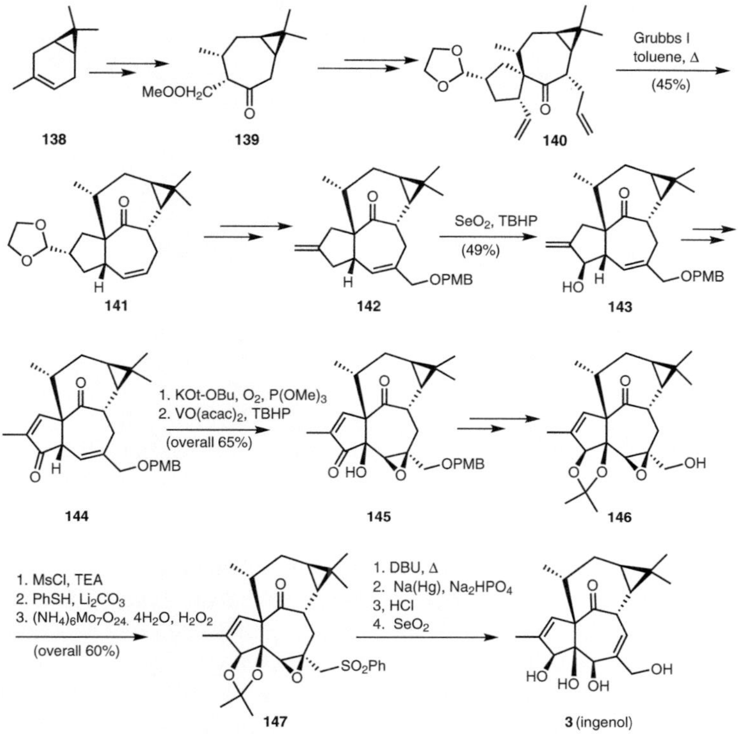


**Fig. 30** The Wood synthesis of ingenol (PMB: *p*-methoxybenzyl, TBHP: *tert*-butyl-hydroperoxide, MsCl: mesyl chloride, DBU: diazabicycloundecene)

desulfurized by treatment with sodium amalgam. After deprotection to 20-deoxyingenol, allylic oxidation delivered the natural product **3** in an optically active form (Fig. 30).

### 3.2.4 The Kigoshi Synthesis of 13-Hydroxyingenol (2012)

This synthesis is conceptually similar to the one of Wood, relying on a ring-closing metathesis to assemble the tricyclic carbon system of the target [115]. The reaction gave the best yields when carried out on the exomethylene enone **148** under the catalysis of the second-generation Hoveyda-Grubbs catalyst. The corresponding alcohols and acetate gave lower yields, suggesting that the presence of an additional $sp^2$-hybridized carbon has a stabilizing effect on the adduct. The Wood and Kigoshi metathesis approaches each generated a $\Delta^5$-ingenoid that could not be directly converted into their $\Delta^6$-natural isomers, suggesting that $\Delta^6$ ingenoids are inherently less stable than their $\Delta^5$ isomers. To overcome this issue, functionalization at C-7 (ingenol numbering) was included in the metathesis precursor. After ring A-modification, reduction of the C-7 ketone to alcohol, change of the protecting

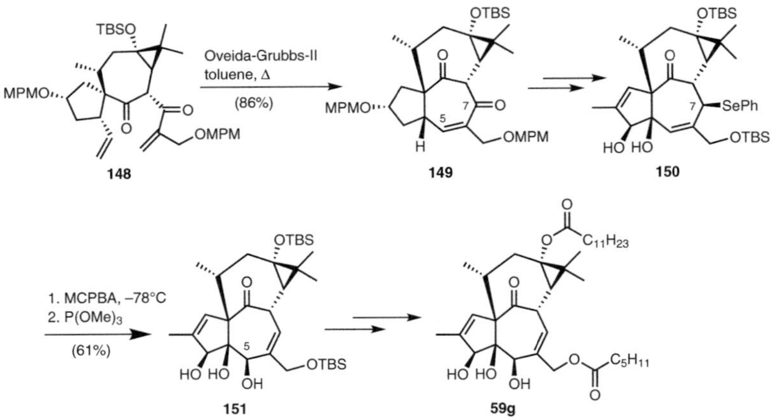


**Fig. 31** The Kigoshi synthesis of the 13-hydroxyingenol ester **59g** (MCPBA: *meta*-chloroperoxybenzoic acid, MPM: 4-methoxyphenylmethyl, TBS: *tert*-butyldimethylsilyl)

group at the 20-hydroxy group site, and oxygen-to selenium exchange, the selenide **150** was chemoselectively oxidized to a selenoxide, that underwent 1,5-sigmatropic rearrangement establishing oxygenation at C-5 and the correct location of the double bond. After protection of the lower-rim hydroxy groups as the diacetonide, the tertiary 13-hydroxy unit was esterified with dodecanoic acid, and, after deprotection, the primary 20-hydroxy group was chemoselectively esterified by treatment with hexanoic anhydride and triethylamine at low temperature, giving the natural 20-hexanoyl-13-dodecanoyl ester (**59g**) occurring in *E. kansui* [104] (Fig. 31).

### 3.2.5 The Baran Synthesis (2013)

Comparison of this synthesis with the previous ones highlights the current trend in natural products synthesis toward brevity and scalability. In a certain sense, this synthesis is a sort of manifesto for a novel era in compound total synthesis, where the virtuosistic duplication of a natural product is aimed at both increasing its availability and exploring the chemical space around its scaffold, and, in general, at improving our knowledge on a specific class of compounds. Unlike all the previous syntheses, the Baran synthesis relies on a biogenetic strategy to assemble the tetracyclic core of the natural product, following the logic of isoprenoid synthesis, where a polycyclic core (cyclase stage) is first assembled from a precursor that can be oxidatively decorated (oxidase stage). This logic is exemplified by the biosynthesis of taxol (paclitaxel) in the yew tree [116], and might be of general relevance for oxygenated terpenoids, if not because oxygenated functionalities can interfere with cyclization reactions, affording oxygen bridges and ether bonds rather that carbon-carbon bonds and polycyclic scaffolds.

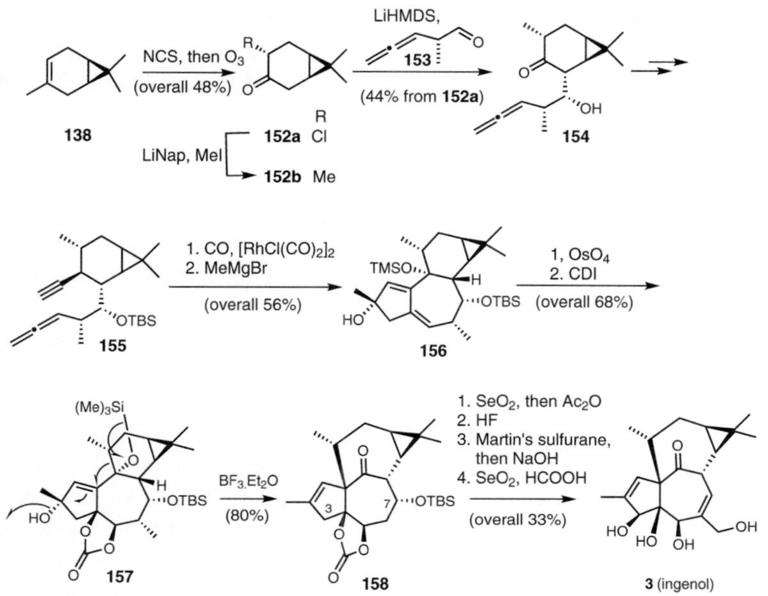


**Fig. 32** The Baran total synthesis of ingenol (CDI carbonyldimidazole, NCS *N*-chloro-succinimide, LiHMDS lithium hexamethyldisylamide, TBS *tert*-butyldimethylsilyl, TMS trimethylsilyl)

The cyclase phase (Fig. 32) begins with a form of configurational inversion of the naturally occurring monoterpene olefin, (+)-3-carene (**138**), a commodity chemical available in crude form from Indian turpentine. (+)-3-Carene is also the starting material for the synthesis of Funk ketone, a precursor for the C,D rings in the Wood synthesis of ingenol [98], as well as of the C,D rings of heterocyclic analogs of phorbol [117]. However, rather than directly expanding it into a seven-membered analog of rings C,D of an ingenoid, the six-membered ring architecture is maintained, and is elaborated into a mimic of the C,D rings of a tigliane precursor of ingenol, that is, a compound capable of generating the C-10 cation that triggers the migration of C-11 from a hydroxylated C-9 carbon. This is accomplished by re-shuffling the location of the methyl substituent of the cyclohexene ring by electrophilic chlorination of the natural product to an exomethylene olefin, which is next ozonized and methylated to **152b**. This compound proved difficult to purify and, in the final process, was only a transient intermediate in the one-pot telescoped process that reductively alkylated the chloroketone **152a** by methyl-chlorine exchange and next aldolized it on the other $\alpha$-side by reaction with the allenyl aldehyde **153**. After installation of the acetylenic group and protection of the hydroxy group functions to allenylacetylene **155**, the tigliane derivative **156** was obtained by treatment of **155** with carbon monoxide and a Rh (I) catalyst. The allenyl Pauson-Khand carbonylation had to be carried out at a high-dilution (5 m*M*), but could, nevertheless, be scaled up at gram level. The availability of the resulting tigliane **156** is per se a remarkable achievement, since it could be used to explore the many areas of the phorbol pharmacophore that are non-accessible from the pool of natural diterpenoids.

The oxidative phase was easier and more straightforward than in the previous syntheses, because of the presence of a hydroxy group at C-7 which, after rearrangement to the ingenane skeleton, served to generate the $\Delta^6$ double bond, a difficult step in the functionalization of the lower rim of ingenol precursors. After dihydroxylation of the $\Delta^4$ double bond and carbonation of the resulting diol, the C-2 allylic hydroxy group of **157** was used to trigger the tigliane-to-ingenane rearrangement discussed in Sect. 4.1, with allylic oxidation at C-3 and C-20 completing the synthesis.

Compared to the previous syntheses, the extraordinary abbreviation achieved by Baran is the result of the generation of the ingenane skeleton in a highly functionalized form that could be converted to the final target in a straightforward way. With the exception of galantamine [118], the total synthesis of complex natural products has been industrialized essentially when no other alternative source existed, as exemplified by the active moiety of the marine polyether halichondrin B (eribulin mesylate) [119]. Given the availability of ingenol from a cheap and renewable agricultural source (the seeds of the caper spurge), the synthesis by Baran probably will never become competitive with isolation. Nevertheless, it might be modulated to generate ingenoid chemotypes not available by modification of the natural polyol, via exploring the chemical space around the ingenol chemotypes through avenues unavailable from the natural polyol. Undoubtedly, it will remain in the history of chemistry to testify how ingenuity and judicious synthetic planning can overcome apparently insurmountable difficulties.

# 4 Bioactivity

## *4.1 Molecular Targets*

The successful introduction to therapy of ingenol-3-acetate (I3A) (**5**, Picato®) in 2012 has fostered studies on all aspects of the pharmacology of ingenol derivatives, revitalizing investigation of phorboids, a field that had stagnated substantially after the great interest in their tumor-promoting properties in the second half of the past century and the substantial deflation of the prostratin anti-HIV hype at the beginning of the present century. Many, but not all, bioactivities identified so far for phorboids are downstream to the activation of PKC, a heterogeneous family of serine/threonine kinases that can be modulated selectively in terms of activation and/or inhibition by these compounds [120]. The ensuing picture is complicated and has still substantially eluded full rationalization, since different phorboid esters show distinct profiles of isoform activation, while compounds apparently devoid of any affinity for PKC still maintain activity in certain bioassays. This suggests that further high-affinity targets might exist, of which some are starting to surface [121].

### 4.1.1 Protein Kinase C (PKC)

This multifamily of lipid-regulated enzymes, consisting of fifteen members in humans, phosphorylates a variety of cellular proteins and plays an essential role in the mechanism of signal transduction [120]. Protein kinases C have been classified into several distinct subfamilies depending on their specific requirements for activation. Classical [(c)PKC] isoforms (α, βI, βII, and γ) require calcium (II) and diacylglycerol (*sn*-1,2-diacylglycerol, DAG); novel [(n)PKC] isoforms (δ-, ε-, η-, and θ) require DAG but not calcium for activation, while activation of the atypical isoforms [(a)PKCs] (Mζ- ι/λ isoforms) is independent from calcium and DAG. The PKC-related kinases (PRK) are the most recent addition to the family. All classical and novel PKC isoforms are sensitive to phorboids, which bind to their regulatory domain acting as ultra-potent biological analogs of DAG [121], eliminating the requirement for DAG for activation, and, for (c)PKC, decreasing the concentration of calcium needed for activation. Considerable uncertainty exists, however, on the binding mode and the selectivity profile of phorboids for the activation of various PKC isoforms, and even the stoichiometry of the interaction is debated [120]. More than a single-binding mode is probably possible within the same phorboid-recognition site of the various PKCs, since phorboids, and especially ingenoids, are remarkably plastic in their interaction with PKC isoforms. A further issue of confusion is that there is incomplete overlapping between biochemical experiments on PKC activation and structure-activity relationships for phenotypic end-points sensitive to PKC modulation, such as cytotoxicity, skin-irritancy, tumor promotion, and reversal of HIV-1 latency. More than three decades after the seminal discovery by Nishizuka that PKC is the target of phorbol esters [122], the details and the implications of this interaction are still largely to be clarified.

The interaction of phorboid polyols with PKC is critically dependent on their type of ester decoration, and requires a combination of optimal hydrogen bondings and hydrophobic contacts for high potency. A lipophilic ester group and a free hydroxy group at a specific position are required for binding, which relies on a network of complementary hydrogen bonding donor and acceptor sites between the two partners. The net result is the replacement of a hydrophilic surface with a hydrophobic one, and this polarity transfer promotes membrane association of cytoplasmatic PKC and implementation of its enzymatic activity. While there is substantial agreement on this general view, the identification of the network of hydrogen bondings that triggers the transfer of polarity has raised considerable controversy, and is still an unsettled issue for both phorbol and ingenol.

Unlike phorbol, ingenol has an intrinsic affinity for PKC, with overall effects in cellular assays in a similar, but less intense, way than those associated with its esters [123]. Activity for this polyol can be demonstrated both at the level of isolated enzymes and of intact cells, with a Ki value around 30 μ*M*, four orders of magnitude higher than phorbol esters like TPA or phorbol dibutyrate, for which the Ki is sub-nanomolar, but not too different from that of the physiological activator DAG ($K_i$ 0.56 μ*M*) [123]. The lipophilic element of phorboids is credited with a dual role, entropic and enthalpic, promoting affinity by increasing the concentration at the membrane interface, and providing an ester carbonyl oxygen function suitably located to be able to act as a hydrogen-bonding acceptor. The surprising activity of ingenol suggests that the lipophic upper part of the molecule can meet the hydrophobic requirements, and that the string of oxygenated functions on rings A and B can somewhat compensate for the lack of a carbonyl ester, providing an extra hydrogen-bonding acceptor site. Phorbol is apparently unable to do so, also because the 12-hydroxy unit is, basically, a deactivating structural element for PKC binding [124].

These observations suggest that ingenol is a more versatile molecule than phorbol for the modulation of PKC activity, but limited information exists on the structure-activity relationships of ingenol esters as PKC ligands. A large body of the literature uses indirect, downstream biochemical end-points of PKC modulation or phenotypical assays (tumor promotion, skin irritation, HIV-1 latency reversal) that are not directly or exclusively related to the activation of PKC. Finally, most studies on the interaction of ingenoids with purified isoforms of PKC have been carried out with I3A (**5**) and it is not clear to what extent the conclusions of these studies can be applied to other esters as well [125].

The most detailed molecular modeling investigation on the interaction of ingenoids with PKC was performed on ingenol 3-benzoate (**159**), an anticancer compound and potent PKC modulator [126]. A combination of complementary

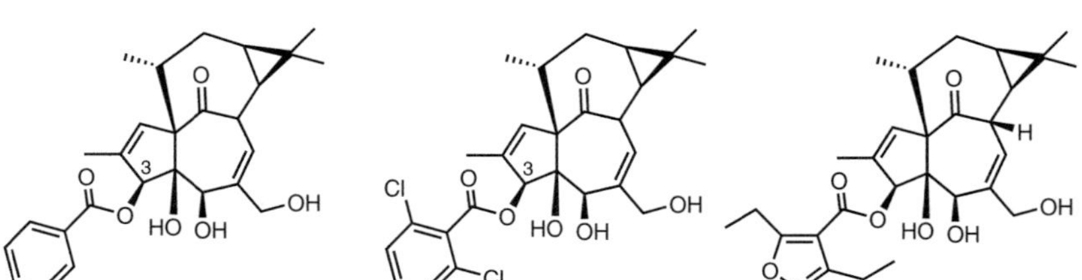


**159** **160** **161**

molecular dynamic and site-directed mutagenesis methods was employed, using as a reference the X-ray data of phorbol-13-acetate bound to one of the two cysteine-rich regulatory domains (C1b) of PKCδ, a classic isozyme, and the Blumberg model of binding derived thereof [127]. When fitted into the binding pocket defined by the interaction of phorbol-13-acetate with the phorboid binding element of PKCδ, the interaction of ingenol-3-benzoate involves a dual role of H-bonding donor and acceptor for the 20-hydroxy group, and as a donor for the 4- and 5-hydroxy groups, and as an acceptor for the ester carbonyl, with lipophilic interactions with the hydrophobic pocket of PKC (Pro11, Leu20, and Leu24) limited to rings C and D, and with the 9-keto group essentially acting as an inert unit (Fig. 33).

This binding mode is very similar to that of the daphnane orthoester thymeleatoxin, a weaker ligand, with the ester carbonyl at position 3 taking the

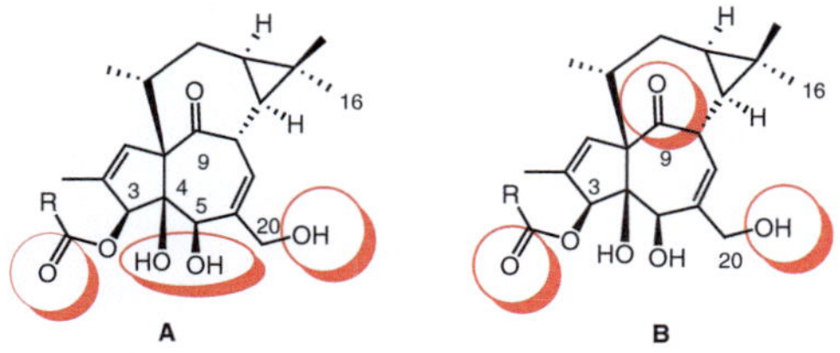


**Fig. 33** Pharmacophoric models for the interaction of ingenol esters with PKC. A: the Blumberg (B) model; B: the Hecker-Rando-Kishi (HRK) model. The red circles represent the oxygen functions involved in binding

place of the carbonyl group at position 3 of phorbol esters as a hydrogen bond acceptor for the amide group of Gly23 of PKC. When compared to the other high-affinity ligands considered in this study (phorbol dibutyrate, thymeleatoxin, and indolactam V), ingenol-3-benzoate was the least sensitive to structural perturbation associated with site-directed mutagenesis, suggesting a noteworthy degree of plasticity to accommodate changes in the phorboid pocket to PKC. This observation might be related to the presence of a closer proximity of the hydroxy groups in ingenol esters compared to phorbol esters, resulting in a better capacity to reorganize the hydrogen bonding network required for PKC binding. For docking calculations, the need to assume that the binding site is rigid and interacts with the various ligands in essentially the same conformation despite the inherent flexibility of the hydrophobic binding domain, was partially compensated for by the adoption of a molecular dynamics technique based on generalized statistic-mechanics theory (q-jumping MD method) [128].

The major limitation of this study is that it relies on the interaction of a very low affinity pseudo-agonist, for which the location in the cysteine-rich domain of PKC might be an artifact, representing a special case of complementarity with little relevance for the higher affinity ligands, or else providing just a hint of how more potent ligands might interact.

A rather different binding mode has emerged from the investigation of the structure-activity relationships of ingenol esters as tumor-promoters [129]. This model, proposed by Hecker, assumes that the ester carbonyl at C-3 and the primary 20-hydroxy group are involved in the interaction, but the 9-carbonyl, and not the hydroxy units at C-4 and C-5, represents the third element of the oxygen triangulation required for binding, as a hydrogen-bonding acceptor (Fig. 33). According to this model, the 9-carbonyl plays the role of hydrogen-bonding acceptor that in phorbol esters is associated with the 4-hydroxy group, and the major lipophilic interactions are associated with the acyl group, and not the hydrophobic moiety of the diterpenoid scaffold, as proposed for phorbol esters. By molecular modeling, it was possible to superimpose the minimum-energy conformation of TPA and ingenol 3-palmitate with *sn*-1,2-diacylglycerol.

The pharmacophore model associate to this binding mode fits with the Rando-Kishi model of PKC activation, which requires the presence of three hydrophilic atoms separated by approximately 6 Å for a significant affinity [130]. The limitation of this approach is that it is based on a phenotypic end-point (tumor-promoting activity) downstream to PKC activation, and therefore sensitive to modulation also by other factors, and, at least for the limited data available, not fully superimposable with the primary molecular event of PKC activation.

The B-model is structure-based, while the HRK-model is ligand-based, and, unsurprisingly, the two models also differ in the identification of the pharmacophore of phorbol esters [131]. Both models emphasize the relevance of the 20-hydroxy group and the ester carbonyl at C-3 of ingenol esters, as well as on the need for a third oxygen functionality and of hydrophobic interactions, but they differ in the location of these elements. Some observations are more in line with the HRK model than with the B-model. Thus, photo-cross linking experiments have highlighted the relevance of the hydrophobic interactions associated to the acyl moiety for the interaction of phorboids with PKC [132], while the HRK model seems better suited to explain the tumor-promoting activity, and therefore PKC affinity, of the 3-esters of 16-hydroxy-20-deoxyingenol and of those of 13-hydroxyingenol, where the hydrophobic moiety of the diterpenoid skeleton is perturbed by the introduction of the oxygen functions [129]. On the other hand, the activity of the 3-esters of both epimers of 9-dihydroingenol can be accommodated in both models [129]. In the B-model, C-9 is located into a space not involved in binding, while the HRK model assumes that both configurations of the 9-hydroxy group can function as a H-bonding acceptor. The similar bioactivity (at least in terms of tumor-promoting properties) of the diastereomeric dihydroingenols parallels a similar observation for the 3-dihydrophorbol esters [129]. The occurrence of ingenol esters based on different oxygenation and acylation patterns from those considered in the modeling studies (16-hydroxy-20-deoxy-, 13,19-dihydroxy-, and 13-hydroxy esters) provides an interesting opportunity to investigate the binding of ingenol esters to PKC, but the biological profile of these compounds is still largely unexplored, apart from information on their tumor-promoting properties [129].

Taken together, these studies evidence an unsettled situation regarding the binding mode of ingenoids to PKC. Unsurprisingly, a confusing situation exists also on their selectivity of isoform activation, with growing evidence that different esters target different isoforms, and that this issue can only be addressed in terms of systems biology, since PKC is essentially a target mix with similar but non-equivalent ligand-binding domains, different phospholipid requirements, and specific intracellular localizations. The largest body of information concerns I3A, a broad-range activator of both classical and novel PKC isoforms first reported as a highly inflammatory and weak tumor-promoting agent [129]. The phenotypic response of I3A (**5**) was, however, basically different from that of PMA, since the inflammatory response induced in mouse skin was transient, and reminiscent of that of the short chain esters of deoxyphorbol, which, although very powerful skin irritants, actually inhibit tumor promotion by long-chain phorbol esters like PMA [129]. Despite the lack of a long-chain hydrophobic moiety and a lower overall lipophilicity compared to PMA, **5** is remarkably potent in the activation of PKC, in line with the suggestion that the hydrophobic diterpenoid core of ingenol can act as a surrogate for a long-chain acyl moiety. On the other hand, the overall profile of activation was functionally distinct from that of PMA and DAG, the physiological activator of PKC, with more emphasis on PKC-δ for **5** and on PKC-α for PMA. Rather than having a different binding mode for the ligands, these differences might be associated with their different polarities, since, with a calculated log P of 3.89, **5** is much less lipophilic than TPA, for which the calculated log P is 6.65. A polarity swap between the hydrophilic surface of PKC and the lipophilic moiety of phorboids is critical for the translocation of PKC to the lipid membrane and its functional activity, rationalizing the conclusion that **5** is not simply another phorbol ester with a different potency. Interestingly, PKC-δ is overexpressed in tumors, and plays an important role in the generation and proliferation of malignant cells, possibly explaining the major emphasis on anticancer therapy for ingenol esters compared to phorbol esters. Selective interaction with PKCβII is also critical for the activation of neutrophils involved in the clinical activity of I3A [133].

Another ingenol derivative with an interesting profile of interaction with PKC is ingenol 3,20-dibenzoate (IDB, **57k**), which has been reported to activate selectively the so-called "novel" forms of PKC, in sharp contrast to the 3-acyl esters of ingenol and phorbol esters, that also activate the classical isoforms [134]. This finding has significant clinical potential for cancer supporting care. Thus, while activation of the classical PKC isoforms inhibits megakaryocyte differentiation, activation of the novel forms is associated with megakaryopoietic and thrombopoietic effects in vivo. The possibility therefore exists to promote early megakaryocyte and platelet reconstitution in conditions associated with therapy-related marrow damage, as in cancer chemotherapy, and IDB was assayed with positive results in a murine model of this condition [134].

### 4.1.2 Topoisomerase II

Many ingenol esters can induce cell cycle arrest in the G2/M phase, potently inhibiting cell proliferation and inducing the DNA-decatenation checkpoint, a typical profile of topoisomerase (topo) inhibitors [135]. Limited structure-activity relationship work has been done, with 3-*O*-((2′*E*,4′*Z*)-decadienoyl)-20-*O*-acetylingenol (3*EZ*,20Ac-ingenol, **57h**) emerging as a lead compound [136], endowed with catalytic (class II) topo-I inhibitory activity similar to that of the thiobarbiturate merbarone [137]. From the database of ingenoids included in these studies, the presence of a 20-ester group seems critical, while a 3-ester group alone is not sufficient for significant activity. In light of these findings, it is tempting to assume that ingenol 3-esters, originally PKC modulators and cytostatic agents, could be transformed by spontaneous acyl migration into cytotoxic compounds.

### 4.1.3 Other Targets

IDB (**57k**) has been shown to directly bind the erythropoietin receptor (EpoR) and mimic the activity of the endogenous hormone, stimulating the proliferation of EPO-dependent cell lines [138]. In animal experiments, IDB can attenuate anemia induced by the cytotoxic drug 5-fluorouracil, showing a synergistic effect on EPO at low concentrations, but an antagonistic effect at higher concentrations. Erythropoietin promotes tumor angiogenesis, and dysregulation of EpoR may affect the growth of certain tumors. Since IDB shows antitumor properties, it qualifies as a potential agent to counteract anemia in cancer patients [138]. Interaction of IDB with tubulin with inhibition of its polymerization, has also been described for ingenol esters, but this preliminary report needs confirmation [139].

### 4.1.4 Target-Based Structure-Activity Relationship Studies

Given the complex pharmacology of ingenoids, it is difficult to summarize their profile in a single molecular- or phenotypic assay and optimize their clinical potential accordingly, as exemplified by the medicinal chemistry of I3A. Aberrant keratinocyte death and lesion-directed immunological responses seem critical for the clinical activity of this compound, and both actions are to a various extent dependent on the activation of PKC isoforms, and especially PKCδ. In terms of cell-based assays, this could be translated into the induction of necrotic cell death in cancer cells and the stimulation of oxidative burst and cytokine release in immune cells and keratinocytes. In practice, cell death was investigated in HeLa cells, and the generation of reactive oxygen species (ROS) in polymorphonuclear leukocytes, and cytokine (IL8 and TNF-α) release in human primary epidermal keratinocytes. Since ingenol-3-esters are not stable toward acyl rearrangement, the bioactivity assays were complemented by stability determinations, evaluating the kinetic of degradation in an aqueous buffer at pH 7.4 and 37°C. Despite the maintenance of the critical angeloylation at C-3 and the presence of a free hydroxy group at C-20, massive loss of activity was observed upon modification of the diterpenoid scaffold (epoxidation of the $\Delta^6$ double bond, methylation of the 5-hydroxy group, fluoridative rearrangement of the 5-hydroxy group) [102]. This suggested that the diterpenoid core of I3A is already substantially optimized for activity, refocusing activities on the nature of the ester group at C-3. Starting from ingenol 5,20-acetonide, various esters were prepared from all of the saturated, unsaturated and aromatic series [102]. Within the set of bioassays investigated, the most critical factor turned out to be the stability to acyl rearrangement. Thus, all compounds that outperformed I3A in the biological end-points also showed an increased chemical stability, suggesting an intrinsically flat SAR, dominated by the concentration of the active species in the assay. These considerations could rationalize the lower potency of ingenol-3-tiglate when compared to ingenol-3-angelate. The carbonyl of tiglic acid is less encumbered by I-strain because of the lack of the *cis*-methyl group, which is replaced by a hydrogen atom, and is therefore more prone to attack from the nearby 4- or 5-hydroxy group and rearrangement. The comparative recovery of the 3-tiglate was, in fact, 30% lower than that of the angelate [102].

These considerations were confirmed in a series of benzoyl derivatives [108]. Ingenol-3-benzoate (**159**) is more easily degraded when compared to I3A (**5**), but stability may be improved by the introduction of substituents at the two *ortho*-positions. On the other hand, this was not accompanied systematically by a substantial increase of potency in the various biological end-points investigated. In particular, a partial dissection between the induction of necrosis, the immune-boosting activity and the affinity for PKCδ was observed. For instance, the 2,6-dichlorobenzoate (**160**) was stable toward rearrangement (>95% recovery), but had less than half of the potency of I3A in terms of cytokine-releasing properties, and was also less potent than I3A in terms of PKC-δ activation ($EC_{50} = 19.8$ n*M* vs. 4.1 n*M*). Nevertheless, it was considerably more potent in terms of necrotic properties than I3A ($LC_{50} = 59$ vs. 230 μ*M*). A dissection between PKC activation and cytotoxicity had already been observed in analogs of ingenol-3-benzoate [109]. A further issue to consider in cell-based studies of bioactivity optimization is the potential increase of tumor-promoting properties with the lipophilicity of the acyl group at C-3, an observation clearly observed in linear esters but of unknown relevance in branched esters. In a parallel series of investigation not yet disclosed, ingenol disoxate (LEO 43204) (**161**) has emerged as an interesting analog of I3A, and this compound is currently undergoing clinical development with promising phase IIa clinical data, where it outperformed I3A without a corresponding increase in side-effects [140–142].

As indicated earlier, the development of an expeditious total synthesis of ingenol has made possible the exploration of areas of its chemical space unavailable by chemical modification of the natural product [133]. This has afforded interesting information on the relationship between the immunostimulating activity and the specificity of PKC isoform activation. Thus, the neutrophil oxidative burst parameter may be better correlated with the activation of PKCβII rather than PKCδ. Since PKCβII is expressed very little in keratinocytes, the availability of ingenol-based selective probes will make it possible to dissect the activity at keratinocytes and at neutrophils, paving the way for new pharmacological applications of these analogs.

Taken together, the results of these medicinal chemistry studies show a complex scenario, where readings in bioassays are dominated by selectivity of PKC isoform expression but are complicated by the need to consider chemical stability. There is little doubt that the clinical potential of I3A could never have been discovered in a mainstream medicinal chemistry campaign.

## 4.2 *Preclinical Pharmacology*

Ingenol was discovered because of the tumor-promoting properties of some of its esters. In general, ingenol derivatives are less potent tumor promoters compared to phorbol esters, but fully retain their potent inflammatory activity. Cytotoxicity was also observed in the early studies, in part paving the way to the eventual development of ingenol 3-angelate (**5**) and Picato®.

### 4.2.1 Tumor Promotion

Despite intense research even since its discovery in 1941 [143], the human relevance of tumor promotion, a mostly rodent process, is still unclear. The process can be demonstrated on mouse skin, but may not applicable to the human situation. Nevertheless, research on ingenol was fueled paradoxically by an interest in these negative properties rather than in the anticancer potential of this class of phorboids. Ingenol 3-hexadecanoate (*Euphorbia* factor $L_5 = I_1$, **2**) is an order of magnitude weaker as a tumor-promoting agent than phorbol-12-tetradecanoate-13-acetate (TPA = PMA), the reference compound for this type of activity [67]. An interesting dissection between irritancy and tumor- promoting properties was observed in the corresponding 2,4,6-dodecatrienoate (*Euphorbia* factor $I_6$, **53q**), that was more irritant but less potent as a tumor-promoter than ingenol-hexadecanoate (**2**), in line with the observation that among phorbol derivatives polyunsaturated and short-chain acyl groups induce strong inflammatory activities, but weak tumor-promoting properties [144]. The tumor-promoting activity of ingenol-3-esters is strongly dependent upon their acyl decoration. Location is critical, since migration of the ester group from the 3-hydroxy to the 20-hydroxy position completely abolishes the tumor-promoting properties [144]. Short chain ($C_2$–$C_6$) esters are weak or very weak promoters, while activity peaks with at with a chain-length of $C_{14}$, with ingenol 3-myristate being almost as potent as PMA [67]. Ingenol 5-monoesters are unstable, and rearrange very quickly to their corresponding and inactive 20-esters, and could not be included in the analysis of the SAR. The presence of both a free 20-hydroxy functionality and of a medium-chain linear acyl moiety are also critical for the tumor-promoting activity of phorbol esters, despite the different topological relationships between these elements. Remarkably, deletion of the 20-hydroxy group can be compensated for by the presence of a hydroxy unit at C-16, since the 3-esters of 20-deoxy-16-hydroxyingenol are potent tumor promoters [144].

### 4.2.2 Anticancer and Immunostimulating Activity

Ingenol-3,20-dibenzoate (IDB, **57k**), a non-tumor promoting ester, was identified as the potent antileukemic principle of *E. esula* by Kupchan in 1976 [145], fostering considerable interest in the anticancer properties of ingenol esters. The same bioactivity-directed fractionation scheme afforded phorbol 12-tiglate-13-decanoate, a moderately potent tumor promoting ester, as the antileukemic principle of croton oil. This early finding suggested that the anticancer and cytotoxic properties of phorboids could, in principle, be dissected from their skin-irritancy and the tumor-promoting properties. On the other hand, ingenol-3-hexadecanoate, the major irritant and tumor-promoting constituent of the caper spurge was also identified as its antitumor constituent [146], suggesting that a clear-cut distinction between these various biological properties is difficult, and that our predictive power in this area is limited.

Some ingenol esters, and I3A (**5**) in particular, show a surprising association of cytotoxicity against leukemia cells with the stimulation of immune cells [147]. This pharmacological oxymoronic profile has been related to the expression of different PKC isoforms in myeloid and immune cells, with apoptosis being induced by activation of PKCδ, and suppressed by activation of PKCθ, an isoform expressed selectively in T cells but absent from myeloid cells. Depending on the cell type, I3A can therefore suppress or, alternatively, promote cell growth, and this might indeed represent one important mechanism of its clinical activity. In addition to PKCδ, I3A can also directly bind to a subset family (RasGRP) of Ras modulators, eventually inducing apoptosis [148]. Both ingenol and phorbol esters can induce senescence in selected cell lines from solid tumors (e.g. melanoma, breast, colon) with permanent growth arrest [149]. This activity was inhibited by the PKC inhibitor bisindoleylmaleimide-I, showing that this type of activity was related to the modulation of these kinases.

A systematic study on the structure-activity relationship of IDB as a potential anticancer agent, using the induction of apoptosis in Jurkat cells as the end point, showed strict structure-activity relationships [150]. Benzoylation of the 20-hydroxy group was required to trigger apoptosis through a pathway that involved caspase-3, which occurred at the specific cell cycle checkpoint that controls the S-M phase transition, and was insensitive to treatment with PKC pan-inhibitors like GF 109130XO. In this series of compounds, activation of PKC was evaluated indirectly, measuring the induction of the transcription factors AP-1 and NF-κB. The caspase-3 and the PKC activation pathway were substantial orthogonal in terms of structure-activity relationships, suggesting that a complete dissection could, in principle, be possible [150].

Taken together, the available biochemical and cellular data suggest that ingenol derivatives have the potential to be developed as anticancer agents. Their activity depends on their profile of PKC isoform modulation, but can also involve other targets unrelated to the PKC pathway. Although the anticancer and tumor-promoting properties may be dissected, phorboids, and ingenol esters in particular, have a complex pharmacology that seems to substantially transcend the current paradigm of anticancer drug discovery, making it difficult to optimize activity, and investigate structure-activity relationships using in vitro and cellular assays that bear a direct relationship with in vivo activity, and overall to be able to qualify a specific compound as a lead based exclusively on cellular and preclinical data.

### 4.2.3 Antiviral Activity

Ingenol triacetate (ITA) potently inhibits the replication of human type 1-immunodeficiency virus (HIV-1) in acutely infected cells [151]. This antiviral activity was observed at nanomolar dosages that are three orders of magnitude lower than the cytotoxic ones, and was associated with the downregulation of CD4, a major HIV-1 receptor on host cells. Although interesting, this therapeutic index is three orders of magnitude lower than that of zidovudine, a clinically useful anti-HIV-1 agent [151]. Activation of PKC and/or its downstream-related transcription factors like NF-κB are known to downregulate the expression of CD4, but ITA showed no affinity for PKC, suggesting that ingenoids have an intrinsic capacity to downregulate the expression of CD4 independently from PKC activation. This was confirmed by the observation that anti-HIV-1 activity was not affected by staurosporine, a PKC inhibitor [151]. Very limited information is, however, available on the structure-activity relationships of ingenol esters as inhibitors of HIV replication. Activity was tolerated by acylation of the 20-hydroxy group, but was substantially lost by methylation of this functionality [152]. From these studies, a diester of 13-hydroxyingenol (3α-methylbutyrate, 13-dodecanoate; RD4-2138) emerged as a sub-nanomolar HIV-1 inhibitor [153]. The potency of this compound was higher than that of the reverse transcriptase and the protease inhibitors used currently in the management of HIV-1, and RD4-2138 was also active against viral strains resistant to reverse transcriptase- and protease inhibitors. Since RD4-2138 was also relatively well tolerated in animals, its lack of further development is surprising, and might be related to the danger of chronic administration of a compound that potentially could affect the expression of oncogenes.

Quite different results were observed in chronically infected cells, where many ingenol derivatives potently enhanced the replication of the HIV virus at nanomolar concentrations, effectively reverting, in a PKC-dependent way, HIV-1 latency [153]. HIV-1 delatentization in T-lymphocytes is an unmet need in the current management of this type of infection, and a major barrier to achieve the complete viral eradication in HIV-1 positive patients on highly active antiretroviral therapy (HAART) [154]. By enhancing the transcriptional activity of latent HIV-1 without inducing the polyclonal activation of non-infected cells, the viral reservoirs potentially could be purged of the virus. Reversal of latency is related to the binding of NF-κB and related transcription factors to HIV-1 LTR (long terminal repeat) promoter, with triggering of viral RNA transcription [154]. Thus, I3A (**5**) [155], IDB (**57k**) [156] and ingenol 3-hexanoate (ingenol-B) [154, 157] showed outstanding reactivation properties, fully comparable to CD3/28 antibody stimulation, and, because of a lower toxicity profile when compared to other PKC modulators like bryostatin, all qualified as excellent latency-reversing agents. A remarkable synergism was observed between ingenol-B and histone deacylase inhibitors like SAHA (suberoylanilidenehydroxamic acid), an effect not observed with IDB [157]. Ingenol-B (**162**) is a semi-synthetic ester developed in Brazil starting from a mixture of ingenol esters isolated from *E. tirucalli*, which is simplified to the parent polyol and then re-esterified with caproic (hexanoic) acid [157]. Surprisingly, while the 3-cinnamate and the 3-hexadecanoate of ingenol showed considerable cytotoxicity, the 3-caproate (= hexanoate) was not cytotoxic [157], and its reactivation properties were coupled to downregulation of the expression of the surface proteins necessary for the replication of HIV (CD4, CXCR4, CCR5), confirming the early reports of a similar activity for ingenol triacetate [151]. In comparison experiments, ingenol-B outperformed a series of HIV-1 latency disruptors of the phorboid type (TPA, IDB) as well as the histone deacylase (HDAC) inhibitor suberoylanilidenehydroxamic acid (SAHA, vorinostat), the PI3K/Akt pathway activator HMBA (hexamethylenebisacetamide), and even TNF-α itself. Ingenol B was well tolerated and efficacious in a simian model of HIV reactivation, with low toxicity upon oral administration in various animal species (rats, mice, dogs, monkeys) [157]. The similarity of activity of IDB, I3A (**5**), and ingenol-B is remarkable, since their profile of PKC activation is not overlapping.

Taken together, these studies show that ingenol esters have the potential to be developed as HIV-1 latency-reversal agents to be used along with retroviral drugs for viral eradication [154]. A surprising diversity between ingenol esters, however, seems to exist in terms of cytotoxicity and synergism with other delatentizing agents [157]. The reasons for the dissection of latency reversal and cytotoxicity are unknown, as are those underlying the synergy with other mechanistically unrelated reversal agents. Undoubtedly, this represents an obstacle for a systematic optimization of pre-clinical profiles, which is thus far based on a very limited database of compounds.

Phorboids have also been at the forefront of research on another RNA virus, the Chikungunya virus (CHIKV), and several phorbol and deoxyphorbol esters show outstanding inhibitory activity on the replication of this RNA virus [158]. However, I3A, IDB, and ingenol itself showed only weak activity, a remarkable example of differentiation between the biological profile of tigliane and ingenane lead compounds [159].

### 4.2.4 Pesticide Activity

Ingenol esters are remarkably potent pesticide agents. It is, however, unclear as to what extent their activity is selective and therefore of practical relevance, while their use could eventually lead to contamination of crops. On the other hand, their substantial potencies could make them interesting probes to discover new mechanisms of action against pests of agricultural relevance and noxious animals, and some pests seem exquisitely sensitive to these compounds. Thus, milliamines and related dipeptide ingenol esters from *E. cornigera* Boiss. [160] and *E. milii* [161, 162] are among the most potent molluscicidal compounds ever reported, outperforming by two orders of magnitude the potency of niclosamide (Bayluscide®) when assayed against the snail *Biomphalaria glabrata*, one important intermediate snail host for the schistosoma trematode (*Schistosoma mansoni*). Milliamines are only mildly irritant and are not tumor-promoting, and, since snails are sensitive to these compounds at nanomolar concentrations non toxic for other animals, the crude latex of *E. milii* was used successfully used in field experiments at a concentration of ca 1 mg/L as part of a control project on schistosomiasis [161]. Of relevance is also the observation that ingenol 3-(2′*E*,4′*Z*)-decadienoate (**53j**), its 20-acetyl derivative, and its 5,20-isomer, show impressive activity (5 μg/cotton ball!) against the pine-wood nematode *Bursaphelenchus xylophilus* [163]. This pest, present in Asia and America and recently discovered also in Europe, is the causal agent of pine wilt disease, and is one of the worst pest for forest trees and the cause of a major economic loss for the timber industry [163].

## *4.3 Clinical Pharmacology*

The only clinical data on ingenol derivatives are those of I3A (**5**), successfully introduced onto the pharmaceutical market as a gel for the management of actinic keratosis (Picato®). The discovery of the clinical potential of I3A is rather unusual and almost at complete odds with the current mainstream paradigm of drug discovery. Rather than emanating from the sophistication of the molecular screening on the cancer receptor "du jour", the discovery of I3A can be traced to the curiosity and motivation of James Aylward, an ex-scientist from the Commonwealth Scientific and Industrial Research Organisation (CSIRO), and to the open-mindness and ingenuity of a small group of Australian researchers based at the Queensland Institute of Medical Research (QIMR) in Brisbane [164]. Overall, the discovery of this agent is a remarkable example of cutting-edge modern "reverse pharmacology", where preliminary clinical validation related to folkloric use was obtained before pharmaceutical development.

### 4.3.1 Discovery of the Anticancer Activity of the Sap from *Euphorbia peplus*

Australia has one of the highest rates of skin cancer in the world, accounting for about 80% of all new cancers diagnosed each year [165]. With the estimate that two in three Australians will be diagnosed with skin cancer by the age of 70 [166], this condition gains a great deal of media attention, and in the 1970s Aylward's mother read in an article that the sap of the petty spurge (*E. peplus* L., known as "radium weed" in Australia) was used by farmers to treat skin cancer and related conditions. She then started to treat her own sun spots, a type of pre-cancerous lesions known more formally as actinic keratosis, and those of other members of her family, with amazing results. In 1996, Aylward contacted Peter Parsons, a Professor at QIMR, with the idea of investigating the potential of the sap to be developed as a mainstream anticancer drug. Parsons evaluated the antiproliferative activity of the sap against five cell lines, documenting a noteworthy degree of selectivity between cancer cells and normal fibroblasts, and discovering a curious effect on the melanoma cell line MM96L, for which the phenotype had lost the polydendritic pattern, back-differentiating into the bipolar shape of the melanocytes from which it had derived [164]. This morphological change was used as the cellular end-point for the activity-guided fractionation of the sap that ultimately led to the discovery of I3A (**5**) as the active anticancer agent in *E. peplus* sap. In light of the confirmation of cytotoxic activity for a wide panel of cancer cell lines, the biotechnology company Peplin was eventually established two years later, in 1998.

Peplin was instrumental in coordinating and funding a network of multidisciplinary collaborations that in the following two years provided further evidence for the anticancer activity of the sap, including the critical observations that activity was fully retained by sap extracts, as well as using in vivo experiments on nude (Foxn1nu) mice injected with melanoma cells [164]. Much of this early stage research was by, at that time, a junior postdoctoral researcher, Steven Ogbourne, who, having won a prestigious National Health and Medical Research, Industry Research Fellowship, was able to focus his research efforts under the mentorship of Parsons at QIMR and Aylward at Peplin on the isolation, anticancer activity determination, and mechanism of action of I3A (**5**). Ogbourne was ultimately employed by Peplin, and was closely involved in all aspects of the development of this compound.

### 4.3.2 Identification of I3A as the Major Active Constituent from the Sap of *Euphorbia peplus*

In the wake of the promising pre-clinical data on the sap, Peplin went on a public offering in September 2000, raising almost five million euros, which fueled studies aimed at clinically validating the activity of the plant sap and initiating a drug development program. The topical administration of the sap showed remarkable activity when evaluated against a small cohort of cancer patients suffering from basal cell-, squamous cell-, and intraepidermal-carcinomas. This study, carried out in 2000–2002 but published later [166], provided a strong rationale for turning the sap into a mainstream drug. Given the lack of information on the molecular mechanism of activity of the sap and its molecular target(s), bioactivity-directed fractionation was carried out using a combination of three phenotypic assays, namely, cytotoxicity, morphological transformation of the melanoma line MM96L to a bipolar state, and mouse xenografts. The active fraction purified from the sap contained a mixture of three ingenoids, all featuring a 3-angeloyl moiety: ingenol-3-angelate (PEP005, **5**) and its 20-deoxy (PEP006, **65a**) and 20-acetyl (PEP008, **57a**) derivatives. Ingenol-3-angelate (**5**) was the most potent of the three, and a patent was filed in 2001 on its uses as an anticancer agent [167], with the first one gram batch being manufactured at QIMR in 2004. Ingenol-3-angelate (**5**) had first been isolated from the petty spurge in 2000 [58], and had already been described in several other species (see Table 2), but not until the discoveries of Aylward and Parsons had **5** been associated with cancer treatment.

### 4.3.3 Development of I3A

A supply chain for the production of the starting biomass (dried aerial parts) of the petty spurge was implemented successfully, and production then moved from QIMR to a custom-built GMP manufacturing facility in the Gold Coast area of Queensland. A deal struck in November 2002 gave the U.S. company Allergan exclusivity of the product for skin diseases in North and South America, but, in the wake of budgetary constraints at Allergan, Peplin re-acquired the product rights in October 2004. Positive phase I results were reported in 2005 [168], and Peplin went on to validate clinically a gel formulation of PEP005 in phase II studies and initiated phase III trials. On September 3, 2009, Peplin was eventually purchased by the Danish company LEO Pharma, a globally recognized dermatology leader, for approximately US $ 287.5 million. The Danish LEO Pharma continued the clinical development of **5** and eventually received U.S. FDA approval for a gel formulation of the compound in January 2012 [169]. Derivative PEP005 was renamed ingenol mebutate, and the formulation received the brand name of Picato®. Approval in the EU was granted in November 2012 [170], and the product is now commercialized worldwide. Ingenol-3-angelate (**5**) is currently obtained by isolation from farmed *E. peplus*. The supply chain directly employs over 60 manufacturing specialists and approximately ten farmers, directly injecting over $30 million into Australia's economy each year.

### 4.3.4 Molecular and Clinical Pharmacology of Picato®

The molecular mechanism of the anticancer activity of I3A (**5**) is unclear, and it has been suggested to be twofold, involving a combination of necrotic and immunostimulating effects [171]. Depending on the concentration and cell type, I3A shows necrotic or apoptotic properties, the latter seemingly dependent upon the activation of PKCδ and its translocation from the cytoplasm into the nuclear membrane. Conversely, interaction with PKCβII is critical for the activation of neutrophils involved in the immunostimulating activity [133]. Ingenol-3-angelate (**5**) also shows immunostimulatory effects, increasing the release of chemokines by a range of cell types inclusive of malignant cells, and promoting, after topical application, the local infiltration of neutrophils, and the development of specific anticancer immune responses by CD8(+) T cells [133].

Ingenol-3-angelate (**5**) shows an impressive potential to be used as a systemic anticancer agent, especially against leukemia [172], but little information exists on its systemic toxicity and therapeutic window, and similar considerations apply for its use as HIV-1 latency-reversal agent [173].

So far, the only clinically approved use of I3A (**5**) is for the treatment of actinic keratosis (AK), a focal area of abnormal keratinocyte proliferation associated with chronic exposure to the sun (sun spots) and considered an early step in a continuum of alterations that leads from normal skin to invasive and metastatic squamous cell carcinoma (SCC). The likelihood of development of an invasive SCC from a single AK lesion is around 0.075–0.096% per lesion per year, meaning that for an individual with an average of 7.7 lesions, the probability of at least one transforming into SCC, within a ten-year period, is almost 10% [174]. The removal of sun spots is therefore, an important form of cancer chemoprevention. The topical treatment of multiple actinic keratosis lesions by application of a pharmaceutical product to a specific area of the skin is referred to in medical terminology as "field therapy", and, before the introduction of **5**, required weeks or months of treatment, alone or in association with surgical operations like laser removal, cryosurgery, electrodesiccation, shave excision, or laser ablation. Before the introduction of **5** the pharmacological treatment was based on 5-fluorouracil, imiquimod, retinoids, or diclofenac [174].

For topical application, two formulations of **5** have been developed. The lower concentration product (0.015%, 150 ppm) is used daily for three consecutive days for actinic keratosis on the face or scalp, while the more concentrated product (0.05%, 500 ppm) is used daily for two consecutive days for actinic keratosis on the trunk or limbs [175]. Complete and sustained clearance is observed in a significant percentage (ca. 50%) of patients, with outstanding quality of life and satisfaction [176]. Ingenol-3-angelate (**5**) has also been proposed as a treatment for genital warts [177], another application in line with the use of the sap of *E. peplus* in folkloric medicine.

## *4.4 Toxicity*

### 4.4.1 Skin Irritancy

Ingenol esters are potent skin irritants, as most tumor promoters are, but they are generally less potent tumor promoters when compared to phorbol esters [178]. Thus, irritancy is a necessary but per se insufficient prerequisite for tumor-promoting activity, and these two types of activities can be dissected as in phorbol 12-retinoate-13-acetate, a compound almost as irritant as TPA but only marginally active as a promoter [179]. The detrimental effect of polyunsaturation on tumor promotion is quite general in phorboids, and has been observed also in ingenol-3-esters. Short-chain esters like I3A are highly irritant for the skin, but, unlike TPA and long-chain esters, induce a transient, and not long lasting, response [7]. From a mechanistic standpoint, co-carcinogenicity is related to the activation of various isoforms of PKC, while it has been suggested that irritancy can be mediated by the activation of specific isoforms of PKC [7] and or by phospholipase A2 and the liberation of arachidonic acid from biological membranes [179].

### 4.4.2 Animal and Human Exposure

Animal exposure to ingenoids is direct, via consumption of their plant food sources or with fodder contaminated with these compounds. Conversely, human exposure is mostly indirect, via the food chain and products (e.g. milk, honey, cheese) obtained from animals exposed to ingenoids, with direct exposure being limited to the medicinal uses of euphorbiaceous species or to gardening activities [180]. Many spurges have a high protein content and low fiber levels, qualifying as a potential highly nutritive food source for animals, comparable to alfalfa, and better than most grasses [87]. However, the presence of ingenol esters triggers aversive responses in cattle after ingestion of even small amounts of ingenoid-containing plants, with a negative impact on the productivity of agricultural areas. The situation is especially severe in the Great Plains of North America, where millions of acres of pasture lands are heavily contaminated by the leafy spurge (*E. esula*), a creeping perennial of Eurasian origin first imported to the U.S. as a seed impurity in the first half of the nineteenth century [87, 181].

Phytochemical studies have established the presence in the leafy spurge of 3-acyl derivatives of ingenol, of which the nature seems to depend on location, due to the presence of distinct chemotypes characterized by the presence of aromatic (benzoyl) [103] or polyunsaturated aliphatic acyl moieties [56], with both chemotypes having been detected in U.S. collections of the plant [87]. Due to their irritancy, these compounds are the most likely candidates as the aversive constituents of the leafy spurge. On the other hand, the concentration of ingenol esters in the leafy spurge is very small, overall in the range of 5–10 ppm, and in vivo experiments in cattle have shown significant activity for ingenol itself and for a compound tentatively identified as 3-acetyl-20-dodecanoylingenol, suggesting a complex situation, not exclusively accounted for by the presence of any single 3-acyl ingenol derivative or by its irritancy [87]. A toxification role for the rumen microbiota has also been suggested, possibly related to the selective hydrolysis of the 3,20-diesters to the more irritant 3-esters, and vegetable tannins (plant polyphenols) might also play a role [182]. Despite a negative economic impact estimated at more than $ 200 million a year [87], the molecular basis for the aversion of cattle to leafy spurge is still not entirely clear. The aggressive diffusion of the plant in the prairies suggests that most wildlife avoids it, and the plant has not only an economic impact but also a negative effect on the overall quality of the natural habitat, replacing native plants and failing to support their associated animal community. Furthermore, the avoidance of the leafy spurge by large herbivores means that non-contaminated area will be intensely browsed, creating additional opportunities for further encroachment by *E. esula*, which can thus spread both by seed dispersal and through its rhizomes. Additionally, other spurges are invasive plants, and they can contaminate fodder, as detected in Egypt for *E. helioscopia* L., a plant containing short- and medium-chain ingenol esters [183, 184].

Lactating animals fed with spurges excrete diterpenoid esters in the milk, rationalizing the observation that milk from animals fed in areas contaminated by these plants is toxic [185]. Contamination from ingenol derivatives was indeed detected in lactating goats fed *E. peplus* [186]. Both the native esters and ingenol itself were isolated from the milk, explaining certain case reports of acute human poisoning from milk contaminated by spurge constituents [186]. The toxicological implications of the chronic consumption of milk contaminated by diterpenoid esters are unknown, but it has been suggested that the unusually high incidence of esophageal cancer observed in certain areas of the Caspian littoral of Iran, where consumption of goat milk is very high, might be related to chronic exposure to diterpenoid esters [187, 188].

In general, rumen microbes cannot degrade phorboids [189], and nor is the rumen biota is affected adversely by them, and it has been suggested that little difference exists in the toxicity of these compounds between ruminants and monogastric animals. Mammary excretion seems therefore an important detoxification pathway for phorboids, and it is interesting to observe that they generally co-occur with macrocyclic diterpenoid esters. These non-irritant compounds are potent inhibitors of transport proteins of the ATP cassette [16], a class of transporters highly expressed in mammary tissues during lactation [190], and it does not seem unreasonable to assume that their evolutionary role might be related to the block of this major detoxification route in herbivorous animals.

Ingenol derivatives have been detected in honey obtained from spurges [191]. Although the pollen of spurges is per se devoid of toxic diterpenoids, pollinators get contaminated during their browsing activity. Transfer of ingenol esters to honey has been documented for *E. seguieriana* Neck. in Iran [192], and honeys from ingenoid-containing plants like *E. marginata* Pursh are generally toxic [192]. Since the flowers of *E. ingens* are a major source of nectar, also the (in) famous toxic and irritant South-African noors honeys might contain ingenol derivatives [193].

Taken together, these studies show that a human dietary exposure to ingenol esters is, overall, possible, but, given the trace amounts involved, poisoning seems limited to the chronic consumption of goat milk contaminated with trace amounts of ingenol esters. Goats, as well as sheep, can feed with impunity on many spurges, and have even been proposed as a "green" alternative to pesticides and insects for the control of invasive spurge like *E. esula* [87].

In Western medicine, the use of spurges in folk medicine is mostly topical, for the removal of warts, and skin cancers, and they are only occasionally used internally as cathartics [180]. Exposure to ingenoids is therefore limited. On the other hand, several *Euphorbia* species are used in traditional Chinese Medicine (TCM), with four of them being included in the Chinese Pharmacopoeia [194]. Among these plants, *E. kansui* is a major source of ingenol esters. This plant is a prolific producer of ingenol and 16-hydroxyingenol esters, and is a popular treatment for edema, ascites, and asthma. However, the roots are processed by stir-baking with vinegar, a procedure that causes a considerable degradation of the ingenol esters [93, 195]. Some homeopathic products are based on ingenoid-containing plants, and, indeed, trace amounts of ingenol and its derivatives were detected in homeopathic tinctures [196]. Concentrations of ingenol in the range of 5–17 mg/L were, in fact, detected in the tincture of *E. resinifera*, while other popular homeopathic tinctures like those from *E. amygdaloides* L. and *E. pilulifera* L. only contained traces amount of this compound. Unfortunately, the presence of deoxyphorbol esters, which often co-occur with ingenols in spurges, was not evaluated [196]. Given the low level of consumption of these products, it seems unlikely that these trace amounts of ingenol ester might be bioactive.

Professional exposure to ingenoids is limited to the species grown for the production of oils or for horticultural purposes. The seeds of *E. lathyris* are mass-cultivated for the production of lamp oil, especially in mainland China, and contain large amounts (>0.1%) of ingenol esters [2]. The oil is detoxified by heating, which degrades the irritant diterpenoids. It is also commonly planted in domestic gardens for an alleged capability of deterring moles (hence its nickname of "mole plant") [2]. *Euphorbia milii* (crown of thorns), a species native to Madagascar, is a popular indoor plant in temperate zones, and is used extensively used as an ornamental in warmer areas [180]. As mentioned earlier, this plant contains peptide conjugates of ingenol based on *N*-methyl anthranilate, a unique type of natural product [78, 79]. Also, *E. marginata* (snow-on-the-mountain), a North American plant of which the type specimen was collected by William Clark during his famous expedition with Lewis, has been reported to contain ingenol esters, but the only phytochemical investigation on this plant actually afforded an ingol ester [197]. Finally, some ingenoid-containing succulent Euphorbias (*E. ingens*, *E. tirucalli*) are used as border plants in warm areas, acting not only as a physical, but also as a veritable chemical barrier [180].

## 5 Conclusions

Phorboids have provided valuable tools to study cell metabolism and neural activity, as exemplified by the tigliane PMA and the daphnane resiniferatoxin (RTX), but ingenoids are the only class of phorboids to have so far been subjected to clinical development. This is surprising not only on account of the instability of ingenol esters to both skeletal and acyl rearrangement, but also because of their low concentrations in plant biomass. Clinical research of ingenol-3-angelate (I3A, **5**) is expected to provide a better comprehension of the biological profile of this class of compounds, clarifying the differences, if any, with the analogous parameters of phorbol esters, while the increased availability of the parent polyol will expand the resultant knowledge on their chemical reactivity and the medicinal chemistry.

Several issues are still unresolved and will be important to address. From the phytochemical standpoint, the structure of two groups of ingenoids, the tetrahydroingenol esters and the 16-hydroxyingenol esters, need confirmation. The nature of the ingenol derivatives occurring in plants from the genus *Elaeophorbia* is still unknown, and the distribution of ingenoids in neo-tropical euphorbiaceous species should be worth a systematic evaluation. Most important, the Jakupovic biogenetic proposal for the formation of phorboids should be rigorously evaluated in labeling experiments. From the biological standpoint, the profiles of several classes of ingenoids substantially transcend the interaction with PKC, and is still incompletely understood in terms of their molecular targets. Thus, the extraordinary systemic toxicity of milliamines ($LD_{50} = <1$ mg/kg in mice for milliamine A hydrochloride (**57o**), a value lower than that of strychnine [198]), with the former compound representing a class of non-irritant and tumor-promoting ingenoids, hints to interference with some basic but yet unknown physiological process. In turn, the kinetics of skin irritancy of several classes of ingenoids differ greatly from the phorbol esters, in being more similar to the daphnane resiniferatoxin, a non-PKC targeting compound. Even within the realm of PKC-activating ingenoids, there is a surprising profligacy of diterpenoid templates (ingenol, 20-deoxyingenol, 5-deoxyingenol, 13-hydroxyingenol), whose acyl decoration is capable of inducing powerful kinase-activating properties. Undoubtedly, ingenoids exemplify vividly the concept that the larger our body of knowledge grows, the longer the interface is extended where current knowledge meets our ignorance of this topic.

**Acknowledgments** This chapter could not have been written without the help of many colleagues and friends. In particular, I am very grateful to Siems Karsten, Fotini Tsichritzis and Sven Jakupovic (Analyticon, Potsdam, Germany) and Franck Jescke for providing me with the material that made it possible to discuss the Jakupovic proposal for the biogenesis of phorboids, to Marco Facini (Sandoz, Munich, Germany) for providing many articles from the early literature on ingenol, and to Louis-Félix Nothias-Scaglia (University of Corsica, France) for his suggestions. A special thanks should also be given to Peter Parson (QIMR Berghofer Medical Research Institute, Brisbane, Australia) and Steven Ogbourne (GeneCology Research Centre, University of the Sunshine Coast, Australia) for carefully reading the biological section of this manuscript and for their suggestions on improving it. Finally, I would like to apologize to my daughter Silvia for the time spent with ingenol and not with her during the summer of 2014, when the first draft of this review was prepared.

My wife Enrica, a medical doctor, medicated me on two occasions when, during my early work on ingenol, I experimented on myself the power of its activity. Were she still with us, she would surely be surprised that an ingenol derivative is now a drug. Jasmin Jakupovic always stimulated me with his ingenuity and inspiring insights, and it is an honor for me to dedicate this work to his memory and to the one of my wife. Paraphrasing a comment made on the premature death of the composer Bellini, *what Jasmin lacked could be learnt. What he had could not.*

## References

1. Flaschenträger B, v Wolffersdorff R (1934) Über den Giftstoff des Crotonöles. 1. Die Säuren des Crotonöles. Helv Chim Acta 17:1444
2. Adolf W, Opferkuch HJ, Hecker E (1968) Diterpene aus dem Samen von *Euphorbia lathyris* and ihre tumorpromovierende Wirkung. Fette Seifen Anstrichmittel Ernaehrungsindustrie 70:825
3. Hecker E (1968) Cocarcinogenic principles from the seed oil of *Croton tiglium* and from other Euphorbiaceae. Cancer Res 28:2338
4. Pettersen RC, Ferguson G, Crombie L, Games ML, Pointer DJ (1967) The structure and stereochemistry of phorbol, diterpene parent of co-carcinogens of croton oil. Chem Commun:716
5. Stout GH, Balkenhol WG, Poling M, Hickernell GL (1970) The isolation and structure of daphnetoxin, the poisonous principle of *Daphne* species. J Am Chem Soc 92:1070
6. Zechmeister K, Brandl F, Hoppe W, Hecker H, Opferkuch HJ, Adolf WA (1970) Structure determination of the new tetracyclic diterpene ingenol-triacetate with triple product methods. Tetrahedron Lett 11:4075
7. Hecker E, Schmidt R (1974) Phorbolesters – the irritant and cocarcinogens of *Croton tiglium* L. Prog Chem Org Nat Prod 31:283
8. Evans FJ, Taylor SE (1983) Pro-inflammatory, tumor promoting and anti-tumour diterpenes of the plant families Euphorbiaceae and Thymelaeaceace. Prog Chem Org Nat Prod 44:17
9. Roe JC, Peirce WEH (1961) Tumor promotion by *Euphorbia* latices. Cancer Res 21:338

10. Adolf W, Hecker E, Balmain A, Lhomme MF, Nakatani Y, Ourisson G, Ponsinet G, Pryce RJ, Santhanakrishna TS, Matyukhina LG, Saltikova IA (1970) "Euphorbiasteroid" (Epoxy-lathyrol). A new tricyclic diterpene from *Euphorbia lathyris* L. Tetrahedron Lett 11:2241
11. Dublyanskaya NF (1937) Chemical characteristics of *E. lathyris* L. as an oleageneous plant. Biokhimiya 2:521
12. Evans FJ (1986) Naturally occurring phorbol esters. CRC, Boca Raton, FL
13. Alder RW, East SP (1996) In/out isomerism. Chem Rev 96:2097
14. Braekman JD, Daloze D, Dupont A, Pasteels J, Tursch B, Declercq JP, Germain G, van Meerssche M (1980) Secotrinervitane, a novel bicyclic diterpene skeleton from a termite soldier. Tetrahedron Lett 21:2761
15. Prestwich GD, Tempesta MS, Turner C (1984) Longipenol, a novel tetracyclic diterpene from the termite soldier *Longipeditermes longipes*. Tetrahedron Lett 25:1531
16. McKerrall SJ, Jørgensen L, Kuttruff CA, Ungeheuer F, Baran PD (2014) Development of a concise synthesis of (+)-ingenol. J Am Chem Soc 136:5799. Correction: ibid, doi: 10.1126/science.aad8935
17. Funk RI, Olmstead TA, Parvez M (1988) A solution to the in, out-bicyclo[4.4.l]undecan-7-one problem inherent in ingenane total synthesis. J Am Chem Soc 110:3298
18. Paquette LA, Ross RJ, Springer JP (1988) Stereocontrolled construction of an ingenol prototype having a complete array of oxygenated and unsaturated centers. J Am Chem Soc 110:1988
19. Winkler JD, Rouse MD, Greaney MF, Harrison SJ, Jeon YT (2002) The first total synthesis of (±)-ingenol. J Am Chem Soc 124:9726
20. Sorg B, Schmidt R, Hecker E (1987) Structure/activity relationships of polyfunctional diterpenes of the ingenane type. I. Tumor-promoting activity of homologous, aliphatic 3-esters of ingenol and of delta 7,8-isoingenol-3-tetradecanoate. Carcinogenesis 8:1
21. Appendino G, Tron GC, Cravotto G, Palmisano P, Jakupovic J (1999) An expeditious procedure for the isolation of ingenol from the seeds of *Euphorbia lathyris*. J Nat Prod 62:76
22. Wu Q-C, Tang Y-P, Ding A-W, You FQ, Xhang L (2009) $^{13}$C-NMR data of three important diterpenes isolated from *Euphorbia* species. Molecules 14:4454
23. Yang M, Lu Z, Yu K, Wang Q, Chen X, Li Y, Liu X, Wu W, Guo D-A (2012) Studies on the fragmentation pathways of ingenol esters isolated from *Euphorbia esula* using IT-MS$^n$ and Q-TOF-MS/MS methods in electrospray ionization mode. Int J Mass Spectrom 323–324:55
24. Ogbourne SM (2011) Crystalline ingenol mebutate. US Patent Appl No 2011/0257262 A1
25. Shi Q-W, Su X-H, Kiyota H (2008) Chemical and pharmacological research of the plants in genus *Euphorbia*. Chem Rev 108:4295
26. Tsang W (1985) The stability of alkyl radicals. J Am Chem Soc 110:7343
27. Nonhebel DC (1993) The chemistry of cyclopropylmethyl and related radicals. Chem Soc Rev:347
28. Appendino G, Tron GC, Jarevang T, Sterner O (2001) Unnatural natural products from the transannular cyclization of lathyrane diterpenes. Org Lett 3:1609
29. Shizuri Y, Ohtsuka J, Kosemura S, Terada Y, Yamamura S (1984) Biomimetic reactions of some macrocyclic diterpenes. Tetrahedron Lett 25:55473
30. Appendino G, Jakupovic J, Varese M, Bombardelli E (1997) Acid and base catalysed rearrangements of 9,10-dioxotaxanes. Tetrahedron Lett 38:727
31. Schmidt RJ (1987) The biosynthesis of tigliane and related diterpenoids: an intriguing problem. In: Jury SL, Reynolds T, Cutler DF, Evans FJ (eds) The Euphorbiales: chemistry, taxonomy and economic botany. Academic Press, London, p 221
32. Jeske F (1996) Naturstoffe aus Euphorbiaceen. Studien zum konformativen Verhalten von Jatrophanen. PhD Thesis, Freie Universität Berlin, p 676
33. Cane DE (1980) The stereochemistry of allyl pyrophosphate metabolism. Tetrahedron 36:1109
34. Micali G, Lacarrubba F, Nasca MR, Schwartz RA (2014) Topical pharmacotherapy for skin cancer. Part 1. Pharmacology. J Am Acad Dermatol 70:965e

35. Fortin JL, Bitar MP, N'Guedia Kenfack F, Jaboura, S, Fortin L, Coste S, Savio C (2014) Toxicité par *Euphorbia lathyris*: Efficacité des solutions polyamphotères de lavage. Clin Toxicol 52:383
36. FDA Drug Safety Communication. http://www.fda.gov/Safety/MedWatch/SafetyInformation/SafetyAlertsforHumanMedicalProducts/ucm459311.htm. Accessed 9 Sept 2015
37. Appendino G. Unpublished observations
38. Goel G, Makkar HPS, Francis G, Becker K (2007) Phorbol esters: structure, biological activity, and toxicity in animals. Int J Toxicol 26:279
39. Lehmanns JF (1966) Nachdruck der 1.ste Aufl. von 1925. Carl-Hanser, Muenchen
40. Friesen JB, McAlpine JB, Chen S-N, Pauli GF (2015) Countercurrent separation of natural products: an update. J Nat Prod 78:1765
41. Bagavathi R, Sorg B, Hecker E (1991) On the chemistry of ingenol, [IV]: ingenol from seeds of *Euphorbia lathyris* L. and preparation of (9*R*)[(9*S*)]-9-deoxo-9-hydroxyingenol with some corresponding 3- and 9-esters. Z Naturforsch 46b:1425
42. Berge JM, Roberts SM (1979) The synthesis of fused and bridged ring systems by free radical carbocyclization. Synthesis 471
43. Herzig J, Nudelman A, Gottlieb HE, Fischer B (1986) Studies in sugar chemistry. 2. A simple method for *O*-deacylation of polyacylated sugars. J Org Chem 51:727
44. Bellido Cabello de Alba ML, Appendino G, Pagani A, Munoz Blanco E (2012) Method of isolating ingenol and synthesis of biologically active ingenol derivatives. WO Patent Appl No 2012-EP69452
45. CITES Appendices I–III (2013) https://www.cites.org/. Accessed 1 Oct 2015
46. Pandeló JD, Bartholomeeusen K, da Cunha RD, Abreu CM, Glinski J, da Costa TB, Bacchi Rabay AF, Pianowski Filho LF, Dudycz LW, Ranga U, Peterlin BM, Pianowski LF, Tanuri A, Aguiar RS (2014) Reactivation of latent HIV-1 by new semi-synthetic ingenol esters. Virology 462–463:328
47. Kuster RM, Calixto MLC, Sabino KCC, da Costa HB, Tose LV, Romão W, Vaz BG, Silvia AG (2015) Identification of malonyl glucans from *Euphorbia tirucalli* by ESI-(−)-FT-ICR MS analyses. Phytochem Lett 12:209
48. Evans FJ, Kinghorn AD (1977) A comparative phytochemical study of the diterpenes of some species of the genus *Euphorbia* and *Elaeophorbia* (Euphorbiaceae). Bot J Linn Soc 74:23
49. Govaerts R, Frodin DG, Radcliffe-Smith A, Carter S (2000) World checklist and bibliography of Euphorbiaceae (with Pandaceae). Royal Botanical Gardens, Kew, Richmond
50. Wang H-B, Wang X-Y, Liu L-P, Qin G-W, Kang T-G (2015) Tigliane diterpenoids from Euphorbiaceae and Thymelaeaceae families. Chem Rev 115:2975
51. Kinghorn AD, Evans FJ (1974) Quantitative gas-liquid chromatographic method for phorbol and related diterpenes as their acetates. J Pharm Pharmacol 26:408
52. Hergenhahn M, Kusumoto S, Hecker E (1974) Diterpene esters from Euphorbium and their irritant and cocarcinogenic activity. Experientia 30:1438
53. Adolf W, Chanai S, Hecker E (1983) 3-*O*-Angeloylingenol, the toxic and skin irritant factor from latex of *Euphorbia antiquorum* L. (Euphorbiaceae) and from a derived Thai purgative and anthelmintic (vermifuge). Drug J Sci Soc Thailand 9:81
54. Marco JA, Sanz-Cervera JF, Yuste A (1997) Ingenane and lathyrane diterpenes from the latex of *Euphorbia canariensis*. Phytochemistry 45:563
55. Hoegberg T, Grue-Sorensen G, Liang X, Horneman AM, Petersen AK (2012) A method of producing ingenol-3-angelate. WO Patent Appl No 2012/010172 A1
56. Seip EH, Hecker E (1982) Skin irritant ingenol esters from *Euphorbia esula*. Planta Med 46:215
57. Rizk A, Hammouda FM, Le-Missiry MM, Williamson EM, Evans FJ (1980) Constituents of Egyptian Euphorbiaceae. IX, Irritant and cytotoxic ingenane esters from *Euphorbia paralias*. Experientia 36:1206
58. Hohmann J, Evanics F, Berta L, Bartok T (2000) Diterpenoids from *Euphorbia peplus*. Planta Med 66:291

59. Jakupovic J, Morgenstern T, Bittner M, Silva M (1998) Diterpenes from *Euphorbia peplus*. Phytochemistry 47:1601
60. Upadhyay RR, Samiyeh R, Tafazuli A (1981) Tumor promoting and skin irritant diterpene esters of *Euphorbia virgata* latex. Neoplasma 28:555
61. Rizk AM, Hammouda FM, El-Missiry MM, Radwan HM, Evans FJ (1985) Constituents of Egyptian Euphorbiaceae. Part 13. Biologically active diterpene esters from *Euphorbia peplus*. Phytochemistry 24:1605
62. Abo KA, Evans F (1982) Ingenol esters from the pro-inflammatory fraction of *Euphorbia kamerunica*. Phytochemistry 31:275
63. Fürstenberger G, Hecker E (1977) New highly irritant *Euphorbia* Factors from the latex of *Euphorbia tirucalli* L. Experientia 33:986
64. Wang L-Y, Wang N-L, Yao X-S, Miyata S, Kitanaka S (2002) Diterpenes from the roots of *Euphorbia kansui* and their in vitro effects on the cell division of *Xenopus*. J Nat Prod 65:1246
65. Tsukasa M, Yamada A, Jung CC, Ohzuka Y (1988) Effects of Kansui extracts on macrophage immune complex binding. Wakan Iyaku Gakkaishi 5:430
66. Matsumoto T, Cyong JC, Yamada H (1992) Stimulatory effects of ingenols from *Euphorbia kansui* on the expression of macrophage Fc receptor. Planta Med 58:255
67. Opferkuch HJ, Hecker E (1982) On the active principles of the spurge family (Euphorbiaceae). IV. Skin irritant and tumor promoting diterpene esters from *Euphorbia ingens* E. Mey. J Cancer Res Clin Oncol 103:255
68. Opferkuch HJ, Hecker E (1974) New diterpenoid irritants from *Euphorbia ingens*. Tetrahedron Lett 3:261
69. Adolf W, Hecker E (1971) Further new diterpene esters from the irritant and cocarcinogenic seed oil and latex of the caper spurge (*Euphorbia lathyris* L.). Experientia 27:1393
70. Adolf W, Hecker E (1975) Active principles of the spurge family. III. Skin irritant and cocarcinogenic factors from the caper spurge. Z Krebsforsch Klin Onkol 84:325
71. Gonzalez Urones J, Basabe Barcala P, Sexmero Cuaradao MJ, Sanchez Marcos I (1988) Diterpenes from the latex of *Euphorbia broteri*. Phytochemistry 27:207
72. Li Y, Suo Y (2005) Ingenane diterpene ester constituents from Tibetan medicine *Euphorbia wallichii*. Zhongcaoyao 36:1763
73. Shi Y-P, Ma B, Joa Z-J, Lahham J, Saleh SA (1996) A new ingenol ester from *Euphorbia petiolata*. Chin Med Lett 7:439
74. Shi YP, Jia ZJ, Ma B, Saleh SD, Lahham J (1996) Ingenane diterpenes from *Euphorbia petiolata*. Planta Med 62:260
75. Wu D, Sorg B, Adolf W, Seip EH, Hecker E (1992) Oligo- and macrocyclic diterpenes in Thymelaeaceae and Euphorbiaceae occurring and utilized in Yunnan (Southwest China). 2. Ingenane diterpene esters from *Euphorbia nematocypha* Hand.-Mazz. Phytother Res 6:237
76. Uemura D, Hirata Y (1973) Isolation and structures of irritant substances obtained from *Euphorbia* species (Euphorbiaceae). Tetrahedron Lett 11:881
77. Upadhyay RR, Ansarin M, Zarintan MH, Shakui P (1976) Tumor promoting constituents of *Euphorbia serrata* L. latex. Experientia 32:1196
78. Uemura D, Hirata Y (1971) The isolation and structure of two new alkaloids, milliamines A and B, obtained from *Euphorbia millii* (sic). Tetrahedron Lett 19:3673
79. Marston A, Hecker E (1983) On the active principles of the Euphorbiaceae. VI. Isolation and biological activity of seven milliamines from *Euphorbia milii*. Planta Med 47:141
80. Brooks G, Evans AT, Markby DP, Harrison ME, Baldwin MA, Evans FJ (1990) An ingenane diterpene from Belizian *Mabea excelsa*. Phytochemistry 29:1615
81. Teng R-W, McManus D, Aylward J, Ogbourne S, Armstrong D, Mau S-L, Johns J, Bacic A (2009) Biotransformation of ingenol-3-angelate in four plant cell suspension cultures. Biocatal Biotransformation 27:186
82. Baloch IB, Baloch MK (2011) Isolation and characterization of ingenol esters from *Euphorbia cornigera*. Chin Chem Lett 22:451

83. Uemura D, Ohwaki H, Hirata Y, Chen YP, Hsu HY (1974) Isolation and structure of 20-deoxyingenol. New diterpene, derivatives and ingenol derivative obtained from "Kansui". Tetrahedron Lett 29:2527
84. Hecker E (1977) New toxic, irritant and cocarcinogenic diterpene esters from Euphorbiaceae and Thymeleaeceae. Pure Appl Chem 49:1423–1431
85. Markby DP, Brooks G, Evans AT, Evans FJ (1989) Checum (poison sap tree) toxin, a potent activator of protein kinase C. Lancet 1320
86. Deng B, Mu SZ, Zhang JX, Hao XJ (2010) New diterpenoids from the roots of *Euphorbia eubracteolata* Hayata. Nat Prod Res 24:1503
87. Halaweish FT, Kronberg S, Hubert MB, Rice JA (2002) Toxic and aversive diterpenoids of *Euphorbia esula*. J Chem Ecol 28:1599
88. Vasas A, Hohmann J (2014) *Euphorbia* diterpenes: isolation, structure, biological activity, and synthesis (2008–2012). Chem Rev 114:8579
89. Corea G, Di Pietro A, Dumontet C, Fattorusso E, Lanzotti V (2009) Jatrophane diterpenes from *Euphorbia* spp. as modulators of multidrug resistance in cancer therapy. Phytochem Rev 8:431
90. Marsh N, Rothshild M, Evans FJ (1984) A new look at Lepidoptera toxins. In: Vane-Wright RI, Ackery PR (eds) The biology of butterflies. Academic, London, p 135
91. Hundsdoerfer AK, Tshibangu JN, Wetterauer B, Wink M (2005) Sequestration of phorbol esters by aposematic larvae of *Hyles euphorbiae* (Lepidoptera: Sphingidae). Chemoecology 15:261
92. Appendino G, Tron GC, Cravotto G, Palmisano G, Annunziata R, Baj G, Surico N (1999) Synthesis of modified ingenol esters. Eur J Org Chem 12:3413–3420
93. Zhang XY, Guo J, Cao Y, Shang E, Tang Y, Ding A, Duan J-A (2014) Processing of Kansui roots stir-baked with vinegar reduces Kansui-induced hepatocyte cytotoxicity by decreasing the contents of toxic terpenoids and regulating the cell apoptosis pathway. Molecules 19:7237
94. Krenske EH, Williams CM (2015) Do anti-Bredt natural products exist? Olefin strain energy as a predictor of isolability. Angew Chem Int Ed Engl 54:10608
95. Mak JYW, Pouwer RH, Williams CM (2014) Natural products with anti-Bredt and bridgehead double bonds. Angew Chem Int Ed Engl 53:13664
96. Nguyen KA, Gordon MS, Wang G-t, Lambert JB (1991) Stabilization of β-positive charge by silicon, germanium, or tin. Organometallics 10:2798
97. Uemura D, Hirata Y, Chen Y-P, Hsu H-Y (1975) 13-Oxingenol derivative, a new diterpene isolated from *Euphorbia kansui*. Tetrahedron Lett 1974:2529
98. Nickel A, Maruyama T, Tang H, Murphy PD, Greene B, Yusuff N, Wood JL (2004) Total synthesis of ingenol. J Am Chem Soc 126:16300
99. Sorg B, Hecker E (1982) On the chemistry of ingenol. III. Synthesis of 3-deoxo-3-oxoingenol, some 5-esters and of ethers and acetals of ingenol. Z Naturforsch 37b:1640
100. Snatzke G (1961) Ueber die Oxydation mit $CrO_3$ in Dimethyulformamid. Chem Ber 94:729
101. Shibuya K (1994) A novel allylic oxidation using a combination of formic acid and selenium dioxide. Synth Commun 24:2923
102. Liang X, Grue-Sørensen G, Månsson K, Vedsø P, Soor A, Stahlhut M, Bertelsen M, Engell KM, Högberg T (2013) Syntheses, biological evaluation and SAR of ingenol mebutate analogues for treatment of actinic keratosis and non-melanoma skin cancer. Bioorg Med Chem Lett 23:5624
103. Uemura D, Hirata Y (1977) The isolation and structure of toxic principles, milliamines A, B, and C, from *Euphorbia millii* (sic). Bull Chem Soc Jpn 50:2005
104. Uemura D, Hirata Y, Chen YP, Hsu HY (1974) New diterpene, 13-oxyingenol, derivative isolated from *Euphorbia kansui* Liou. Tetrahedron Lett 29:2529
105. Lin LJ, Marshall GT, Kinghorn AD (1983) The dermatitis-producing constituents of *Euphorbia hermentiana* latex. J Nat Prod 46:723
106. Liang X, Grue-Sørensen G, Petersen AK, Högberg T (2012) Semisynthesis of ingenol-3-angelate (PEP005): efficient stereoconservative angeloylation of alcohols. Synlett 23:2647

107. Brown MB, Crothers MED (2007) Therapeutic compositions comprising ingenol-3-angelate. WO Patent Appl No 2007/068963 A3
108. Bundgaard H, Hansen J (1981) Studies on the stability of corticosteroids. VI. Kinetics of the rearrangement of betamethasone-17-valerate to the 21-valerate esters in aqueous solution. Int J Pharm 7:197
109. Grue-Sørensen G, Liang X, Månsson K, Vedsø P, Dahl Sørensen M, Soor A, Stahlhut M, Bertelsen M, Engell KM, Högberg T (2014) Synthesis, biological evaluation and SAR of 3-benzoates of ingenol for treatment of actinic keratosis and non-melanoma skin cancer. Bioorg Med Chem Lett 24:54
110. Kim S, Winkler JD (1997) Approaches to the synthesis of ingenol. Chem Soc Rev 26:387
111. Kuwajima I, Tanino K (2005) Total synthesis of ingenol. Chem Rev 105:4661
112. Cha JK, Epstein OL (2006) Synthetic approaches to ingenol. Tetrahedron 62:1329
113. Watanabe K, Suzuki Y, Aoki K, Sakakura A, Suenaga K, Kigoshi H (2004) Formal synthesis of optically active ingenol via ring-closing olefin metathesis. J Org Chem 69:7802
114. Tanino K, Onuki K, Asano K, Miyashita M, Nakamura T, Takahashi Y, Kuwajima I (2003) Total synthesis of ingenol. J Am Chem Soc 125:1499
115. Ohyoshi T, Funakubo S, Miyazawa Y, Niida K, Hayakawa I, Kigoshi H (2012) Total synthesis of (–)-13-oxyingenol and its natural derivative. Angew Chem Int Ed Engl 124:5056
116. Croteau R, Ketchum REB, Long RM, Kaspera R, Wildung MR (2006) Taxol biosynthesis and molecular genetics. Phytochem Rev 5:75
117. Appendino G (2014) *Omnia Praeclara Rara*. The quest for ingenol heats up. Angew Chem Int Ed 53:927
118. Marco-Contelles J, do Carmo Carreiras M, Rodríguez C, Villarroya M, García AG (2006) Synthesis and pharmacology of galantamine. Chem Rev 106:116
119. Yu MJ, Zheng W, Seletsky BM (2013) From micrograms to grams: scale-up synthesis of eribulin mesylate. Nat Prod Rep 30:1158
120. Newton AC (2001) Protein kinase C: structural and spatial regulation by phosphorylation, cofactors, and macromolecular interactions. Chem Rev 101:2353
121. Ron D, Kazanietz MG (1999) New insights into the regulation of protein kinase C and novel phorbol ester receptors. FASEB J 13:1658
122. Castagna M, Takai Y, Kaibuchi K, Sano K, Kikkawa U, Nishizuka Y (1982) Direct activation of calcium-activated, phospholipid-dependent protein kinase by tumor-promoting phorbol esters. J Biol Chem 257:7847
123. Hasler CM, Geza A, Blumberg PM (1992) Specific binding to protein kinase C by ingenol and its induction of biological responses. Cancer Res 52:202
124. Yeh E, Sharkey NA, Blumberg PM (1987) Influence of side chain on phorbol ester binding to protein kinase C. Phytother Res 1:135
125. Kedei N, Dundberg DJ, Toth A, Welburn P, Garfield S, Blumberg PM (2004) Characterization of the interaction of ingenol-3-angelate with protein kinase C. Cancer Res 64:3243
126. Pak Y, Enyedy IJ, Varady J, Kun JW, Lorenzo PS, Blumberg PM, Wang S (2001) Structural basis of binding of high-affinity ligands to protein kinase C: prediction of the binding modes through a new molecular dynamics method and evaluation by site-directed mutagenesis. J Med Chem 44:1690
127. Zhang GG, Kazanietz MG, Blumberg PM, Hurley JH (1995) Crystal structure of the cys2 activator-binding domain of protein kinase C delta in complex with phorbol ester. Cell 81:917
128. Pak Y, Wang S (1999) Folding of a 16-residue helical peptide using molecular dynamics simulation with Tsallis effective potential. J Chem Phys 111:4359
129. Krauter G, von der Liet CW, Schmidt R, Hecker E (1996) Structure/activity relationships of polyfunctional diterpenes of the tigliane type. Eur J Biochem 242:417
130. Kishi Y, Rando RR (1998) Structural basis of protein kinase C activation by tumor promoters. Acc Chem Res 31:163

131. Kazanietz MG, Caloca MJ, Eroles P, Fujii T, Garcia-Bermejo ML, Reilly M, Wang H (2000) Pharmacology of the receptors for the phorbol esters tumor promoters. Biochem Pharmacol 60:1417
132. Sodeoka M, Uotsu K, Shibasaki M (1995) Photoaffinity labelling of PKC with a phorbol derivative: importance of the 13-acyl group in phorbol ester-PKC interactions. Tetrahedron Lett 36:879599
133. Jin Y, Yeh CH, Kuttruff CA, Jørgensen L, Dünstl G, Felding J, Natarajan SR, Baran PS (2015) C-H oxidation of ingenanes enables potent and selective protein kinase C isoform activation. Angew Chem Int Ed 54:14044
134. Racke FK, Baird M, Barth RF, Huo T, Yang W, Gupta N, Weldon M, Rutledge H (2012) Unique in vitro and in vivo thrombopoietic activities of ingenol 3,20-dibenzoate, a $Ca^{++}$-independent protein kinase C isoform agonist. PLoS One 7:e51059
135. Yoshida C, Hishiyama K, Miyazaki K, Watanabe M, Kanbe M, Yamada Y, Matsuzaki K, Miyashita K, Kitanaka S, Miyata S (2010) Analysis of inhibition of topoisomerase II-alpha and cancer cell proliferation by ingenol EZ. Cancer Sci 101:374
136. Fukuda Y, Kanbe M, Watanabe M, Dan K, Matsuzaki K, Kitanaka S, Miyata S (2013) 3*EZ*,20Ac-Ingenol, a catalytic inhibitor of topoisomerases, downregulates p-Akt and induces DSBs and apoptosis of DT40 cells. Arch Pharm Res 36:1029
137. Miyata S, Fukuda Y, Tojima H, Matsuzaki K, Kitanaka S, Sawada H (2015) Mechanism of the inhibition of leukemia cell growth and induction of apoptosis through the activation of ATR and PTEN by the topoisomerase inhibitor 3*EZ*, 20Ac-ingenol. Leuk Res 15:30331
138. Oh JG, Chin YW, Kim SJ, Choi JM, Kim SK, Kang HE, Heo TH (2015) Biphasic effects of ingenol 3,20-dibenzoate on the erythropoietin receptor: synergism at low doses and antagonism at high doses. Mol Pharmacol 88:392
139. Khaleghian A, Riazi GH, Ghafari M, Rezaie M, Takahashi A, Nakaya Y, Nazari H (2010) Effect of inganen (sic) anticancer properties on microtubule organization. Pak J Pharm Sci 23:273
140. Selmer J, Knudsen KM (2015) A method for topically treating actinic keratosis with ingenol 3-(3,5-diethylisoxazole-4-carboxylate). WO Patent Appl No 2015059572
141. Sinnya S, Tan JM, Prow TW, Primiero C, McEniery E, Selmer J, Østerdal ML, Soyer HP (2016) A randomized, phase IIa exploratory trial to assess the safety and preliminary efficacy of LEO 43204 in patients with actinic keratosis. Br J Dermatol doi 174:305
142. Clinical Trials.Gov. A study of increasing strengths, safety and efficacy of two formulas of LEO 43204 on the face or the chest in patients with actinic keratosis. https://clinicaltrials.gov/ct2/show/NCT01922050
143. Berenblum I (1941) The co-carcinogenic action of *Croton* resin. Cancer Res 1:44
144. Furstenberger G, Berry DL, Bertsch S, Sorg B, Marks F (1981) Skin tumor promotion by phorbol esters is a two-stage process. Proc Natl Acad Sci USA 78:7722
145. Kupchan SM, Uchida I, Braufman AR, Dailey RG, Fei BY (1976) Antileukemic principles isolated from Euphorbiaceous plants. Science 191:571
146. Itokawa H, Ichihara Y, Watanabe K, Takeya K (1989) An antitumor principle from *Euphorbia lathyris*. Planta Med 55:271
147. Lee W-Y, Hampson P, Coulthard L, Ali F, Salmon M, Lord JM, Scheel-Toellner D (2010) Novel antileukemic compound ingenol 3-angelate inhibits T cell apoptosis by activating protein kinase C theta. J Biol Chem 285:23889
148. Song X, Lopez-Campistrous A, Sun L, Dower NA, Kedei N, Yang J, Kelsey JS, Lewin NE, Esch TE, Blumberg PM, Stone JC (2013) GRPs are targets of the anti-cancer agent ingenol-3-angelate. PLoS One 8:e72331
149. Mason SA, Cozzi SJ, Pierce CJ, Pavey SJ, Parsons PG, Boyle GM (2010) The induction of senescence-like growth arrest by protein kinase C-activating diterpene esters in solid tumor cells. Invest New Drugs 28:575

150. Blanco-Molina M, Tron GC, Macho A, Lucena C, Calzado MA, Muñoz E, Appendino G (2001) Ingenol esters induce apoptosis in Jurkat cells through an AP-1 and NF-KappaB independent pathway. Chem Biol 8:767
151. Fujiwara M, Ijichi K, Katsuura K, Shigeta S, Konno K, Wang GY, Uemura D, Yokota T, Baba M (1996) Mechanism of selective inhibition of human immunodeficiency virus by ingenol triacetate. Antimicrob Agents Chemother 40:271
152. Fujiwara M, Ijichi K, Tokuhisa K, Katsuura K, Wang G-YS, Uemura D, Shigeta S, Konno K, Yokota T, Baba M (1997) Ingenol derivatives are highly potent and selective inhibitors of HIV replication in vitro. Antivir Chem Chemother 7:230
153. Fujiwara M, Okamoto M, Ijichi K, Tokuhisa K, Anasaki Y, Katsuura K, Uemura D, Shigeta S, Konno K, Yokota T (1998) Upregulation of HIV-1 replication in chronically infected cells by ingenol derivatives. Arch Virol 143:2003
154. Proveda E (2014) Ingenol derivatives promising for HIV eradication. AIDS Rev 16:246
155. Jiang G, Mendes EA, Kaiser P, Wong DP, Tang Y, Cai I, Fenton A, Melcher GP, Hildreth JE, Thompson GR, Wong JK, Dandekar S (2015) Synergistic reactivation of latent HIV expression by ingenol-3-angelate, PEP005, targeted NF-κB signaling in combination with JQ1 induced p-TEFb activation. PLoS Pathog 11:e1005066
156. Spivak AM, Bosque A, Balch AH, Smyth D, Martins L, Planelles V (2015) Ex vivo bioactivity and HIV-1 latency reversal by ingenol dibenzoate and panobinostat in resting CD4+ T cells from aviremic patients. Antimicrob Agents Chemother 59:5984
157. Darcis G, Kula A, Bouchat S, Fujinaga K, Corazza F, Ait-Ammar A, Delacourt N, Melard A, Kabeya K, Vanhulle C, Van Driessche B, Gatot JS, Cherrier T, Pianowski LF, Gama L, Schwartz C, Vila J, Burny A, Clumeck N, Moutschen M, De Wit S, Peterlin BM, Rouzioux C, Rohr O, Van Lint C (2015) An in-depth comparison of latency-reversing agent combinations in various in vitro and ex vivo HIV-1 latency models identified bryostatin-1 + JQ1 and ingenol-B + JQ1 to potently reactivate viral gene expression. PLoS Pathog 11:e1005063
158. Nothias-Scaglia LF, Pannecouque C, Renucci F, Delang L, Neyts J, Roussi F, Costa J, Leyssen P, Litaudon M, Paolini J (2015) Antiviral activity of diterpene esters on Chikungunya virus and HIV replication. J Nat Prod 78:1277
159. Nothias-Scaglia L-F, Dumontet V, Neyts J, Roussi F, Costa J, Leyssen P, Litaudon M, Paolini J (2015) LC-$MS^2$-Based dereplication of *Euphorbia* extracts with anti-Chikungunya virus activity. Fitoterapia 105:202
160. Baloch LB, Baloch MK, Baloch AK (2009) Bio-active compounds from *Euphorbia cornigera* Boiss. Eur J Med Chem 44:3188
161. Nascimento BA, Zani CL (1999) A simple high pressure liquid chromatographic method for the quantitative determination of the molluscicide milliamine L in the latex of *Euphorbia splendens*. Phytochem Anal 10:93
162. Zani CL, Marston A, Hamburger M, Hostettmann K (1993) Molluscicidal milliamines from *Euphorbia milii* var. *hislopii*. Phytochemistry 34:89
163. Shi JX, Li ZX, Nitoda T, Izumi M, Kanzaki H, Baba N, Kawazu K, Nakajima S (2007) Three antinematodal diterpenes from *Euphorbia kansui*. Biosci Biotechnol Biochem 71:1086
164. Ogbourne SM, Parson PG (2014) The value of Nature's natural product library for the discovery of new chemical entities: the discovery of ingenol mebutate. Fitoterapia 98:36
165. Staples MP, Elwood M, Burton RC, Williams JL, Marks R, Giles GG (2006) Non-melanoma skin cancer in Australia: the 2002 National Survey and trends since 1985. Med J Aust 184:6
166. Ramsay JR, Suhrbier A, Aylward JH, Ogbourne SM, Cozzi S-J, Poulsen MGG (2011) The sap from *Euphorbia peplus* is effective against human nonmelanoma skin cancers. Br J Dermatol 164:633
167. Aylward JH (2001) Therapeutic agents-1. PCT Patent Appl No AU2001/000679
168. ASX and Media Release (2005) http://files.shareholder.com/downloads/PEPL/0x0x206196/9fe6aefa-aa82-4bfc-8dfd-05a178142a71/206196.pdf. Accessed 11 Aug 2015

169. Center Watch for Information on FDA-Approved Drugs (2015) https://www.centerwatch.com/drug-information/fda-approved-drugs/drug/1181/picato-ingenol-mebutate-gel. Accessed 11 Aug 2015
170. European Medicinal Agency (2012) http://www.ema.europa.eu/ema/index.jsp?curl=pages/medicines/human/medicines/002275/human_med_001600.jsp&mid=WC0b01ac058001d124. Accessed 11 Aug 2015
171. Rosen RH, Gupta AK, Tyring SK (2012) Dual mechanism of action of ingenol mebutate gel for topical treatment of actinic keratoses: rapid lesion necrosis followed by lesion-specific immune response. J Am Acad Dermatol 66:486
172. Hampson P, Chahal H, Khanim F, Hayden R, Mulder A, Assi LK, Bunce CM, Lord JM (2005) PEP005, a selective small-molecule activator of protein kinase C, has potent antileukemic activity mediated via the delta isoform of PKC. Blood 106:1362
173. Berman B (2015) Safety and tolerability of ingenol mebutate in the treatment of actinic keratosis. Expert Opin Drug Saf 14:1969
174. Dodson JM, DeSpain J, Hewett JE, Clark DP (1991) Malignant potential of actinic keratoses and the controversy over treatment. Arch Dermatol 127:1029
175. Augustin M, Tu JH, Knudsen KM, Erntoft S, Larsson T, Hanke CW (2015) Ingenol mebutate gel for actinic keratosis: the link between quality of life, treatment satisfaction, and clinical outcomes. J Am Acad Dermatol 72:816
176. Lebwohl MG, Shumack S, Stein Gold L, Melgaard A, Larsson T, Tyring SK (2013) Long-term follow-up study of ingenol mebutate gel for the treatment of actinic keratoses. JAMA Dermatol 149:666
177. Schopf RE (2015) Ingenol mebutate gel is effective against anogenital warts — a case series in 17 patients. J Eur Acad Dermatol Venereol. doi:10.1111/jdv.13097
178. Fürstenberger G, Hecker E (1986) On the active principles of the Euphorbiaceae XII. Highly unsaturated irritant diterpene esters from *Euphorbia tirucalli* originating from Madagascar. J Nat Prod 49:386
179. Tremp GL, Hecker E (1985) Stability of the "second stage" promoter 12-*O*-retinoylphorbol-13-acetate. Cancer Res 45:2390
180. Turner R (1995) Euphorbias. A gardener's guide. BT Batsford, London
181. Colorado State University. http://www.ext.colostate.edu/pubs/natres/03107.html. Accessed 10 Aug 2015
182. Kronberg SL, Halaweish FT, Hubert MB, Weimer PJ (2006) Interactions between *Euphorbia esula* toxins and bovine ruminal microbes. J Chem Ecol 32:15
183. Zayed SM, Farghaly M, Soliman SM, Gotta H, Sorg B, Hecker E (2001) Dietary cancer risk from conditional cancerogens (tumor promoters) in produce of livestock fed on species of spurge (Euphorbiaceae). V. Skin irritant and tumor-promoting diterpene ester toxins of the tigliane and ingenane type in the herbs *Euphorbia nubica* and *Euphorbia helioscopia* contaminating fodder of livestock. J Cancer Res Clin Oncol 127:40
184. Nawito M, Ahmed YF, Shalaby SI, Nada A, Zayed SM, Hecker E (2001) Dietary cancer risk from conditional cancerogens (tumor promoters) in products of livestock fed on species of spurge (Euphorbiaceae). IV. Toxicologic and pathophysiologic observations in lactating goats and their suckling kids fed on the irritant herbs *Euphorbia nubica* and *Euphorbia helioscopia*: an etiologic model for investigations on the putative risk of cancer by consumption of food. J Cancer Res Clin Oncol 127:34
185. Forsyth AA (1968) British poisonous plants. Ministry of Agriculture, Fisheries, and Food, Bulletin 161. HMSO, London
186. Zayed SM, Farghaly M, Taha H, Gminski R, Hecker E (1998) Milk of lactating goats fed on the skin irritant herb *Euphorbia peplus* is polluted by tumor promoters of the ingenane diterpene ester type. J Cancer Res Clin Oncol 124:301
187. Upadhyay RR, Bakhtavar F, Ghaisarzadeh M, Tilabri J (1978) Cocarcinogenic and irritant factors of *Euphorbia esula* L. latex. Tumori 64:99

188. Nawito M, Ahmed YF, Zayed SM, Hecker E (1998) Dietary cancer risk from conditional cancerogens in produce of livestock fed on species of spurge (Euphorbiaceae). II. Pathophysiological investigations in lactating goats fed on the skin irritant herb *Euphorbia peplus* and in their milk-raised kids. J Cancer Res Clin Oncol 124:179
189. Makkar HPS, Becker K (2010) Are *Jatropha curcas* phorbol esters degraded by rumen microbes? J Sci Food Agric 90:1562
190. van Herwaarden AE, Wagenaar E, Merino G, Jonker JW, Rosing H, Beijnen JH, Schinkel AH (2007) Multidrug transporter ABCG2/breast cancer resistance protein secretes riboflavin (Vitamin $B_2$) into milk. Mol Cell Biol 27:1247
191. Upadhyay RR, Islampanah S, Davoodi A (1980) Presence of a tumor-promoting factor in honey. Gann 71:557
192. Sosath S, Ott HH, Hecker E. (1988) Irritant principles of the spurge family (Euphorbiaceae) XIII. Oligocyclic and macrocyclic diterpene esters from latices of some *Euphorbia* species utilized as source plants of honey. J Nat Prod 51:1062
193. Prota: Medicinal plants/Plantes médicinales. http://database.prota.org/PROTAhtml/Euphorbia%20ingens_En.htm. Accessed 10 Aug 2015
194. Quoted in Ref. 120
195. Liang A (1995) A review of the processing of *Euphorbia kansui* Liou. Zhongguo Zhong Yao Zhi 20:3
196. Girin MA, Paphassarang S, David-Eteve C, Chaboud A, Raynaud J (1993) Determination of ingenol in homeopathic tinctures of *Euphorbia* species by high-performance liquid chromatography. J Chromatogr 637:206
197. Branch SK, Rowan MG (1992) Concerted use of homo- and heteronuclear 2D NMR in the elucidation of the structure of a novel pentaester of 19-hydroxyingol from *Euphorbia marginata*. Magn Reson Chem 30:632
198. INCHEM: Chemical safety information from intergovernmental organizations: Strychnine. http://www.inchem.org/documents/pims/chemical/pim507.htm. Accessed 14 Oct
199. Jakupovic J, Jeske F, Morgenstern T, Tsichritzis F, Marco JA, Berendohn W (1998) Diterpene from *Euphorbia segetalis*. Phytochemistry 47:1583
200. Betancur-Galvis L, Roldan J, Morales G, Sierra J, Blanco-Molina M, Macho A, Marco JA (2002) Ingenols isolated from *Euphorbia cotinifolia* induce apoptosis in Jurkat cells. Rev Latinoam Quim 30:68
201. Cao D, Su Y, Yang J (1992) Chemical constituents of Dalangdu (*Euphorbia nematocypha*). II. Isolation and identification of the aromatic compounds and diterpenoid. Zhongcaoyao 23:625
202. Mbwambo ZH, Lee SK, Mshiu EN, Pezzuto JM, Kinghorn AD (1996) Constituents from the stem wood of *Euphorbia quinquecostata* with phorbol dibutyrate receptor-binding inhibitory activity. J Nat Prod 59:1051
203. Jia Z, Ding Y (1991) New diterpene from *Euphorbia sieboldiana*. Planta Med 57:569
204. Sevil O, Ulubelen A, Barla A, Kohlbau H-J, Voelter W (1999) Triterpenoids and a diterpene from *Euphorbia iberica*. Planta Med 65:475
205. Hirota M, Ohigashi H, Oki Y, Koshimizu K (1980) New ingenol-esters as piscicidal constituents from *Euphorbia cotinifolia* L. Agric Biol Chem 44:1351
206. Tada M, Seki H (1989) Toxic diterpenes from *Euphorbia trigona*. Agric Biol Chem 53:425
207. Wang Y-B, Li Y-Y, Wang H-B, Qin G-W (2007) Chemical constituents from the roots of *Euphorbia kansui*. Zhongguo Tianran Yaowu 5:182
208. Falsone G, Crea AE, Noack EA (1982) Constituents of Euphorbiaceae. VII. 20-Deoxyingenol monoesters and ingenol diesters from *Euphorbia biglandulosa* Desf. Arkiv Pharm 315:1026
209. Vogg G, Mattes E, Rothenburger J, Hertkorn N, Achatz S, Sandermann H Jr (1999) Tumor promoting diterpenes from *Euphorbia leuconeura* L. Phytochemistry 51:289
210. Marco JA, Sanz-Cervera JF, Ropero FJ, Checa J, Fraga BM (1998) Ingenane and lathyrane diterpenes from the latex of *Euphorbia acrurensis*. Phytochemistry 49:1095

211. Shi JS, Li ZX, Nitoda T, Minoru I, Hiroshi K, Naomichi B, Kazuyoshi K, Shuhei N (2008) Antinematodal activities of ingenane diterpenes from *Euphorbia kansui* and their derivatives against the pine wood nematode. Z Naturforsh C Biosci 23:59
212. Lu Z-H, Yang M, Zhang J-Q, Chen G-T, Huang H-L, Guan S-H, Ma C, Liu X, Guo D-A (2008) Ingenane diterpenoids from *Euphorbia esula*. Phytochemistry 69:812
213. Ott HH, Hecker E (1981) Highly irritant ingenane type diterpene esters from *Euphorbia cyparissias* L. Experientia 37:88
214. Pan Q, Min ZD (2003) Studies on ingenol-type diterpene esters in root tuber of *Euphorbia kansui*. Chin Trad Herb Drugs 34:489
215. Lin LJ, Kinghorn AD (1983) Three new ingenane derivatives from the latex of *Euphorbia canariensis* L. J Agric Food Chem 31:396
216. Baloch IB, Baloch MK, Us Saqib QN (2005) Tumor promoting diterpene esters from latex of *Euphorbia caducifolia* L. Helv Chim Acta 88:3145
217. Connolly JD, Facunle CO, Rycroft DS (1984) Five ingol esters and a 17-hydroxyingenol ester from the latex of *Euphorbia kamerunica*. Assignment of esters using $^{13}$C-NMR methods. Tetrahedron Lett 25:3773
218. Zarei SM, Ayatollahi AM, Ghanadian M, Kobarfard F, Aghaei M, Choudhary MI, Fallahian F (2013) Unusual ingenoids from *Euphorbia erythradenia* Boiss. with pro-apoptotic effects. Fitoterapia 91:87
219. Pan DJ, Hu CQ, Chang JJ, Lee TTY, Chen YP, Hsu HY, McPhail DR, McPhail AT, Lee KH (1991) Kansuiphorin-C and -D, cytotoxic diterpenes from *Euphorbia kansui*. Phytochemistry 30:1018
220. Gotta H, Adolf W, Opferkuch HJ, Hecker E (1984) On the active principles of the Euphorbiaceae, IXa. Ingenane type diterpene esters from five *Euphorbia* species. Z Naturforsch B 39B:683
221. Zhao J-X, Liu C-P, Qi W-Y, Han M-L, Han Y-S, Wainberg MA, Yue J-M (2014) Eurifoloids A-R, structurally diverse diterpenoids from *Euphorbia neriifolia*. J Nat Prod 77:2224

**Giovanni Appendino** is Professor of Organic Chemistry at the Università del Piemonte Orientale, Department of Pharmaceutical Sciences, Novara, Italy. He graduated from the University of Torino (Italy) and did postdoctoral work in organic synthesis at the University of Gent (Belgium) with Prof. Pierre De Clercq. He is Editor-in-Chief of the journal *Fitoterapia*, and is a member of the Advisory Board of *Progress in the Chemistry of Organic Natural Products*. For his work on the chemistry of bioactive natural products, he was awarded the Quilico Medal of the Italian Chemical Society in 2009 and the Bruker Prize of the Phytochemical Society of Europe in 2014.

# The *Laurencia* Paradox: An Endless Source of Chemodiversity

**Maria Harizani, Efstathia Ioannou, and Vassilios Roussis**

## Contents



---

M. Harizani • E. Ioannou (✉) • V. Roussis (✉)
Department of Pharmacognosy and Chemistry of Natural Products, Faculty of Pharmacy, National and Kapodistrian University of Athens, Panepistimiopolis Zografou, Athens 15771, Greece
e-mail: mariachariz@pharm.uoa.gr; eioannou@pharm.uoa.gr; roussis@pharm.uoa.gr



A.D. Kinghorn, H. Falk, S. Gibbons, J. Kobayashi (eds.), *Progress in the Chemistry of Organic Natural Products, Vol. 102*, DOI 10.1007/978-3-319-33172-0_2

# 1 Introduction

Nature, the most prolific source of biological and chemical diversity, has provided mankind with remedies for health problems since ancient times, and continues to be the most promising reservoir of bioactive chemicals for the development of modern drugs. The terrestrial flora and fauna has served as the basis of local pharmacopoeias from as early as 2600 BC and some of the earliest documentation comes from inscriptions found in Mesopotamia. Records from the ancient Egyptians and Chinese show that plants were used for the formulation of a large number of life-saving drugs.

As early as 300 BC, the Greek philosopher and naturalist Theophrastus compiled the botanical characteristics and medicinal properties of herbs in a series of nine books named "History of Plants". Four centuries later, the physician Dioscorides developed and recorded a wealth of complex prescriptions and formulas using medicinal herbs.

Even though the biomedical and pharmaceutical sciences, through the development of technology, have led to many milestones and hundreds of efficient drugs have become available in the medical armamentum, plant-based remedies are still essential components of the healthcare systems of many communities in developing countries. According to the World Health Organization approximately 80% of the world's population uses traditional medicines in its primary healthcare [1, 2]. Nature is equally important for the healthcare systems of developed countries, since almost 25% of the prescribed drugs contain plant extracts or metabolites and an additional significant percentage of the commercial drugs has been developed through studies employing natural products as the lead molecules [3, 4].

In addition to the terrestrial organisms that still remain a promising source of new bioactive metabolites, the marine environment, covering approximately 70% of the Earth's surface and hosting a largely unexplored biodiversity, offers an enormous resource for the discovery of novel compounds. Due to the wide range of competitive ecosystems they survive in, marine organisms have developed unique defense strategies and bioactive compounds that, in some cases, are unparalleled by their terrestrial counterparts.

Systematic investigations of marine organisms were initiated 60 years ago with the pioneering work of Bergmann and Burke, initially on bivalves and later on sponges, especially those of the genus *Cryptotethya* that paved the way to the development of the antiviral and antileukemic drugs Ara-A and Ara-C [5].

The development of SCUBA (Self-Contained Underwater Breathing Apparatus) and underwater technology made the marine biota relatively easily accessible to marine biologists and chemists and the number of investigated organisms increased exponentially in the last 50 years. According to *MarinLit* [6], more than 27,000 metabolites from marine macro- and microorganisms have been isolated to date providing material and key structures for the development of new products in the pharmaceutical, food, cosmeceutical, chemical, and agrochemical sectors.

Algae, as photosynthetic organisms, thrive in the euphotic zone and were among the first marine organisms that were investigated as sources of food, nutrition supplements, soil fertilizers, and bioactive metabolites. The early attempts to distinguish separate groups of algae relied on pigmentation and characteristics, such as the presence or absence of flagella, cell wall composition, and type of photosynthetic products. It is evident though that at least ultrastructural features, especially details of chloroplast structure, are indispensable taxonomic characters for the safe identification of algae [7].

Seaweeds, as macroscopic algae are frequently called, are divided into three classes known as green (Chlorophyceae), brown (Phaeophyceae), and red (Rhodophyceae) algae. Most of the red algae share the surface of the coastal rocks with green and brown algae and are usually smaller and more delicate than brown algae. Red algae tend to be more abundant in subtropical and tropical waters and their populations in deeper waters are denser than those of green and brown algae [8]. Among phycologists, red algae are notable for the complexity of their life cycles and the difficulties encountered in their taxonomy.

Among all marine macroalgae, red algae are the most prolific producers of secondary metabolites. Although these molecules are not involved in the basic machinery of life and often account for only a very small fraction of the total biomass of the organism, their contribution to survival is frequently comparable to constituents resulting from primary metabolism.

Red algae of the genus *Laurencia* are accepted unanimously as one of the richest sources of new secondary metabolites [9]. The plasticity of the morphoanatomical characters of *Laurencia* species has resulted in frequent taxonomic revisions of the genus that along with the genera *Chondrophycus*, *Osmundea*, *Palisada*, *Yuzurua*, and *Laurenciella*, are now included in the so called "*Laurencia* complex". The cosmopolitan distribution, along with the chemical variation influenced to a significant degree by environmental and genetic factors, have resulted in an almost endless array of metabolites, often featuring multiple halogenation sites.

Comprehensive and copious reviews have been reported previously, focusing on the chemistry of *Laurencia*, on specific chemical classes of secondary metabolites isolated from the family Rhodomelaceae or on halogenated compounds from natural sources [10–14].

The present contribution, including the chemical diversity and the bioactivity of the secondary metabolites isolated from species of the genus *Laurencia*, covers the literature until August 2015 and offers a comprehensive view of the chemical wealth and the taxonomic problems currently impeding chemical and biological investigations. Mollusks feeding on *Laurencia* are, in many cases, bioaccumulating, and utilize algal metabolites as chemical weaponry against natural enemies. For this reason, metabolites from *Laurencia*-feeding sea hares with such a postulated dietary origin are also included in the present review. In some cases, the frequently unique structures of *Laurencia* metabolites, originally proposed on the basis of their spectroscopic data, were later proven to be incorrect and were revised utilizing additional data provided by chemical synthesis, X-ray crystallographic analysis, or sophisticated spectroscopic techniques. In the present literature review, when chemical structures have been revised, only the currently accepted structure is depicted.

## 2 The Genus *Laurencia*

The genus *Laurencia*, currently including 146 taxonomically accepted species, belongs to the order Ceramiales, family Rhodomelaceae, tribe Laurencieae and is mainly found in tropical, subtropical, and temperate coastal waters [15]. It is considered one of the most prolific genera and even though it has been investigated intensively over the last 50 years, many new molecules are being reported continuously [6, 9, 16].

The Rhodomelaceae is estimated to be the largest marine red algal family, with approximately 125 genera and 700 species recognized worldwide [15]. According to *MarinLit* [6], more than 1200 naturally occurring compounds have been isolated and characterized from species belonging to this family, with the vast majority of them being halogenated [12].

The tribe Laurencieae is comprised of eight genera, six of which, under the current taxonomic status, are included in the "*Laurencia* complex" (or *Laurencia* sensu lato). These genera are *Chondrophycus* (Tokida et Saito) Garbary et J.T. Harper, *Laurencia* J.V. Lamouroux, *Laurenciella* Cassano, *Osmundea* Stackhouse, *Palisada* (Yamada) K.W. Nam, and *Yuzurua* (K.W. Nam) Martin-Lescanne [17].

## 2.1 The "*Laurencia* Complex"

The frequently observed color morphs within single interbreeding populations, the color adaptation mainly affected from the levels of irradiation and the color change through aging of individual specimens, co-existing in most cases on the same substrate, renders species discrimination very difficult for the untrained eye (see Fig. 1).

The genus *Laurencia*, as classified by Lamouroux in 1813, included eight species in its original description, but in the subsequent years the taxonomic hierarchy of the genus underwent a number of revisions [18].

**Fig. 1** Different color morphs co-occuring within a *Laurencia* population

Thorough anatomical studies in the last few decades have revealed that *Laurencia* is a highly diverse genus, encompassing species that display distinctive features, usually diagnostic at the generic level [19–23]. Saito in 1967 was the first to divide *Laurencia* into the subgenera, *Laurencia* and *Chondrophycus* and the genus has thereafter been referred to as the "*Laurencia* complex" [19]. Successive detailed studies on the vegetative and reproductive structures of the included species led to the addition of the genera *Osmundea*, *Chondrophycus*, and *Palisada* to the "*Laurencia* complex" to reflect its morphological diversity [21, 22]. These genera share the typical and superficial Rhodomelacean morphology with apical cells sunk in apical pits of branchlets, a recognizable axial cell row only near the apical cell and an extensive cortex [24]. *Osmundea* is easily distinguished from *Laurencia* and *Chondrophycus* as it has a filament-type, rather than a trichoblast-type, spermatangial development and tetrasporangial production from random epidermal cells, rather than particular pericentral cells [20, 22]. *Laurencia* is delimited from *Chondrophycus* by the vegetative axial segment feature of four instead of two pericentral cells [25]. In 1999, Nam proposed an infrageneric classification scheme for *Chondrophycus* that included the subgenus *Yuzurua*. Subsequently, Martin-Lescanne and co-workers in 2010 proposed the elevation of this subgenus to generic rank and *Yuzurua* was recognized as the fifth member of "*Laurencia* complex" [18, 22]. More recent studies on the basis of morphological characters, especially the reproductive traits, and molecular sequences of the plastid-encoded gene *rbcL* (RuBisCO, Ribulose-1,5-Bisphosphate Carboxylase Oxygenase, Large chain) have led to a further division of the "*Laurencia* complex". This phylogenetic analysis by Cassano and co-workers identified the genus *Laurenciella*, which is morphologically identical with the genus *Laurencia*, but differs in the *rbcL* sequence [17].

Of these genera, *Laurencia* has a worldwide distribution (see Fig. 2), although the majority of species are found in the Southern Hemisphere [25]. *Osmundea* occurs in the North American Pacific Ocean, Brazil, the Mediterranean Sea, the

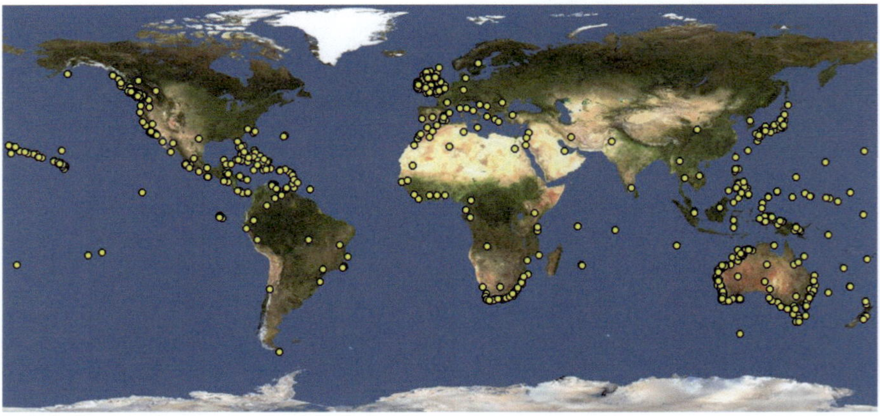

**Fig. 2** Global distribution map for the genus *Laurencia*

European Atlantic Ocean, India, Australia, and North Africa [26–29]. *Chondrophycus* has been reported mainly in regions of the Pacific Ocean [21, 22]. Species of *Palisada* have been collected from Asia, Europe, America, Africa, the Atlantic islands, the Caribbean islands, the Indian Ocean islands, Australia, New Zealand, the Pacific islands, and the Western Atlantic Ocean [15]. Reports on the occurrence of *Yuzurua* include the Caribbean islands, the Atlantic islands, North America, Africa, Queensland (Australia), Indonesia, the Western Atlantic, the Arabian Gulf, and the Philippines [15]. The range of *Laurenciella* distribution is relatively narrower, including the Canary Islands, the Azores, Mexico, Brazil, and the Western Atlantic Ocean [15].

The vegetative and reproductive characters used for the distinction of the genera in the "*Laurencia* complex" are the: (1) number of pericentral cells per vegetative axis; (2) position of the first pericentral cell relative to the trichoblast; (3) origin of the tetrasporangia; (4) presence or absence of fertility in the second pericentral cells; (5) number of sterile pericentral cells in the tetrasporangial axis; (6) origin of the spermatangial branches; (7) formation pattern of the spermatangial branches on trichoblasts; (8) number of pericentral cells in procarp-bearing segments of female trichoblasts; and (9) post-fertilization features associated with the timing of auxiliary cell formation [22, 23]. It is worth noting that among the genera comprising the complex, *Laurencia* is the only genus that possesses four pericentral cells per vegetative axial segment [17].

The extensive morphological plasticity, in combination with the difficulty in defining diacritical characters for the morphology and anatomy makes species discrimination within the complex extremely difficult when utilizing only traditional characters. It is now evident that the morphological, anatomical, and reproductive characters have a diagnostic value only at the generic level in the taxonomy of the "*Laurencia* complex".

In an alternative approach, as species separation in "*Laurencia* complex" is complicated by the high degree of morphological variation within the genus, the secondary metabolites have been considered as a useful taxonomic tool [30, 31]. However, the plasticity in morphoanatomical characters has led frequently in the past to the wrong species identification and therefore to serious confusion when recognition of chemotaxonomic markers was attempted. For example, in a chemosystematic study performed by Barrow and co-workers on *Laurencia* species, digeneaside, which was previously suggested as a characteristic photoassimilatory product found only in representatives of the order Ceramiales, was not detected in any of the investigated species, including five species in which it was identified in previous studies. Accordingly, the proposed high diagnostic value of digeneaside was negated and its occurrence in these species is questioned [32].

Therefore, the use of molecular markers seems indispensable in delimiting taxa and inferring their phylogenetic relationships. The promises and perils of DNA barcoding are now well-known, but no systematic investigations have been performed to reveal the extent of taxonomic misidentification of algal specimens available in primary DNA sequence repositories. An attempt towards the reassignment of the taxonomic status of chemically investigated *Laurencia* species was reported in 2011 by Fujii and co-workers, but unfortunately included only *Laurencia* species that had been collected in Brazilian waters [33].

## 2.2 Botanical Description of *Laurencia*

Specimens of *Laurencia* species (see Fig. 3) display erect thalli that in some cases can reach 40 cm and are attached by discoidal to rhizoidal appendages on encrusting holdfasts. Branching of thalli can extend to a different degree either bilaterally or in all directions. The branches can be either cylindrical or flat and the branchlets, which are not much narrower than the branches, vary between being blunt and truncate and in some cases are claviform.

Many species appear iridescent when alive. Individuals are polysiphonous, but soon form several layers of pseudoparenchymatous cortication masking the polysiphonous axes. The apical cells are always immersed at the bottom of the apical pit, from which dichotomously branched colorless trichoblasts develop. Stoloniferous propagation is common in some species [15]. The surface cortical cells can display secondary pit connections between them and this character is used for separating groups of species. In fresh material of some species, refringent intracellular "corps en cerise" with discoid to spheroid inclusions are present in the cortical and trichoblast cells [15]. The cortical cells contain photosynthetic pigments arranged in the thylakoids, while other internal cells are colorless [34].

In some species, unequal lenticular thickening of medullary cell walls is observed. The reproductive structures are developed usually on the branchlets, but sometimes are observed on the main branches as well. The procarps are formed in apical pits and the cystocarps protrude from the surface of the thallus. The

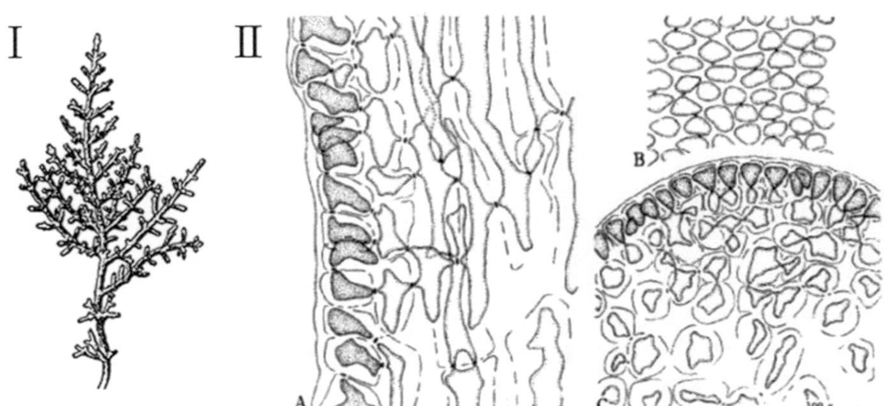


**Fig. 3** (I) Drawing of a *Laurencia obtusa* specimen; (II) (A) Longitudinal section through a branch, (B) Surface view of the cortical cell arrangement in the upper part of the ultimate branchlet, (C) Transverse section through a branchlet [18]

spermatangial capitula are cylindrical with a cup-shaped to a more deeper shape exhibiting conceptacle-like pockets at lateral or terminal apices. The tetrasporangia on subcortical cells, products of divisions of elongated fertile pericentral cells [19], occur parallel to the central axis or at right angles to the central axis. This character is important for the separation of groups within species. The tetrasporangial initially is cut off adaxially or abaxially from the pericentral cell product. This characteristic is also used to separate groups of species [15].

## 2.3 Growth and Population Dynamics of Laurencia

*Laurencia* species are found in variable population densities in temperate to tropical shores of the world, inhabiting littoral to sublittoral zones, spreading to depths reaching 65 m. The growth of *Laurencia* and frequently the content of the secondary metabolites are influenced by various factors. Nutrients and temperature exert the most significant impact on the growth rate and maturity of the algae, whereas changes in the day length and salinity influence the growth rate to a smaller degree. For *Laurencia okamurae* (syn. *L. okamurai*) the optimum growth was observed at 25°C regardless of the day length. The algae were able to grow at salinities of 14–50‰ with maximum growth observed at 26‰. Variation in temperature between 15 and 25°C did not result in any obvious change in the content of the major metabolites laurinterol and debromolaurinterol [35]. The lower survival temperatures for *Laurencia cartilaginea* and temperature requirements for growth, ranging between 15 and 30°C with an optimum of 25–30°C, are mostly similar to those of tropical west Atlantic and amphi-Atlantic sub-tropical macroalgae [36]. A recent study on Brazilian algae reported that climate change associated with variations in the frequency and intensity of extreme temperatures and precipitation events on the local and regional scale can lead to the disappearance or to expansion of the distribution range of major algal groups [37].

The distribution and abundance of 87 species of macrophytes, including *Laurencia* species, on the intertidal reef flat at Agatti atoll (Lakshadweep, India) was recorded showing that species diversity increased with distance from the shore, while the percentage coverage and species richness exhibited the reverse trend. Herbivory with sea urchins as the major grazers appeared to significantly affect the population density [38].

Comparative studies on *Laurencia cruciata* growing in Indian Ocean sites affected from variable industrial and urban pollution indicated significant adaptability for this particular species [39]. Contradictory results from the Gulf of Antalya (north-eastern Mediterranean), showed that the highest values of biomass were found within the unpolluted stations, while a progressive decrease was observed as the effects of the anthropogenic pressure were more marked [40]. The intertidal vegetation of the "Abra de Bilbao" (on the Basque coast in northern Spain) was studied, following a pollution gradient to show that the richness of the species and the algal abundance suffer a parallel reduction, easily observed under the most degraded environmental conditions. Pollution altered the zonation pattern, reducing the number of vegetation belts, and algae were gradually replaced by filter-feeders (mussels and barnacles). *Laurencia obtusa*, which requires high quality environmental characteristics, disappeared in the most polluted stations [41].

## 2.4 Ecological Relations of Laurencia

*Laurencia* algae, having multiple ecological roles, can grow in dense populations and form "forests" where several animals seek refuge. They host microorganisms and parasitic algae, such as species of *Janczewskia* (belonging to the same tribe), but can also grow as parasites on other marine organisms [42]. Even though they are in general rejected by the majority of grazing organisms due to their deterrent/toxic chemical defense, there is a number of marine invertebrates selectively feeding on *Laurencia* that have developed mechanisms for the detoxification of the bioactive molecules.

Chemical defenses can vary dramatically among geographic regions, habitats, individuals within a local habitat, and within different portions of the same individual. Factors affecting this variance include physical stresses and induction due to previous attack. Further studies are needed though to assess which consumers induce prey defenses, how responses vary in environments with differing physical characteristics and whether the "induced" responses are a direct response to consumer attack or a defense against microbial pathogens invading via feeding wounds [43].

Crabs of the genus *Microphrys* use *Laurencia papillosa* as a food source and as a camouflage [44, 45], whereas the queen conch *Strombus gigas* consumes *Laurencia* algae not only as a source of nutrients, but also for the provision of chemical cues that induce the metamorphosis of its larvae [46, 47].

Competition for space is also an imperative necessity for species survival. In field experiments conducted on declined coral reefs showing predominance of macroalgae, the presence of *Laurencia* species resulted in inhibition of coral larvae recruitment and increased coral larvae mortality [48].

Shell-less sea hares, which are slow-moving marine invertebrates, in many cases exhibit surprising survival rates, even in the most competitive habitats. These animals are postulated to have developed chemical defense mechanisms based on a wide variety of bioactive compounds [49]. Is well known that marine mollusks obtain and bioaccumulate secondary metabolites from their food sources and for this reason are investigated intensely as a rich source of new natural products [50]. Among the invertebrates feeding on *Laurencia*, sea hares of the genus *Aplysia* are the most frequent grazers. They obtain their means of chemical defense from the algae in forms that can be used for their protection either directly or usually after minor structural modifications.

In a study focusing on the patterns and dynamics of algal secondary metabolites in mollusks, *L. obtusa* metabolites were detected in the mucous and opaline secretions of *Aplysia parvula* feeding on this alga species. The loss of *L. obtusa* metabolites from *A. parvula* was found to be faster than predicted for growth alone, suggesting that metabolites were actively metabolized or excreted. The deployment of sequestered metabolites in the secretions of *A. parvula* may explain the more rapid loss of these compounds and is consistent with a possible defensive role for the acquired metabolites [51].

Since in many cases the bioaccumulated molecules are found in higher quantities in the consumers than in the producing organisms, many minor algal metabolites have been isolated only from the grazing mollusks. It is noteworthy that the brominated aromatic sesquiterpene aplysin, the first *Laurencia* metabolite reported, was isolated in 1963 from *Aplysia kurodai* mollusks and later traced to the *Laurencia* substrates [52].

Symbiotic or parasitic host-microbe interaction is a well-recognized driving force for co-evolution in the terrestrial and marine environment. Nonetheless, molecular studies of seaweed-associated microbial communities are still rare. Despite the high number of research studies describing the chemical composition of *Laurencia* species, the genetic knowledge regarding this genus is currently restricted to taxonomic markers and general genome features. In recent work, de Oliveira and co-workers analyzed the transcriptomic profile of *Laurencia dendroidea*, aiming to unveil the genes involved in the biosynthesis of terpenoid compounds in this seaweed and the interactions between this host and its associated microbiome, and showed bacteria to be the dominant active group in its microbiome [53].

## 2.5 "Corps en cerise"

*Laurencia* is renowned for its unique ability to biosynthesize a large variety of secondary metabolites, frequently featuring uncommon carbocycles and a high degree of halogenation. The phenotypic plasticity of the genus is to a large degree reflected on the insurmountable chemical diversity and the endless number of new natural products constantly reported, especially when the investigated specimens are collected from different geographic regions. The extended chemodiversity can also be seen as a consequence of the halogenated nature of *Laurencia* metabolites, a characteristic permitting intramolecular reactions and generation of new derivatives.

The most striking feature of *Laurencia* metabolites and marine natural products in general is the frequency with which halogenated compounds are described. The majority of the halogenated *Laurencia* metabolites contain one or more bromine atoms, with some of them containing both bromine and chlorine and in some cases even iodine. Since seawater is a halogen-rich environment, it is not surprising that marine organisms are capable of biosynthesizing halogenated metabolites. It is curious, however, that brominated compounds are more abundant than chlorinated ones, even though seawater contains a much higher concentration of chloride (559 m*M*) than bromide (0.86 m*M*) ions. A plausible explanation can be the relative ease of oxidation of bromide ions resulting in bromonium ions that can undergo electrophilic addition to alkenes and aromatic systems [54]. Despite the fact that halogenated metabolites represent the prevalent group of natural products isolated from red algae, their biosynthesis is still poorly understood, even though biomimetic syntheses and enzyme-catalyzed halogenation reactions have provided useful insights into the biogenesis of these molecules.

The abundance of halogenated metabolites and their important biological activities have intrigued marine natural product chemists, who more than 30 years ago predicted the presence of haloperoxidase enzymes. Since that time, FeHeme- and vanadium-haloperoxidases have been discovered and characterized in many marine organisms [55, 56].

Vanadium-dependent haloperoxidases contain a vanadate prosthetic group and utilize hydrogen peroxide to oxidize a halide ion into a reactive electrophilic intermediate. These metalloenzymes, having been exclusively characterized in eukaryotes, show a large distribution in Nature, including macroalgae, fungi, and bacteria [57].

The production of labeled brominated metabolites with radioactive $^{82}Br$ in *Laurencia* was investigated by TLC and TLC–autoradioluminography, when several *Laurencia* species were cultured in artificial seawater medium ($ASP_{12}NTA$ including $Na^{82}Br$) under 16:8 h light:dark illumination cycles to show the production of specific $^{82}Br$-containing metabolites. Furthermore, when *Laurencia* spp. specimens were cultured in the dark, the production rates of brominated metabolites were diminished [58].

In *Laurencia*, the halogenated metabolites, following their biosynthesis in the chloroplasts by haloperoxidases are stored in the intensively osmiophilic intracellular organelles, called “corps en cerise” (or cherry body) that transport them to the outer surface of the thallus [55, 59]. These organelles occur as intracellular membrane-bound vesicles in trichoblasts and in the outermost cortical cells and have a spherical, reniform to claviform shape [34]. The cellular distribution of halogenated compounds has been demonstrated by fluorescence microscopy that showed them to be stored mainly in the “corps en cerise” and to a lesser degree in vesicles found along the cytoplasm and within the chloroplasts [59].

When these organelles are not observed in the algae specimens, then halogenated metabolites are not detected [60–62]. For chemical defense at the surface level, these metabolites are intracellularly mobilized through vesicle transport from the “corps en cerise” to the cell periphery for their posterior exocytosis. The cell structures involved in this specific vesicle traffic, as well as the cellular structures related to the positioning and anchoring of the “corps en cerise” within the cell are not well known [63].

The transportation of the halogenated metabolites from the "corps en cerise" is affected by changes in temperature and irradiation, desiccation and bacterial fouling. Traffic of vesicles was induced by exposing *L. obtusa* to low temperatures and variations in irradiance. Under high temperatures and desiccation, the membranous tubular connections were lost and transport of vesicles was not observed [64]. Degradation of the "corps en cerise" and release of the halogenated compounds leads to cell death [59]. It has been reported that the number and size of the "corps en cerise" are constant within a species, independent of the geographical distribution, corroborating their use for taxonomical purposes to differentiate groups of species that present a lower number from those that have a higher number of these vesicles [15].

Investigation of *L. obtusa* specimens has revealed that the concentrations of elatol at the algal surface is not sufficient for inhibition of herbivory (sea urchins) and fouling (settlement of barnacle larvae and mussel attachment), whereas the natural within-thallus concentration is sufficient for exhibition of activity. Therefore, transport between the "corps en cerise", where elatol is probably stored, and the cell wall of *L. obtusa* was proposed through channel-like membranous connections that shift vesicles from the "corps en cerise" to the cell wall region. Finally, it was suggested that the natural concentration of defensive chemicals on the surface of *L. obtusa* is probably not absolute, but may be variable according to environmental conditions [65–67].

## 2.6 Chemical Variation in Laurencia

The frequently observed marked chemical variation of *Laurencia*, even within morphologically homogeneous species and populations, has been the subject of numerous studies that have not reached a unanimous conclusion. Possible explanations include the effects of environmental factors and/or genetic variables. Cultivation of *Laurencia pacifica* under different physical environmental factors has had little or no effect on the chemical profile of the extracts. In parallel, the interfertility between four sympatric populations of *L. pacifica*, which were morphologically similar, but chemically distinct, was investigated. Crossbreeding experiments showed that sympatric, but chemically distinct populations of *L. pacifica* did not interbreed in culture. From these results, it seems reasonable that the unique chemical profiles of the populations reflect genetic differences that are sufficient to cause reproductive isolation [68].

Another research group two decades later investigated a number of morphologically similar, but chemically distinct populations of *Laurencia nipponica* growing in Japan. The selected chemical types were characterized by specific halogenated secondary metabolites that remained the same in the wild and under various cultivation conditions. Experimental crosses between representative individuals of the chemical types showed that such chemically distinct populations are not reproductively isolated. Thus, the diverse chemical types can be referred to as races (chemical races) [69].

A recent study on *Laurencia dendroidea* showed high inter- and intrapopulation variability in the amount of the sesquiterpene elatol, caused by genetic variation, as well as environmental factors. The concentration of elatol varied from two to ten times among the *L. dendroidea* individuals, but was approximately three orders of magnitude broader among the different populations ranging from 0.001% to 1.24% of the dry weight. The results indicated that abiotic factors, as well as genetic factors clearly take part in fostering chemical variations observed in natural populations and can promote differential susceptibility of algae specimens to natural enemies [70, 71].

The majority of *Laurencia* species from which diterpenes have been isolated occurs in the Northern Hemisphere and the greatest diversity is found in the Pacific Ocean and Sea of Japan (Japan, China). This observation indicates that diterpene-producing species emerged in the Pacific Ocean and were later radiated to other regions [72].

# 3 Structures and Occurrence

The majority of secondary metabolites isolated from species of the genus *Laurencia*, as well as from mollusks grazing on them, fall into the chemical classes of sesquiterpenes, diterpenes, triterpenes, and $C_{15}$ acetogenins. Moreover, there are a number of metabolites that belong to the relatively small classes of indoles, aromatic compounds and steroids, as well as compounds that either constitute a rather small class or cannot readily be classified and in the present review are included as miscellaneous metabolites.

Besides the chemical structures of the metabolites, detailed information in tabular form, including the name of compound and the corresponding references for isolation sources or synthetic efforts, are provided.

It should be emphasized that extensive analysis of NMR spectroscopic data is sometimes not enough to establish unequivocally the chemical structure of a given compound, especially in regard to the placement of heteroatoms or the assignment of the relative or absolute configurations of stereocenters. Indeed, there are a number of cases in the genus *Laurencia* where the initial structure elucidation was incorrect, e.g. in the positioning of halogen substituents. In this regard, it is to be noted that efforts should be paid to support the structure proposed either with an X-ray crystallographic analysis when possible or with the use of chemical derivatization, partial or total synthesis, or NMR data calculation and comparison with experimental results.

### *3.1 Sesquiterpenes*

Sesquiterpenes, $C_{15}$ compounds consisting of three isoprene units, arise from the common precursor farnesyl pyrophosphate by various modes of cyclizations, followed in many cases by skeletal rearrangement, such as methyl group migration. Sesquiterpenes are by far the largest group of secondary metabolites isolated from species of the genus *Laurencia*, as well as mollusks grazing on them. Chemical investigations have yielded up to now 512 sesquiterpenes and related metabolites (**1–512**, Table 1) belonging to more than 50 carbon frameworks (see Figs. 4 and 5). Most of these feature mono-, bi-, or tri-carbocyclic skeletons, often consisting of fused- or *spiro*-ring systems.

**Table 1** Sesquiterpenes from *Laurencia* spp.

| No. | Name | Source of isolation/synthesis |
|---|---|---|
| **1** | (–)-(10*R*)-10-Bromo-α-chamigrene | *L. chondrioides* [73], *L. mariannensis* [74], *L. nipponica* [75, 76[d], 77, 78], *L. okamurai* [79], *L. pacifica* [80[a]], *L. rigida* [81]; Synthesis [82] |
| ***ent*-1** | (+)-(10*S*)-10-Bromo-α-chamigrene | *L. obtusa* [83[d]] |
| **2** | Laurecomin C | *L. composita* [84] |
| **3** | 2,10*β*-Dibromochamigra-2,7-dien-9*α*-ol | *L. saitoi* [85] |
| **4** | – | *A. dactylomela* [86] |
| **5** | 2,10-Dibromo-3-chloro-α-chamigrene | *L. claviformis* [87], *L. composita* [69, 88], *L. filiformis* [89], *L. glomerata* [90], *L. implicata* [91], *L. japonensis* [92, 93], *L. nangii* [94], *L. nidifica* [95], *L. nipponica* [76[d], 78, 96, 97], *L. obtusa* [98], *L. okamurai* [99, 100], *L. papillosa* [101], *L. scoparia* [102], *Laurencia* sp. cf. *L. gracilis* [103], *Laurencia* sp. [104[a], 105, 106], *A. parvula* [89] |
| **6** | 9-Hydroxy-4,10-dibromo-3-chloro-α-chamigrene | *L. claviformis* [87, 107], *L. composita* [69, 88], *L. glomerata* [90], *L. nidifica* [95], *L. nipponica* [108[c,d]], *L. okamurai* [99, 100, 109], *L. pacifica* [110[a]], *L. saitoi* [85] |
| **7** | 2,10-Dibromo-3-chloro-7-chamigren-9-ol acetate | *L. composita* [84[d]] |
| **8** | 2,10-Dibromo-3-hydroxy-α-chamigrene | *L. nipponica* [97, 111[a,d]] |
| **9** | Glanduliferol | *L. nipponica* [76[d], 112[a]] |
| **10** | Okamurene E | *L. okamurai* [109] |
| **11** | 10-Bromo-α-chamigren-4-one | *L. nipponica* [76[d], 113[a]] |
| **12** | Laurocamin C | *L. okamurai* [79] |
| **13** | 10-Bromo-α-chamigren-3,4-epoxide | *L. nipponica* [75, 76[c,d], 113[a]] |

(continued)

**Table 1** (continued)

| No. | Name | Source of isolation/synthesis |
|---|---|---|
| **14** | 10*α*-Bromo-9*β*-hydroxy-α-chamigrene | *L. rigida* [81] |
| **15** | – | *L. implicata* [114] |
| **16** | 9-Acetoxy-3-chloro-4,10-dibromo-α-chamigrene | *A. dactylomela* grazing on *L. claviformis* [115] |
| **17** | (6*S*)-2,10-Dibromo-3-chloro-7,9-chamigradiene | *L. okamurai* [100] |
| **18** | 2,10-Dibromo-3-chloro-chamigrane-7,9-diene-5-ol | *L. composita* [88], *L. nidifica* [95[a]] |
| **19** | (6*R*)-2,10-Dibromo-3-chloro-7,9-chamigrediene | *L. okamurai* [100] |
| **20** | – | *L. majuscula* [116[a], 117[d], 118], *A. dactylomela* [118]; Synthesis [119[f], 120[f], 121[f]] |
| **21** | – | *L. majuscula* [122]; Synthesis [121[f]] |
| **22** | Laurencenone C | *L. obtusa* [102]; Synthesis [123[f]] |
| **23** | Laurencenone B | *L. obtusa* [102] |
| **24** | Laurencenone A | *L. obtusa* [102] |
| **25** | Laurencenone D | *L. obtusa* [102] |
| **26** | 2-Chloro-3-hydroxy-α-chamigren-9-one | *L. obtusa* [124] |
| **27** | 2,10-Dibromo-3-chloro-8-hydroxy-β-chamigrene | *L. nipponica* [111[a,d]], *A. dactylomela* [125] |
| **28** | Obtusol | *L. dendroidea* [126], *L. majuscula* [127, 128], *L. obtusa* [129[a], 130[b,d], 131–133, 136[c]], *L. scoparia* [134], *A. dactylomela* [50, 135] |
| **29** | Nidificene | *L. composita* [88], *L. mariannensis* [74], *L. nidifica* [95, 137[a]], *L. nipponica* [111[d], 138], *L. obtusa* [139], *L. okamurai* [79, 99], *A. dactylomela* [140] |
| **30** | 1*β*,4*β*,10*β*-Tribromo-3-chloro-7(14)-en-α-chamigrene | *L. pinnatifida* [141] |
| **31** | Cartilagineol (*allo*-Isoobtusol) | *L. cartilaginea* [83[b], 142[a]], *L. dendroidea* [126], *Laurencia* sp. [143[b,c,d]], *A. dactylomela* [50] |
| **32** | – | *A. dactylomela* [144[d]] |
| **33** | Hurgadol | *L. obtusa* [145[c,d]] |
| **34** | Dendroidiol | *L. dendroidea* [146[c,d]] |
| **35** | Isorigidol | *L. scoparia* [134[a], 147[c,d]] |
| **36** | Ma'iliohydrin | *Laurencia* sp. [148] |
| **37** | Laurecomin B | *L. composita* [84] |
| **38** | 4-Bromo-β-chamigren-8-one | *L. flexilis* [149], *L. nipponica* [113[a]], *L. snackeyi* [50] |
| **39** | Dactylone | *A. dactylomela* [151[a,c,d]] |
| **40** | Dendroidone | *L. dendroidea* [146] |
| **41** | 10-Bromo-β-chamigren-8-ol | *L. okamurai* [152] |
| **42** | (1*Z*,8*R**,9*R**)-8-Bromo-chamigra-1,11(12)-dien-9-ol | *L. majuscula* [114[a], 153], *L. scoparia* [134] |

(continued)

**Table 1** (continued)

| No. | Name | Source of isolation/synthesis |
|---|---|---|
| **43** | (+)-(3*Z*)-Bromomethylidiene-10*β*-bromo-β-chamigrane | *L. scoparia* [34] |
| **44** | (*Z*)-10,15-Dibromo-9-hydroxy-chamigra-1,3(15),7(14)-triene | *L. cartilaginea* [142], *L. majuscula* [62, 69, 116[a], 117[c,d], 154, 155], *L. scoparia* [134], *Laurencia* sp. [143], *A. dactylomela* [118, 156] |
| ***ent*-44** | – | *L. chondrioides* [157], *L. majuscula* [153, 158[a,d]] |
| **45** | (–)-(3*E*)-Bromomethylidiene-10*β*-bromo-β-chamigrane | *L. scoparia* [134] |
| **46** | (*E*)-10,15-Dibromo-9-hydroxy-chamigra-1,3(15),7(14)-triene | *L. cartilaginea* [142], *L. majuscula* [62, 122, 153–155, 158[d]], *L. scoparia* [134], *Laurencia* sp. [143], *A. dactylomela* [118, 156[a]] |
| **47** | (–)-Obtusane | *L. nipponica* [159[d]] |
| ***ent*-47** | Obtusane | *L. composita* [88], *L. decumbens* [160], *L. dendroidea* [126, 146], *L. majuscula* [127], *L. mariannensis* [74], *L. obtusa* [132, 145, 161[a]], *L. okamurai* [79], *L. scoparia* [134] |
| **48** | Rogiolol | *L. majuscula* [118, 158[a], 162], *L. microcladia* [163, 164[b,d]], *L. obtusa* [83], *A. dactylomela* [118] |
| **49** | Isoobtusol | *L. majuscula* [127, 153, 162, 165], *L. obtusa* [102, 129[a], 130[b,c,d], 132], *A. dactylomela* [50, 135], *A. parvula* [166] |
| **50** | Isoobtusol acetate | *A. dactylomela* [50, 156[a,d]] |
| **51** | Epiobtusane | *A. oculifera* [167] |
| **52** | (2*S**,6*R**,8*R**,9*R**)-2,8-Dibromo-9-hydroxy-β-chamigrene | *L. flexilis* [149]; Synthesis [168[g]] |
| **53** | Rigidol | *L. rigida* [81] |
| **54** | – | *L. obtusa* [169[c,d]] |
| **55** | (+)-(10*S*)-10-Bromo-β-chamigrene | *L. mariannensis* [74], *L. nipponica* [97], *L. obtusa* [83], *L. okamurai* [79], *L. pacifica* [10[a]], *L. rigida* [81, 170[d]]; Synthesis [171[g]] |
| **56** | Deschlorelatol | *L. mariannensis* [74], *L. obtusa* [102[a,d]], *L. rigida* [81], *A. dactylomela* [86] |
| ***ent*-56** | (6*R*,9*R*,10*S*)-10-Bromo-9-hydroxy-chamigra-2,7(14)-diene | *L. majuscula* [62] |
| **57** | Acetyldeschlorelatol | *A. dactylomela* [86[d]] |
| **58** | 9-Deoxyelatol | *L. mariannensis* [74] |
| **59** | Elatol | *L. cartilaginea* [142], *L. chondrioides* [157], *L. decumbens* [160], *L. dendroidea* [172, 173], *L. elata* [174[a,c,d]], *L. majuscula* [118, 127, 128, 162, 165], *L. mariannensis* [74], *L. microcladia* [175], *L. obtusa* [102, 124, 129, 131, 132], *L. pacifica* [30], *L. rigida* [81], *L. scoparia* [102], *L. similis* [176], *Laurencia* sp. [165, 177], *A. dactylomela* [50, 86, 118, 135, 156, 178], *A. parvula* [166] |

(continued)

**Table 1** (continued)

| No. | Name | Source of isolation/synthesis |
|---|---|---|
| ***ent*-59** | (–)-Elatol | *L. dendroidea* [126[a,d]] |
| **60** | Acetylelatol | *A. dactylomela* [86[d]] |
| **61** | – | *A. dactylomela* [144[d]] |
| **62** | – | *A. dactylomela* [144[d]] |
| **63** | Obtusadiene | *L. obtusa* [131[d]] |
| **64** | Isoobtusadiene | *L. cartilaginea* [142], *L. majuscula* [153, 154, 158], *L. obtusa* [131[a]] |
| **65** | 2-Bromo-3-hydroxy-11-acetoxy-β-chamigrene | *L. obtusa* [132] |
| **66** | 2-Bromo-3,11-diacetoxy-β-chamigrene | *L. obtusa* [132] |
| **67** | – | *L. mariannensis* [74] |
| **68** | – | *L. nipponica* [179] |
| **69** | (2*S*,3*S*,8*R*)-2,10-Dibromo-3-chlorochamigra-7(14),9-dien-8-ol | *L. nipponica* [179] |
| **70** | – | *L. obtusa* [129[a], 130[b,d]] |
| **71** | – | *L. obtusa* [129[a], 130[b,d]] |
| **72** | Debromoelatol | *L. dendroidea* [146], *L. obtusa* [129[a,d]] |
| **73** | Nidifidienol | *L. nidifica* [95, 180[a]] |
| **74** | Nidifidiene | *L. nidifica* [137] |
| **75** | – | *L. majuscula* [181], *A. dactylomela* [182[a,c,d]] |
| **76** | – | *L. majuscula* [181], *L. mariannensis* [62], *A. dactylomela* [182[a,c,d]] |
| **77** | (2*Z*,6*S**,9*R**)-Chamigra-2,5(14)-dien-8-one | *L. flexilis* [149[a]], *L. snackeyi* [150] |
| **78** | Acetoxy-intricatol | *L. intricata* [183[c,d]] |
| **79** | 4,10-Dibromo-3-chloro-7*β*,8*β*-epoxy-α-chamigrene | *L. glomerata* [90], *L. nidifica* [95], *L. pinnatifida* [141, 185], *L. scoparia* [102], *Laurencia* sp. [104[a], 184[c,d]] |
| **80** | 4,10-Dibromo-3-chloro-7,8-epoxy-9-hydroxy-α-chamigrane | *L. glomerata* [90] |
| **81** | 4,10-Dibromo-3-chloro-7*α*,8*α*-epoxy-α-chamigrene | *L. composita* [88], *L. nipponica* [96[a], 186[c,d]], *L. okamurai* [99, 100], *L. pinnatifida* [141], *Laurencia* sp. [187] |
| **82** | 10-Bromo-7*α*,8*α*-epoxy-chamigr-1-en-3-ol | *L. okamurai* [152] |
| **83** | 5,12-Epoxy-obtusol | *L. obtusa* [132[c,d]] |
| **84** | – | *L. flagellifera* [188[a]], *L. scoparia* [102] |
| **85** | Dehydroxyprepacifenol epoxide | *L. nidifica* [95], *A. dactylomela* [140, 189[a]] |
| **86** | Prepacifenol epoxide | *L. composita* [31, 69], *L. johnstonii* [190[a,d], 191], *L. nidifica* [95], *L. okamurai* [99, 192], *Laurencia* sp. [193], *A. californica* [190[a,d], 191], *A. dactylomela* [140, 189, 194] |

(continued)

**Table 1** (continued)

| No. | Name | Source of isolation/synthesis |
|---|---|---|
| **87** | Deoxyprepacifenol | *L. caduciramulosa* [195], *L. claviformis* [87, 107], *L. composita* [84], *L. elata* [196], *L. filiformis* [89], *L. majuscula* [197], *L. mariannensis* [198], *L. nangii* [94], *L. nidifica* [95], *L. nipponica* [78, 97, 199, 200], *L. okamurai* [99, 100, 109, 192, 201], *L. pacifica* [68], *Laurencia* sp. cf. *L. gracilis* [103], *A. californica* [191[a,c,d]], *A. parvula* [89] |
| **88** | Prepacifenol | *L. claviformis* [87, 107], *L. composita* [31], *L. filiformis* [202[a,d]], *L. nidifica* [95], *L. nipponica* [97, 203], *L. okamurai* [99, 204], *L. pacifica* [30], *Laurencia* sp. cf. *L. gracilis* [103], *Laurencia* sp. [193] |
| **89** | 5-Acetoxy-2,10-dibromo-3-chloro-7,8-epoxy-α-chamigrene | *L. filiformis* [89[c,d]], *A. parvula* [89] |
| **90** | – | *L. nipponica* [205[c,d]] |
| **91** | Dehydrochloroprepacifenol | *L. majuscula* [206[c,d]] |
| **92** | – | *L. microcladia* [163], *A. dactylomela* [50, 207[a,c,d]] |
| **93** | – | *L. microcladia* [163], *A. dactylomela* [207[a]] |
| **94** | Laurecomin D | *L. composita* [84] |
| **95** | – | *L. scoparia* [102] |
| **96** | Pinnatifinone | *L. pinnatifida* [141, 208[a]] |
| **97** | Oxachamigrene | *L. obtusa* [139] |
| **98** | 5-Acetoxy-oxachamigrene | *L. obtusa* [139] |
| **99** | Cyclodebromoacetoxyintricatol | *L. intricata* [209] |
| **100** | – | *A. californica* [210] |
| **101** | Pacifidiene | *L. pacifica* [211], *A. californica* [210, 212[a,d]], *A. dactylomela* [194, 213] |
| **102** | Pacifinediol | *L. composita* [214], *L. nidifica* [95], *A. californica* [190[a,d], 191], *A. dactylomela* [194, 213] |
| **103** | Nidifocene | *L. nidifica* [215[a], 216[b,c,d]]; Synthesis [217[f], 218[f]] |
| **104** | Pinnatifenol | *L. pinnatifida* [141, 219[a]] |
| **105** | Pacifenol | *L. caduciramulosa* [220], *L. claviformis* [87, 107], *L. composita* [31, 69, 88], *L. elata* [196, 221], *L. filiformis* [89], *L. majuscula* [222, 223[c,d]], *L. mariannensis* [198], *L. nidifica* [95, 137], *L. nipponica* [75, 97, 200, 224], *L. okamurai* [99, 192], *L. pacifica* [68, 211, 225[a,c,d]], *L. tasmanica* [202], *Laurencia* sp. cf. *L. gracilis* [103], *Laurencia* sp. [193], *A. californica* [212], *A. dactylomela asymmetrica* [226, 227], *A. dactylomela* [194, 213], *A. parvula* [89] |
| **106** | 4,10-Dibromo-1,10-epoxy-chamigra-8-ene-4,7-diol | *L. nipponica* [199, 224[a]] |
| **107** | 10-Bromo-1,10-epoxy-chamigra-8-ene-7-ol | *L. nipponica* [224[d]] |

(continued)

**Table 1** (continued)

| No. | Name | Source of isolation/synthesis |
|---|---|---|
| **108** | – | *L. nipponica* [205[c,d]] |
| **109** | – | *L. nipponica* [199, 203[a,c,d]] |
| **110** | Bromochlorohydroxy-oxidochamigradiene | *A. dactylomela* [228] |
| **111** | Yicterterpene A | *L. composita* [229[d]] |
| **112** | Yicterterpene B | *L. composita* [229[d]] |
| **113** | Johnstonol | *L. composita* [31, 69, 88], *L. johnstonii* [230[a,c,d]], *L. nidifica* [95], *L. okamurai* [99, 192, 231], *L. pacifica* [230], *L. tristicha* [232], *A. californica* [190], *A. dactylomela* [194, 213] |
| **114** | Pinnatazane | *L. pinnatifida* [141, 233[a,c]] |
| **115** | Almadioxide | *L. intricata* [234[c,d]] |
| **116** | Pinnatifidone | *L. pinnatifida* [141, 235[a,d]] |
| **117** | – | *L. nipponica* [199[d]] |
| **118** | Cycloelatanene B | *L. elata* [221[a]], *L. nangii* [94] |
| **119** | Cycloelatanene A | *L. elata* [221] |
| **120** | Laurokamin B | *L. okamurai* [79] |
| **121** | 10-Bromo-3-chloro-2,7-epoxy-chamigr-9-en-8$\beta$-ol | *L. composita* [214] |
| **122** | 10-Bromo-3-chloro-2,7-epoxy-chamigr-9-en-8$\alpha$-ol | *L. okamurai* [109], *L. saitoi* [85[a]] |
| **123** | – | *L. obtusa* [124] |
| **124** | Claviol | *L. claviformis* [87] |
| **125** | Gomerolactone A | *L. majuscula* [128] |
| **126** | Gomerolactone B | *L. majuscula* [128] |
| **127** | Gomerolactone C | *L. majuscula* [128] |
| **128** | Gomerolactone D | *L. majuscula* [128] |
| **129** | Chamigrenelactone | *L. obtusa* [133] |
| **130** | Majusculone | *L. majuscula* [122[a], 128], *L. scoparia* [134]; Synthesis [121[f], 236[d,g]] |
| **131** | Ma'ilione | *L. cartilaginea* [142[a]], *L. majuscula* [118], *L. scoparia* [134, 147[c,d]], *Laurencia* sp. [143], *A. dactylomela* [118] |
| **132** | – | *A. dactylomela* [144[d]] |
| **133** | – | *A. dactylomela* [182]; Synthesis [237[b,f]] |
| **134** | – | *A. dactylomela* [182]; Synthesis [237[b,f]] |
| **135** | Laurencomposidiene | *L. composita* [88] |
| **136** | Scopariol | *L. scoparia* [134] |
| **137** | Pannosane | *L. pannosa* [238] |
| **138** | Laurenokamurin | *L. okamurai* [239] |
| **139** | Laurokamin A | *L. okamurai* [79] |
| **140** | Laurecomin A | *L. composita* [84] |
| **141** | – | *A. dactylomela* [240[d]] |
| **142** | Pannosanol | *L. pannosa* [238] |

(continued)

**Table 1** (continued)

| No. | Name | Source of isolation/synthesis |
|---|---|---|
| **143** | – | *L. flagellifera* [188[a]], *L. pinnatifida* [141, 219], *L. scoparia* [102] |
| **144** | Pinnatifate | *L. pinnatifida* [241] |
| **145** | Spironippol | *L. nipponica* [97, 242[a]] |
| **146** | 2-Bromo-spironippol | *L. composita* [88] |
| **147** | Kylinone | *L. pacifica* [211] |
| **148** | Laurene | *L. elata* [196], *L. filiformis* f. *heteroclada* [243], *L. glandulifera* [244[a], 245[d]], *L. nipponica* [77, 78, 96, 138, 246], *L. okamurai* [109, 152, 247], *L. subopposita* [68, 248], *Laurencia* sp. cf. *L. gracilis* [103], *A. punctata* [249]; Synthesis [250[h], 251[f], 252[f], 253[f], 254[g]] |
| **149** | Debromoallolaurinterol (7-Hydroxy-laurene) | *L. caduciramulosa* [220], *L. nipponica* [138], *L. obtusa* [255], *L. okamurai* [109], *L. subopposita* [248[a]], *A. punctata* [249] |
| **150** | Debromoallolaurinterol acetate | *L. okamurai* [152] |
| **151** | Allolaurinterol (10-Bromo-7-hydroxy-laurene) | *L. caduciramulosa* [220], *L. caraibica* [256], *L. filiformis* f. *heteroclada* [127, 243[a,d], 257], *L. glandulifera* [258], *L. majuscula* [69], *L. obtusa* [259], *L. subopposita* [248], *A. dactylomela* [178, 260]; Synthesis [261[f]] |
| **152** | Allolaurinterol acetate | *L. okamurai* [152], *A. dactylomela* [156[a,d], 260] |
| **153** | 10,11-Dibromo-7-hydroxy-laurene | *L. majuscula* [69] |
| **154** | 10-Bromo-7-hydroxy-11-iodolaurene | *L. caraibica* [256] |
| **155** | Laurenol | *L. nipponica* [96[d]] |
| **156** | Laurenol acetate | *L. nipponica* [96[d]] |
| **157** | Laurenal | *L. nipponica* [96[d]] |
| **158** | – | *L. microcladia* [262] |
| **159** | Debromoisolaurinterol | *L. decidua* [60, 263], *L. distichophylla* [264], *L. majuscula* [181], *L. okamurai* [60[d], 192, 201, 247, 265[a]], *L. pacifica* [68]; Synthesis [266[f]] |
| **160** | Debromoisolaurinterol acetate | *L. okamurai* [267[d]] |
| **161** | Isolaurinterol | *L. decidua* [60, 263], *L. filiformis* f. *heteroclada* [127], *L. nidifica* [62], *L. okamurai* [60, 99, 192, 201, 247, 265[a,d], 268], *L. pacifica* [60, 68], *L. tristicha* [269], *A. dactylomela* [270]; Synthesis [266[f]] |
| **162** | Isolaurinterol acetate | *L. okamurai* [267[d]] |
| **163** | Laur-11-en-10-ol | *L. tristicha* [232] |
| **164** | Laur-11-en-2,10-diol | *L. tristicha* [232] |
| **165** | Laur-11-en-1,10-diol | *L. tristicha* [232] |

(continued)

**Table 1** (continued)

| No. | Name | Source of isolation/synthesis |
|---|---|---|
| **166** | 4-Bromo-laur-11-en-1,10$\beta$-diol | *L. tristicha* [269] |
| **167** | Laur-2-ene-3,12-diol | *L. obtusa* [271] |
| **168** | Laur-11-en-1,10$\alpha$-diol | *L. tristicha* [269] |
| **169** | 4-Bromo-laur-11-en-1,10$\alpha$-diol | *L. tristicha* [269] |
| **170** | – | *L. pinnatifida* [272] |
| **171** | Laurenisol | *L. claviformis* [107], *L. filiformis* f. *heteroclada* [127], *L. glandulifera* [258], *L. nipponica* [273[a,d]], *A. punctata* [249] |
| **172** | Laurenisol acetate | *A. dactylomela* grazing on *L. claviformis* [115] |
| **173** | – | *L. glandulifera* [258[d]] |
| **174** | *iso*-Laurenisol | *L. chondrioides* [73], *L. obtusa* [255[a]], *L. saitoi* [85], *Laurencia* sp. cf. *L. gracilis* [103[a]] |
| **175** | Bromolaurenisol | *L. filiformis* f. *heteroclada* [127[a], 257], *L. microcladia* [255] |
| **176** | Isolaurenisol | *L. distichophylla* [264[a]], *A. dactylomela* [260] |
| **177** | Isolaurenisol acetate | *A. dactylomela* [260] |
| **178** | Isoallolaurinterol | *L. filiformis* f. *heteroclada* [257] |
| **179** | 10-Bromo-7,12-dihydroxy-$\Delta^{2,3}$-laurene | *L. caraibica* [256] |
| **180** | Isolaurene | *L. elata* [196], *L. glandulifera* [258], *L. nipponica* [60], *L. okamurai* [60, 192, 247[a]], *L. tristicha* [269]; Synthesis [244[d], 245, 246] |
| **181** | 10-Hydroxy-isolaurene | *L. microcladia* [274[a]], *L. okamurai* [267[a]] |
| **182** | 12-Hydroxy-isolaurene | *L. obtusa* [98] |
| **183** | Isolauraldehyde | *L. obtusa* [98] |
| **184** | 3,7-Dihydroxy-dihydrolaurene | *L. obtusa* [255] |
| **185** | Laurepoxyene | *L. okamurai* [275] |
| **186** | Laurol | *L. pinnatifida* [276] |
| **187** | Isodihydrolaurene | *L. nipponica* [78, 96[a,d], 97, 138] |
| **188** | Isodihydrolaurenol | *L. nipponica* [96[d]] |
| **189** | Isodihydrolaurenol acetate | *L. nipponica* [96[d]] |
| **190** | – | *L. pinnatifida* [272] |
| **191** | 8,11-Dihydro-12-hydroxy-isolaurene | *L. obtusa* [98] |
| **192** | Dihydrolaurene | *L. filiformis* f. *heteroclada* [243[a,d]], *Laurencia* sp. cf. *L. gracilis* [103] |
| **193** | 16,16-Dichlorohomolaurane | *A. dactylomela* [277] |
| **194** | Debromoaplysin | *L. distichophylla* [264], *L. nidifica* [278], *L. okamurai* [99, 231, 279], *L. pacifica* [211], *L. tristicha* [269], *A. californica* [212], *A. kurodai* [52[a,d]]; Synthesis [266[f], 280[f], 281[h], 282[f], 283[f], 284[f], 285[g], 286[d,g], 287[g], 288[g], 289[g]] |

(continued)

**Table 1** (continued)

| No. | Name | Source of isolation/synthesis |
|---|---|---|
| **195** | Aplysin | *L. decidua* [263], *L. nidifica* [137, 278], *L. okamurai* [99, 231, 268, 275], *L. pacifica* [211, 263], *L. tristicha* [232, 290], *A. californica* [212], *A. kurodai* [52[a], 292[d], 293], *A. parvula* [291]; Synthesis [266[f], 280[f], 281[h], 282[f], 283[f], 284[f], 285[g], 286[g], 287[g], 288[g], 289[g]] |
| **196** | 10-Hydroxy-aplysin | *L. okamurai* [275, 279], *L. tristicha* [290, 294[a]] |
| **197** | 3*β*-Hydroperoxyaplysin | *L. okamurai* [275[d]] |
| **198** | 10-Hydroxy-debromo-*epi*-aplysin | *L. okamurai* [279], *L. tristicha* [290, 294[a]] |
| **199** | 10-Hydroxy-*epi*-aplysin | *L. okamurai* [275], *L. tristicha* [290, 294[a]] |
| **200** | 3*α*-Hydroperoxy-3-*epi*-aplysin | *L. okamurai* [275[d]] |
| **201** | Aplysin-9-ene | *L. tristicha* [290] |
| **202** | Debromoaplysinol | *L. okamurai* [60[a], 201], *L. tristicha* [269, 295]; Synthesis [266[f], 283[f]] |
| **203** | Isoaplysin | *L. okamurai* [60, 247[a,d], 267, 296]; Synthesis [266[f], 283[f]] |
| **204** | Aplysinol | *L. decidua* [263, 297[c,d]], *L. okamurai* [60, 231, 267, 298], *L. pacifica* [263], *L. tristicha* [232, 269, 290], *A. kurodai* [52[a], 292[d], 293]; Synthesis [266[f], 283[f]] |
| **205** | Aplysinol acetate | *L. okamurai* [201[d]] |
| **206** | Laureperoxide | *L. okamurai* [267] |
| **207** | 10-Bromoisoaplysin | *L. okamurai* [267[a], 275] |
| **208** | 8,10-Dibromoisoaplysin | *L. okamurai* [275] |
| **209** | Aplysinal | *L. okamurai* [268[d]]; Synthesis [266[f]] |
| **210** | Debromo-*epi*-aplysinol | *L. tristicha* [290] |
| **211** | *epi*-Aplysinol | *L. tristicha* [290] |
| **212** | Ibhayinol | *A. dactylomela* [140, 299[c,d]] |
| **213** | Debromofiliformin | *L. caduciramulosa* [220[a]], *L. okamurai* [109]; Synthesis [284[f]] |
| **214** | Filiformin | *L. caduciramulosa* [220], *L. caraibica* [256], *L. filiformis* f. *heteroclada* [127, 243[a], 257], *L. okamurai* [109, 201]; Synthesis [284[f], 300[d,g]] |
| **215** | Isofiliformin | *L. glandulifera* [258[d], 301[a]], *L. pinnatifida* [272], *A. punctata* [249] |
| **216** | 12-Bromo-filiformin | *L. caraibica* [256], *L. glandulifera* [258[d], 301[a]], *L. okamurai* [109], *L. pinnatifida* [272]; Synthesis [302[f]] |
| **217** | – | *L. microcladia* [262] |
| **218** | Filiforminol | *L. filiformis* f. *heteroclada* [243[a], 257]; Synthesis [302[f]] |
| **219** | 8,10-Dibromo-3,7-epoxy-laur-13-ol | *Laurencia* sp. [303] |
| **220** | – | *L. caraibica* [256] |
| **221** | – | *L. glandulifera* [258, 301[a]] |

(continued)

**Table 1** (continued)

| No. | Name | Source of isolation/synthesis |
|---|---|---|
| **222** | 4-Bromo-1,10-epoxy-laur-11-ene | *L. tristicha* [232] |
| **223** | Caraibical | *L. caraibica* [304] |
| **224** | Cycloisoallolaurinterol | *L. filiformis* f. *heteroclada* [257] |
| **225** | Cyclolaurene | *A. dactylomela* [305[d]]; Synthesis [306[f]] |
| **226** | Debromolaurinterol | *L. decidua* [263], *L. flexilis* [149], *L. nidifica* [278], *L. okamurai* [31, 60, 192, 247, 265, 307[a,d]], *L. pacifica* [68, 211, 263], *L. tristicha* [269], *A. californica* [212], *A. kurodai* [308], *A. parvula* [291]; Synthesis [281[h]] |
| **227** | Debromolaurinterol acetate | *L. okamurai* [267], *A. kurodai* [309[a,d]]; Synthesis [281[f]] |
| **228** | Cyclolauren-2-ol | *L. tristicha* [232] |
| **229** | Isodebromolaurinterol | *L. okamurai* [267] |
| **230** | Neolaurinterol | *L. okamurai* [247[d]] |
| **231** | Laurinterol | *L. decidua* [263], *L. microcladia* [255], *L. nidifica* [62, 137, 278], *L. okamurai* [31, 60, 99, 192, 247, 265, 268, 279, 292[c,d], 293, 298, 307[a]], *L. pacifica* [68, 225, 263], *L. tristicha* [269], *Laurencia* sp. [105], *A. californica* [212], *A. dactylomela* [270], *A. kurodai* [308], *A. parvula* [291], *A. punctata* [249]; Synthesis [281[h]] |
| **232** | Laurinterol acetate | *L. okamurai* [267], *A. kurodai* [292[c,d], 293, 309[a]]; Synthesis [281[h]] |
| **233** | Cyclolaurenol | *L. calliclada* [310], *L. venusta* [69], *A. dactylomela* [305[a,d]] |
| **234** | Cyclolaurenol acetate | *A. dactylomela* [305[d]] |
| **235** | Bromolaurinterol | *L. decidua* [263], *L. microcladia* [262], *L. okamurai* [60[a,d]], *L. tristicha* [269] |
| **236** | 8-Iodolaurinterol | *L. microcladia* [262] |
| **237** | Laurequinone | *L. nidifica* [278[a,d]], *L. okamurai* [192]; Synthesis [311[g]] |
| **238** | – | *L. microcladia* [262] |
| **239** | Laurebiphenyl | *L. nidifica* [312[a,d]], *L. tristicha* [232] |
| **240** | – | *L. microcladia* [255[a]], *L. tristicha* [269] |
| **241** | Laurokamurene B | *L. okamurai* [279]; Synthesis [313[f], 314[d,g]] |
| **242** | Laurokamurene A | *L. microcladia* [274[a]], *L. okamurai* [275, 279], *Laurencia* sp. [118], *A. dactylomela* [118] |
| **243** | Laurokamurene C | *L. okamurai* [201[a], 275] |
| **244** | Laurokamurene D | *L. okamurai* [275] |
| **245** | 8,11-Dihydro-1-methoxy-laurokamuren-12-ol | *L. obtusa* [271] |
| **246** | Laurentristich-4-ol | *L. tristicha* [232[a], 315[b]]; Synthesis [316[b]] |
| **247** | α-Snyderol | *L. nangii* [94], *L. obtusa* [317[a], 318, 319] |
| **248** | α-Snyderol acetate | *L. obtusa* [319, 320[a]] |

(continued)

**Table 1** (continued)

| No. | Name | Source of isolation/synthesis |
|---|---|---|
| **249** | β-Snyderol | *L. obtusa* [145, 321], *L. snyderae* [317[a]], *A. punctata* [249]; Synthesis [322[f]] |
| **250** | β-Snyderol acetate | *L. obtusa* [145] |
| **251** | Luzonensin | *L. luzonensis* [323] |
| **252** | Luzonensol | *L. luzonensis* [323[a], 324[b,d]], *L. similis* [325] |
| **253** | Luzonensol acetate | *L. luzonensis* [323[d]] |
| **254** | – | *L. similis* [325] |
| **255** | 6-*epi*-β-Snyderol | *A. fasciata* [326] |
| **256** | (8*S**)-8-Bromo-β-snyderol | *L. obtusa* [319] |
| **257** | 8-Hydroxy-β-snyderol | *L. caespitosa* [327[d]] |
| **258** | (8*R**)-8-Bromo-10-*epi*-β-snyderol | *L. obtusa* [319] |
| **259** | – | *L. implicata* [91], *L. snackeyi* [118], *A. dactylomela* [118] |
| **260** | Palisol | *L. flexilis* [149], *L. implicata* [91, 114], *L. luzonensis* [323], *L. snackeyi* [118, 328[a,d], 329, 330], *A. dactylomela* [118] |
| **261** | Isopalisol | *L. luzonensis* [323[a]], *L. similis* [325] |
| **262** | 8-Oxo-10-dehydrobromo-β-snyderol | *L. caespitosa* [327[d]] |
| **263** | 8-Hydroxy-γ-snyderol | *L. caespitosa* [327[d]] |
| **264** | – | *L. obtusa* [319] |
| **265** | – | *L. obtusa* [319] |
| **266** | Palisadin B | *L. flexilis* [331, 332], *L. implicata* [91, 114], *L. karlae* [333], *L. luzonensis* [323, 334], *L. saitoi* [335], *L. similis* [325], *L. snackeyi* [118, 150, 328[a], 336], *A. dactylomela* [118, 337]; Synthesis [338[d,g]] |
| **267** | 5*β*-Hydroxy-palisadin B | *L. flexilis* [331[a]], *L. snackeyi* [150] |
| **268** | 5*β*-Acetoxy-palisadin B | *L. flexilis* [331], *L. karlae* [333], *L. saitoi* [335], *L. similis* [325], *L. snackeyi* [118, 150, 328[a], 329, 330, 339], *A. dactylomela* [118, 337, 339] |
| **269** | 12-Hydroxy-palisadin B | *L. flexilis* [331], *L. karlae* [333], *L. snackeyi* [328[a,d]], *A. dactylomela* [337]; Synthesis [338[g]] |
| **270** | 12-Acetoxy-palisadin B | *L. snackeyi* [118], *A. dactylomela* [118] |
| **271** | 12-Bromopalisadin B | *L. flexilis* [331] |
| **272** | 4-Hydroxy-palisadin C | *L. saitoi* [335[a]], *L. similis* [325] |
| **273** | Palisadin C | *L. flexilis* [331] |
| **274** | 3,4-Epoxy-palisadin B | *L. luzonensis* [324[a], 334, 340] |
| **275** | 1,2-Dehydro-3,4-epoxy-palisadin B | *L. luzonensis* [324] |
| **276** | Aplysistatin | *L. filiformis* [341], *L. flexilis* [331], *L. implicata* [91, 114], *L. karlae* [333], *L. luzonensis* [334], *L. saitoi* [335], *L. similis* [325], *L. snackeyi* [118, 150, 328–330, 339], *C. papillosus* [342], *A. angasi* [343[a,c,d]], *A. dactylomela* [118, 337, 339]; Synthesis [338[g], 344[f], 345[f], 346[f], 347[g]] |

(continued)

**Table 1** (continued)

| No. | Name | Source of isolation/synthesis |
|---|---|---|
| **277** | 6*β*-Hydroxy-aplysistatin | *L. filiformis* [341[a,c,d]], *L. implicata* [91, 114] |
| **278** | Palisadin A | *L. flexilis* [331], *L. implicata* [91, 114], *L. karlae* [333], *L. luzonensis* [323, 334], *L. saitoi* [335], *L. similis* [325], *L. snackeyi* [118, 150, 328[a,d], 329, 330, 339], *A. dactylomela* [118, 337, 339], *A. parvula* [166]; Synthesis [338[g]] |
| **279** | 15-Hydroxy-palisadin A | *L. luzonensis* [324[a], 340] |
| **280** | 5*β*-Acetoxy-palisadin A | *L. flexilis* [331] |
| **281** | 3,4-Epoxy-palisadin A | *L. flexilis* [331, 332[c,d]] |
| **282** | 3*β*-Bromo-8-epicaparrapi oxide | *L. obtusa* [145, 348[a], 349] |
| **283** | Luzonenone | *L. luzonensis* [324[a], 340] |
| **284** | (3*R**,4*S**)-Luzonolone | *L. luzonensis* [334] |
| **285** | (3*S**,4*R**)-Luzonolone | *L. luzonensis* [334] |
| **286** | Luzondiol | *L. luzonensis* [334] |
| **287** | Luzofuran | *L. luzonensis* [324[a], 340]; Synthesis [350[f]] |
| **288** | 2-Hydroxy-luzofuranone | *L. saitoi* [335] |
| **289** | 2-Hydroxy-luzofuranone B | *L. saitoi* [335] |
| **290** | Bicyclolaurencenol | *L. intricata* [351] |
| **291** | Dactylenol | *L. intricata* [351], *A. dactylomela* [352[a]] |
| **292** | Dactylenol acetate | *A. dactylomela* [352] |
| **293** | – | *L. nangii* [94], *A. punctata* [249[a]] |
| **294** | – | *L. nangii* [94], *A. punctata* [249[a]] |
| **295** | – | *L. intricata* [351], *L. nidifica* [353[a]], *L. snyderae* [354]; Synthesis [355[f]] |
| **296** | – | *L. intricata* [351], *L. nidifica* [353[a]] |
| **297** | – | *L. luzonensis* [334] |
| **298** | – | *L. luzonensis* [334] |
| **299** | Dactyloxene B | *A. dactylomela* [352, 356[a]]; Synthesis [357[b], 358[d,g], 359[g], 360[g], 361[h]] |
| **300** | Dactyloxene C | *A. dactylomela* [352[a]]; Synthesis [357[b], 358[d,g], 359[g], 360[g], 361[h]] |
| **301** | Dactyloxene A | *L. nangii* [94], *A. dactylomela* [352[a]] |
| **302** | Isodactyloxene A | *L. mariannensis* [74] |
| **303** | Dactyloxene D | *L. obtusa* [362] |
| **304** | (*E*)-γ-Bisabolene | *L. nipponica* [363] |
| **305** | 3-Acetoxy-(*E*)-γ-bisabolene | *L. rigida* [81] |
| **306** | – | *L. microcladia* [262] |
| **307** | (5*S*)-5-Acetoxy-β-bisabolene | *L. okamurai* [275[d]] |
| **308** | (*E*)-γ-Bisabolene 8,9-epoxide | *L. nipponica* [363]; Synthesis [171[d,g], 364[f]] |
| **309** | Preintricatol | *L. intricata* [209[a]], *Laurencia* sp. cf. *L. gracilis* [103] |
| **310** | Puertitol A | *L. obtusa* [365[d]] |
| **311** | Puertitol B | *L. obtusa* [365[d]] |
| **312** | Puertitol B acetate | *A. dactylomela* [50[d]] |

(continued)

**Table 1** (continued)

| No. | Name | Source of isolation/synthesis |
|---|---|---|
| **313** | – | *L. scoparia* [366[c,d]] |
| **314** | 2-Bromo-3-chloro-bisabola-7(14),11-diene-6,10-diol | *L. composita* [214] |
| **315** | – | *L. scoparia* [366] |
| **316** | – | *L. scoparia* [366] |
| **317** | – | *L. caespitosa* [367] |
| **318** | – | *L. caespitosa* [367] |
| **319** | Gossonorol | *L. tristicha* [232] |
| **320** | α-Bisabol-9-en-7,11-diol | *L. tristicha* [232] |
| **321** | Okamurene D | *L. okamurai* [109] |
| **322** | Okamurene C | *L. okamurai* [109] |
| **323** | (9*R*)-2-Bromo-3-chloro-6,9-epoxy-bisabola-7(14),10-diene | *L. saitoi* [85] |
| **324** | (9*S*)-2-Bromo-3-chloro-6,9-epoxy-bisabola-7(14),10-diene | *L. saitoi* [85] |
| **325** | – | *L. scoparia* [366] |
| **326** | – | *L. scoparia* [366] |
| **327** | 7,10-Epoxy-α-bisabol-11-ol | *L. tristicha* [232] |
| **328** | 10-*epi*-7,10-Epoxy-α-bisabol-11-ol | *L. tristicha* [232] |
| **329** | Deschlorobromocaespitol | *A. dactylomela* [368[d]] |
| **330** | Deschlorobromocaespitenone | *A. dactylomela* [368[d]] |
| **331** | Okamurene A | *L. okamurai* [109] |
| **332** | Okamurene B | *L. okamurai* [109] |
| **333** | Caespitane (Deoxycaespitol) | *L. caespitosa* [369[a,c,d]], *L. catarinensis* [370], *A. dactylomela* [50, 135, 368] |
| **334** | Deodactol | *A. dactylomela* [371[c,d]] |
| **335** | (5*S*)-5-Acetoxy-deoxycaespitol | *L. catarinensis* [370] |
| **336** | Isodeodactol | *A. dactylomela* [372] |
| **337** | Caespitol (Aldingenin C) | *L. aldingensis* [373, 374[b]], *L. caespitosa* [375[a], 376[b]], *L. caespitosa* [369[d]], *L. catarinensis* [370], *L. intricata* [377], *L. obtusa* [129, 318], *A. dactylomela* [50, 368] |
| **338** | 8-Acetylcaespitol | *L. catarinensis* [370], *A. dactylomela* [50[a,d], 368] |
| **339** | 6-Hydroxy-caespitol | *L. caespitosa* [369[c,d]] |
| **340** | Aldingenin A | *L. aldingensis* [378, 379[b]] |
| **341** | (5*S*)-5-Hydroxy-acetylcaespitol | *L. catarinensis* [370] |
| **342** | (5*S*)-5-Acetoxy-caespitol (aldingenin D) | *L. aldingensis* [373[a], 374[b], 379], *L. catarinensis* [370[d]] |
| **343** | (5*S*)-5-Acetoxy-acetylcaespitol | *L. catarinensis* [370] |
| **344** | Dihydroxy-deodactol monoacetate | *A. dactylomela* [380] |
| **345** | 8-Desoxy-isocaespitol | *L. caespitosa* [381[a]], *L. catarinensis* [370]; Synthesis [381[f]] |
| **346** | Isocaespitol | *L. caespitosa* [369[d], 376[b], 382[a], *L. catarinensis* [370], *L. obtusa* [318]; Synthesis [383[g]] |

(continued)

**Table 1** (continued)

| No. | Name | Source of isolation/synthesis |
|---|---|---|
| **347** | Isocaespitol acetate | *A. dactylomela* [384] |
| **348** | 10-*epi*-Deoxycaespitol | *L. catarinensis* [370] |
| **349** | Caespitenone | *L. catarinensis* [370], *A. dactylomela* [50[a], 135, 368] |
| **350** | Aldingenin B | *L. aldingensis* [373, 379[b]]; Synthesis [385[e], 386[b]] |
| **351** | – | *L. caespitosa* [387[c,d]] |
| **352** | Furocaespitane | *L. caespitosa* [161, 387[b,d], 388[a]], *A. dactylomela* [50, 368] |
| **353** | Isofurocaespitane | *L. caespitosa* [161] |
| **354** | Furocaespitanelactol | *L. catarinensis* [370], *A. dactylomela* [368[a]] |
| **355** | (10*R**)-10-*O*-Methyl-furocaespitanelactol | *L. catarinensis* [370] |
| **356** | (10*S**)-10-*O*-Methyl-furocaespitanelactol | *L. catarinensis* [370] |
| **357** | – | *L. caespitosa* [387[c,d]] |
| **358** | – | *L. majuscula* [153] |
| **359** | – | *L. majuscula* [153] |
| **360** | – | *L. majuscula* [153] |
| **361** | – | *L. obtusa* [389[c,d]] |
| **362** | Majapolene A | *L. caraibica* [390], *L. majuscula* [390[a]], *Laurencia* sp. [193] |
| **363** | Acetylmajapolene A | *Laurencia* sp. [193] |
| **364** | Majapolene B | *L. majuscula* [390[a], 391[d]], *Laurencia* sp. [193] |
| **365** | Acetylmajapolene B | *Laurencia* sp. [193, 391[d]] |
| **366** | Tiomanene | *Laurencia* sp. [193] |
| **367** | Majapolone | *L. majuscula* [390] |
| **368** | Majapol A | *L. majuscula* [390] |
| **369** | Majapol B | *L. majuscula* [390] |
| **370** | Majapol C | *L. majuscula* [390] |
| **371** | Majapol D | *L. majuscula* [390] |
| **372** | Laucapyranoid A | *L. caespitosa* [369[a]], *A. dactylomela* [50, 368] |
| **373** | Laucapyranoid B | *L. caespitosa* [369[c,d]] |
| **374** | Laucapyranoid C | *L. caespitosa* [369[c,d]] |
| **375** | Aplysiadactydiol | *A. dactylomela* [368[a], 392] |
| **376** | Perforenone | *L. obtusa* [393], *L. perforata* [394[a]], *L. snyderae* var. *guadalupensis* [395], *A. punctata* [249]; Synthesis [396[f], 397[f]] |
| **377** | Perforenone A | *L. obtusa* [255], *L. perforata* [398[a]] |
| **378** | Perforenone B | *L. perforata* [398] |
| **379** | Perforenone C (3-*epi*-Perforenone A) | *L. obtusa* [255], *L. perforata* [394[a], 399] |
| **380** | Perforenone D | *L. obtusa* [393] |
| **381** | 9-Hydroxy-3-*epi*-perforenone A | *L. perforata* [399] |
| **382** | Perforenol | *L. obtusa* [255], *L. perforata* [400[a,c,d]] |
| **383** | Guadalupol | *L. snyderae* var. *guadalupensis* [395] |

(continued)

**Table 1** (continued)

| No. | Name | Source of isolation/synthesis |
|---|---|---|
| **384** | Perforenol B | *L. obtusa* [255] |
| **385** | Epiguadalupol | *L. snyderae* var. *guadalupensis* [395] |
| **386** | Perforene | *L. perforata* [401[a]] |
| **387** | Debromoperforatone | *L. obtusa* [393[b]]; Synthesis [396[f], 398[f]] |
| **388** | Perforatone | *L. obtusa* [393], *L. perforata* [398[a]] |
| **389** | – | *L. tenera* [402[c,d]] |
| **390** | – | *L. obtusa* [393] |
| **391** | – | *L. obtusa* [393] |
| **392** | – | *L. obtusa* [393] |
| **393** | – | *L. obtusa* [255] |
| **394** | Perforatol | *A. punctata* [249] |
| **395** | Tenerol acetate | *L. tenera* [403] |
| **396** | – | *L. tenera* [402] |
| **397** | 4-Hydroxy-1,8-*epi*-isotenerone | *L. perforata* [399] |
| **398** | Brasilenol | *L. implicata* [114], *L. obtusa* [320, 404, 405], *A. brasiliana* [406[a]], *A. dactylomela* [407], *A. fasciata* [326]; Synthesis [408[d,g], 409[f]] |
| **399** | Brasilenol acetate | *A. brasiliana* [406[a,d]], *A. fasciata* [326] |
| **400** | Epibrasilenol | *L. obtusa* [320, 405, 410, 411], *A. brasiliana* [406[a,d]], *A. fasciata* [326] |
| **401** | Epibrasilenol acetate | *A. fasciata* [326[d]] |
| **402** | 4-Hydroxy-5-brasilene | *L. implicata* [91], *L. obtusa* [410[a], 411], *A. fasciata* [326] |
| **403** | 4-Acetoxy-5-brasilene | *A. fasciata* [326] |
| **404** | – | *L. implicata* [114] |
| **405** | – | *L. implicata* [91, 412[d]] |
| **406** | 6-Hydroxy-1-brasilene | *A. fasciata* [326] |
| **407** | *epi*-5,9-Dihydroxy-brasil-1(6)-en-7-one | *L. obtusa* [410] |
| **408** | 5,9-Dihydroxy-brasil-1(6)-en-7-one | *L. obtusa* [404] |
| **409** | 9-Hydroxy-brasil-1(6),4-dien-7-one | *L. obtusa* [410] |
| **410** | – | *L. obtusa* [413] |
| **411** | – | *L. obtusa* [405] |
| **412** | – | *L. obtusa* [405] |
| **413** | – | *L. obtusa* [405] |
| **414** | 2,5-Epoxy-brasil-1(6)-en-9-ol | *L. obtusa* [404] |
| **415** | (+)-$\alpha$-Bromocuparene | *L. caraibica* [256, 304], *L. chondrioides* [73], *L. distichophylla* [264], *L. flexilis* [149], *L. glandulifera* [76, 414[a,d]], *L. okamurai* [60, 192, 247], *A. punctata* [249] |
| ***ent*-415** | (–)-$\alpha$-Bromocuparene | *L. filiformis* f. *heteroclada* [127[a,d]], *L. microcladia* [262], *Laurencia* sp. [83] |
| **416** | – | *L. implicata* [415] |

(continued)

**Table 1** (continued)

| No. | Name | Source of isolation/synthesis |
|---|---|---|
| **417** | $\alpha$-Isobromocuparene | *L. glandulifera* [76, 414[a,d]], *L. majuscula* [416], *L. microcladia* [262], *L. nipponica* [414[a,d]], *L. okamurai* [60, 109, 201, 247], *Laurencia* sp. [83], *A. punctata* [249] |
| **418** | Cupalaurenol | *L. venusta* [69], *Laurencia* sp. [187], *A. dactylomela* [305[a,d]] |
| **419** | Cupalaurenol acetate | *A. dactylomela* [305[d]] |
| **420** | – | *L. obtusa* [271] |
| **421** | – | *L. distichophylla* [264], *L. microcladia* [255], *L. okamurai* [60, 201, 247[a]] |
| **422** | – | *L. microcladia* [274] |
| **423** | – | *L. majuscula* [416] |
| **424** | 10-Bromo-3-chloro-cupar-5-en-2-ol | *L. okamurai* [152] |
| **425** | 1-Deacetoxy-8-deoxyalgoane | *A. dactylomela* [140] |
| **426** | 1-Deacetoxy-algoane | *A. dactylomela* [140] |
| **427** | Algoane | *A. dactylomela* [140[c,d]] |
| **428** | Oculiferane | *A. oculifera* [167] |
| **429** | (+)-Selin-4,7(11)-diene | *L. nidifica* [417[a,d]], *L. nipponica* [363] |
| **430** | (4*S*,5*R*,10*R*)-Selin-6(7)-en-11-ol | *L. nipponica* [418[d]] |
| **431** | (5*S*,7*R*,10*R*)-Selin-4(14)-en-5$\alpha$-ol | *L. nipponica* [78[a,d], 418] |
| **432** | Brasudol | *A. brasiliana* [419[d]] |
| **433** | Isobrasudol | *A. brasiliana* [419[d]] |
| **434** | 1-Bromo-selin-4(14),11-diene | *L. composita* [84] |
| **435** | 9-Bromo-selin-4(14),11-diene | *L. composita* [84] |
| **436** | – | *A. dactylomela* [277] |
| **437** | Austradiol | *L. filiformis* f. *heteroclada* [420[a,d], 421] |
| **438** | Austradiol acetate | *L. filiformis* f. *heteroclada* [420[a,d], 421] |
| **439** | Itomanol | *L. intricata* [422] |
| **440** | Eudesma-4(15),7-diene-5,11-diol | *L. obtusa* [423] |
| **441** | (1*S*)-Bromo-(4*R*)-hydroxy-7-chloroselin-7-ene | *Laurencia* sp. [177[a,d], 424[a,d]] |
| **442** | (1*S*,4*R*,7*R*)-1-Bromo-4-hydroxy-7-chloroselinane | *Laurencia* sp. [425[c,d]] |
| **443** | Heterocladol | *L. filiformis* f. *heteroclada* [426[c,d]] |
| **444** | Lankalapuol A | *A. dactylomela* [427[c,d]] |
| **445** | Lankalapuol B | *A. dactylomela* [427] |
| **446** | Cycloeudesmol | *L. nipponica* [77, 199, 363, 428[b,d]], *Laurencia* sp. [429[a]] |
| **447** | – | *L. microcladia* [430] |
| **448** | Chabrolidione B | *L. obtusa* [423] |
| **449** | Teuhetenone A | *L. obtusa* [423] |
| **450** | 1(10)-Aristolene | *L. decumbens* [160] |

(continued)

**Table 1** (continued)

| No. | Name | Source of isolation/synthesis |
|---|---|---|
| **451** | Aristol-1(10)-en-9*β*-ol | *L. similis* [176[a], 325, 431], *Laurencia* sp. [303], *C. papillosus* [342] |
| ***ent*-451** | – | *L. similis* [432] |
| **452** | Aristol-1(10)-en-9-one | *L. similis* [176, 325] |
| **453** | – | *L. similis* [432] |
| **454** | Aristolone | *L. similis* [325, 432], *Laurencia* sp. [303], *C. papillosus* [342] |
| **455** | Axinysone B | *L. similis* [432] |
| **456** | Debilone | *L. complanata* [433] |
| **457** | Aristol-9-en-1-one | *L. similis* [176[a], 325, 431] |
| **458** | Aristol-8-en-1-one | *L. similis* [176[a], 325, 431, 434] |
| **459** | Aristol-1(10),8-diene | *L. similis* [176, 325] |
| **460** | 1(10),8-Aristoladien-2-one | *C. papillosus* [342] |
| **461** | Aristol-1,9-diene | *L. similis* [176, 325] |
| **462** | 10*β*-Hydroxy-aristolan-9-one | *L. similis* [176[a], 325, 434] |
| **463** | Aristolan-1*α*-bromo-9*β*,10*β*-epoxide | *L. similis* [431] |
| **464** | – | *L. luzonensis* [323] |
| **465** | Obtusenol | *L. obtusa* [413, 435[a]], *A. punctata* [249]; Synthesis [436[b,f]] |
| **466** | – | *L. subopposita* [248] |
| **467** | Humulene | *L. saitoi* [437] |
| **468** | – | *L. saitoi* [437[c,d]] |
| ***ent*-468** | – | *L. saitoi* [437[c,d]] |
| **469** | *seco*-Laurokamurone | *L. okamurai* [275[d]] |
| **470** | Lauracetal A | *L. nipponica* [75[a], 199] |
| **471** | Lauracetal B | *L. nipponica* [199, 438[a]] |
| **472** | Lauracetal D | *L. nipponica* [439[c,d]] |
| **473** | Lauracetal E | *L. nipponica* [439[c,d]] |
| **474** | Lauracetal C | *L. nipponica* [440] |
| **475** | (–)-δ-Cadinene | *L. microcladia* [430] |
| **476** | (+)-α-Cadinol | *L. microcladia* [430] |
| **477** | Oplopanone | *L. subopposita* [248] |
| **478** | Dactylol | *L. poitei* [441[d]], *A. dactylomela* [442[a,d]]; Synthesis [443[f], 444[f], 445[g], 446[g]] |
| **479** | Poitediol | *L. poitei* [441[c]]; Synthesis [447[f]] |
| **480** | Guaiazulene | *L. obtusa* [320] |
| **481** | Oppositol | *L. subopposita* [448[c,d]]; Synthesis [449[f], 450[f]] |
| **482** | – | *L. subopposita* [248] |
| **483** | – | *L. subopposita* [248] |
| **484** | – | *L. mariannensis* [198] |

(continued)

**Table 1** (continued)

| No. | Name | Source of isolation/synthesis |
|---|---|---|
| **485** | Pacifigorgiol | *L. flexilis* [149], *L. luzonensis* [323], *L. saitoi* [335], *L. snackeyi* [118, 451], *A. dactylomela* [118] |
| **486** | Laurencial | *L. nipponica* [452[c,d]]; Synthesis [453[f]] |
| **487** | Spirolaurenone | *L. nipponica* [78, 199, 454[a], 455[d]]; Synthesis [456[f,d]] |
| **488** | Laurenone A | *L. nipponica* [457[c,d]] |
| **489** | Laurenone B | *L. nipponica* [457[d]] |
| **490** | – | *L. dendroidea* [126, 146, 458], *L. majuscula* [153, 158[a]], *L. rigida* [81], *L. scoparia* [366] |
| **491** | – | *L. majuscula* [114], *L. scoparia* [366] |
| **492** | – | *L. dendroidea* [458] |
| **493** | Viridianol | *L. viridis* [459] |
| **494** | Majusin | *L. majuscula* [460] |
| **495** | Isoafricanol | *L. mariannensis* [74] |
| **496** | (–)-Aromadendrene | *L. filiformis* [421] |
| **497** | 1-Hydroxy-alloaromadendrene | *L. subopposita* [248[d]] |
| **498** | 10$\beta$-Hydroxy-$\Delta^{1(2)}$-aromadendrene | *L. subopposita* [248] |
| **499** | Bromocyclococanol | *L. obtusa* [461] |
| **500** | Calenzanol | *L. microcladia* [462, 463] |
| **501** | Debromoisocalenzanol | *L. microcladia* [463] |
| **502** | Laurobtusol | *L. obtusa* [464]; Synthesis [465[b,f]] |
| **503** | Guimarediol | *L. obtusa* [132], *Laurencia* sp. [466[a,c,d]] |
| **504** | Dactylomelatriol | *A. dactylomela* [135] |
| **505** | Gomerone A | *L. majuscula* [467] |
| **506** | Gomerone B | *L. majuscula* [467]; Synthesis [468[b]] |
| **507** | Gomerone C | *L. majuscula* [467]; Synthesis [468[b,f]] |
| **508** | Rhodolauradiol | *L. obtusa* [132], *Laurencia* sp. [469[a,c]]; Synthesis [470] |
| **509** | Rhodolaureol | *L. obtusa* [132], *Laurencia* sp. [469[a,c]]; Synthesis [470] |
| **510** | Isorhodolaureol | *L. majuscula* [158] |
| **511** | – | *L. mariannensis* [74]; Synthesis [470] |
| **512** | Aplydactone | *A. dactylomela* [471[c,d]] |

[a]The isolation and structure elucidation of this compound was reported for the first time
[b]The structure of this compound has been revised
[c]The structure of this compound or of a derivative has been confirmed by X-ray crystallographic analysis
[d]The absolute stereochemistry of this compound has been established
[e]Partial synthesis
[f]Stereoselective total synthesis
[g]Enantioselective total synthesis
[h]Formal synthesis

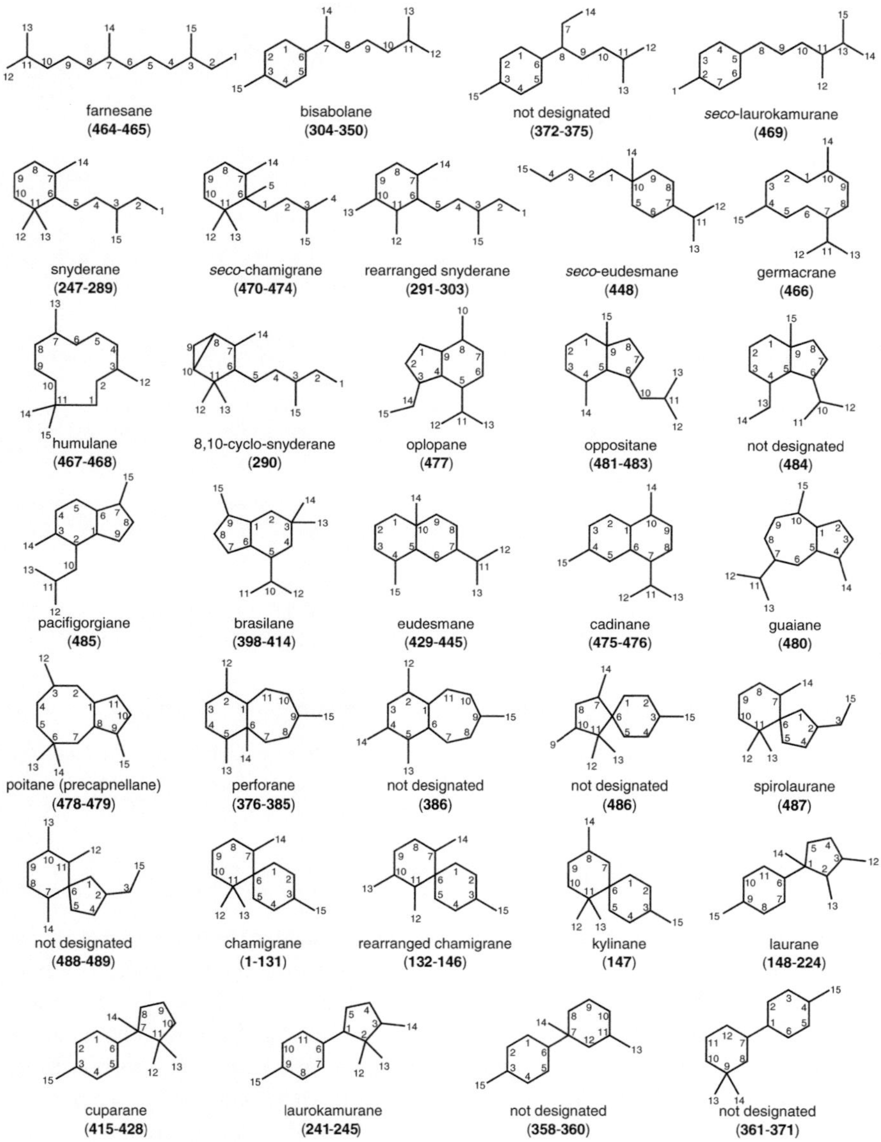


**Fig. 4** Acyclic, monocyclic, and bicyclic carbon skeletons of sesquiterpenes isolated from *Laurencia* species

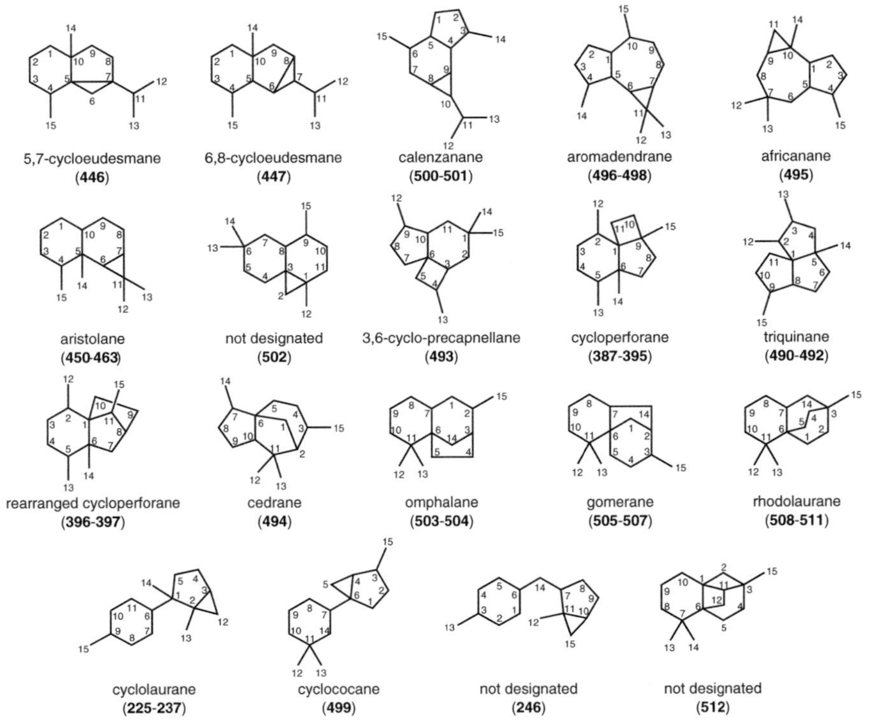


**Fig. 5** Tricyclic and tetracyclic carbon skeletons of sesquiterpenes isolated from *Laurencia* species

### 3.1.1 Chamigranes and Related Sesquiterpenes

Chamigranes and related metabolites are the most frequently occurring class of sesquiterpenes isolated from *Laurencia* species or sea hares feeding on them, with a total of 147 (**1–147**) having been reported in the literature up to now. Structurally, sesquiterpenes of the chamigrane class (**1–131**) are characterized by a *spiro*[5.5]undecane skeleton bearing a stereogenic quaternary carbon at the spirocycle junction (C-6). The first six-membered ring, which is moderately to densely functionalized, contains a second non-stereogenic quaternary carbon (C-11) that is vicinal to the spirocycle junction and bears a *gem*-dimethyl group, whereas the second six-membered ring often possesses a tri- or tetrasubstituted olefin or presents a dihalogenation pattern at C-2 and C-3 usually in a 2,3-*trans*-halide arrangement. They can be divided roughly into four subclasses, namely: (1) those belonging to the α-chamigrene series possessing a trisubstituted double bond between C-7 and C-8 (**1–26**); (2) those belonging to the β-chamigrene series possessing an exomethylene group at C-7 (**27–77**); (3) those featuring an oxygenation site at C-7 (**78–99**); and (4) those with either a 5,10-epoxide ring system (**100–117**) or another epoxide (**118–124**) or a lactone ring connecting the two six-membered rings (**125–129**). Most chamigranes possess various halogen and oxygen substitutions, with C-10 being a favored position for a Br substitution. Often, a bromine atom is located at the secondary carbon C-2 and a chlorine atom is attached to the tertiary carbon C-3 on the opposite face of the former, while seldom a *cis*-arrangement of the two halogens can be observed. Moreover, less frequently halogenation on a double bond or a conjugated bromodiene system is incorporated into the second six-membered ring. In the 1990s, Guella and co-workers undertook thorough conformational studies of α- and β-chamigrenes using temperature-dependent NMR spectra [164, 472, 473]. Majusculone (**130**) and mailione (**131**) are the only two known nor-chamigranes isolated from a number of different *Laurencia* species [118, 122, 128, 134, 142, 143, 147]. Compounds **132–146** possess a rearranged chamigrane carbocycle that arises from a methyl migration from C-11 to C-10 of the chamigrane skeleton, whereas kylinone (**147**), isolated as a constituent of *L. pacifica* from the Monterey Peninsula, California [211], is the sole representative of another rearranged chamigrane framework, in all probability derived from a methyl migration from C-7 to C-8.

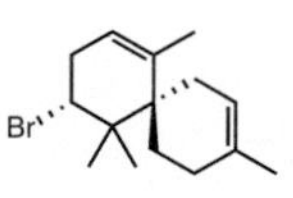

**1**

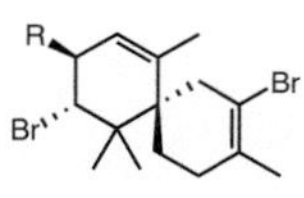

**2** R = H
**3** R = OH

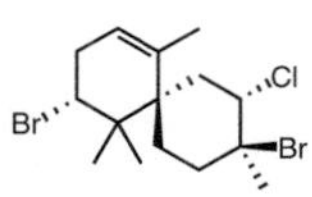

**4**

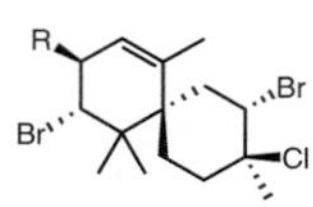

**5** R = H
**6** R = OH
**7** R = OAc

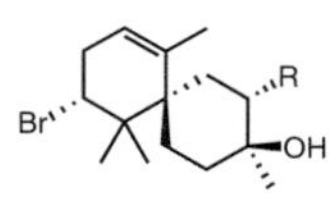

**8** R = Br
**9** R = Cl

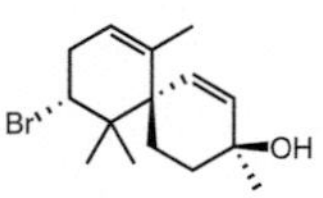

**10**

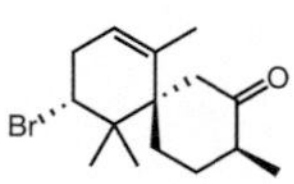

**11**

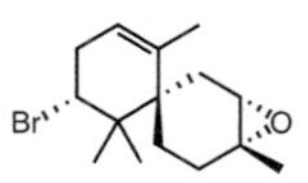

**12**

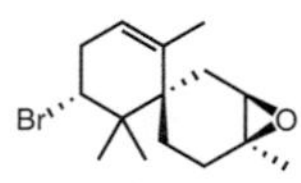

**13**

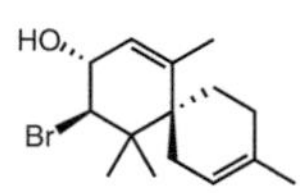

**14**

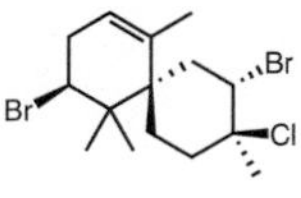

**15**

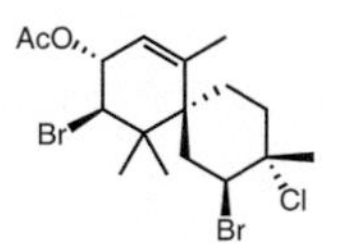

**16**

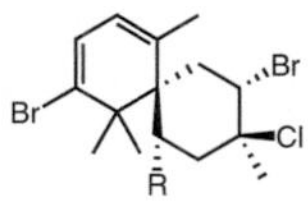

**17** R = H
**18** R = OH

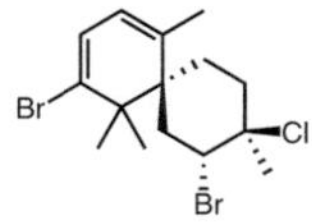

**19**

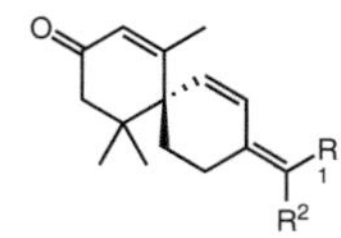

**20** $R^1$ = Br, $R^2$ = H
**21** $R^1$ = H, $R^2$ = Br

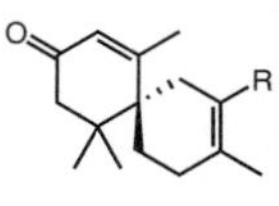

**22** R = H
**23** R = Cl

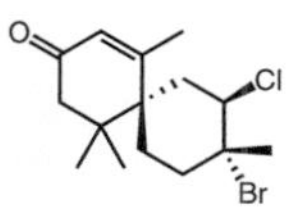

**24**

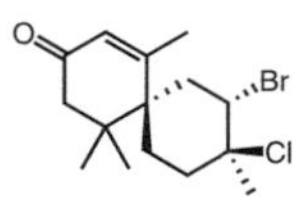

**25**

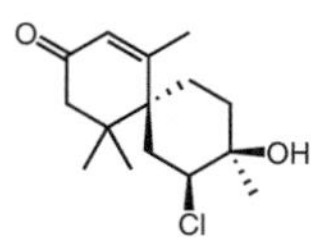

**26**

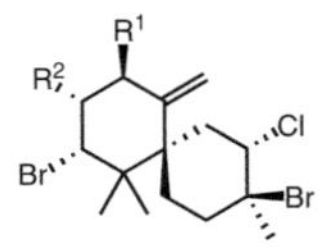

**27** $R^1$ = OH, $R^2$ = H
**28** $R^1$ = H, $R^2$ = OH

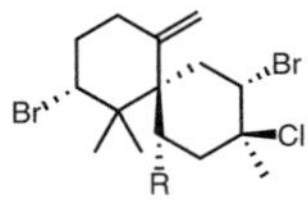

**29** R = H
**30** R = Br

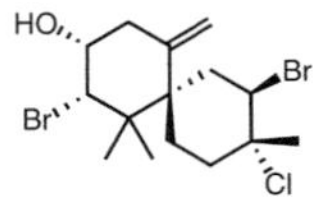

**31**

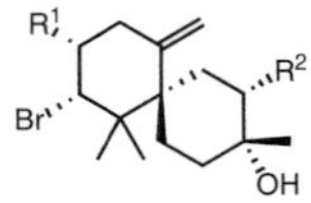

**32** $R^1$ = H, $R^2$ = OH
**33** $R^1$ = H, $R^2$ = Cl
**34** $R^1$ = OH, $R^2$ = Cl

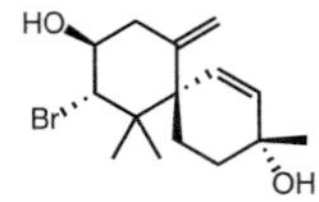

**35**

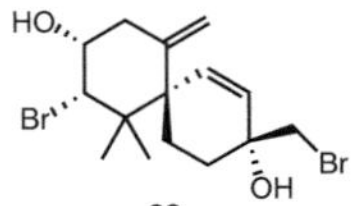

**36**

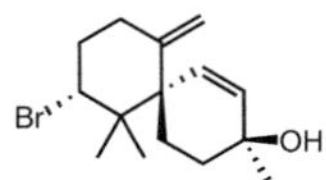

**37**

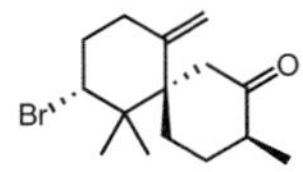

**38**

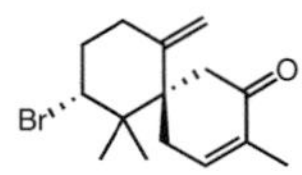

**39**

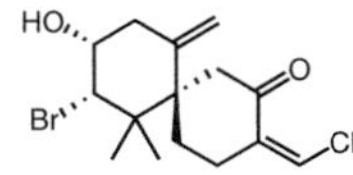

**40**

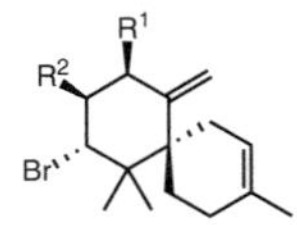

**41** $R^1$ = OH, $R^2$ = H
**42** $R^1$ = H, $R^2$ = OH

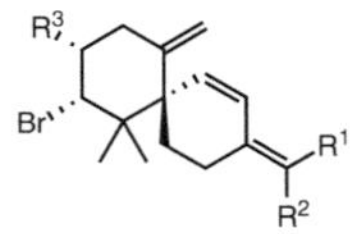

**43** $R^1$ = Br, $R^2$ = $R^3$ = H
**44** $R^1$ = Br, $R^2$ = H, $R^3$ = OH
**45** $R^1$ = $R^3$ = H, $R^2$ = Br
**46** $R^1$ = H, $R^2$ = Br, $R^3$ = OH

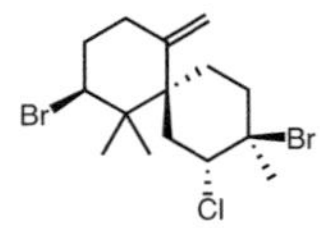

**47**

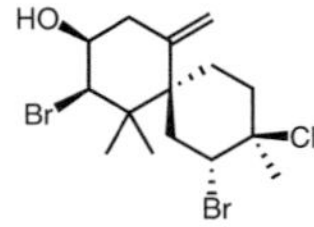

**48**

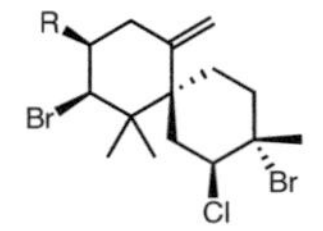

**49** R = OH
**50** R = OAc

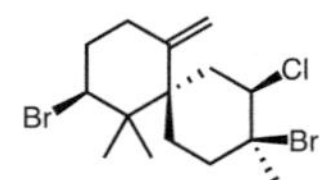

**51**

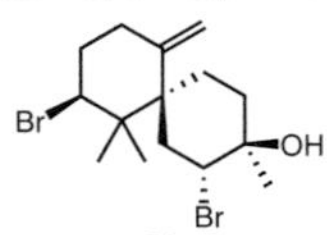

**52**

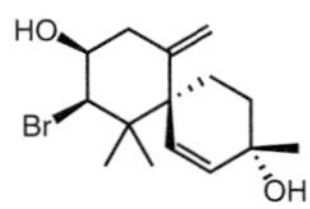

**53**

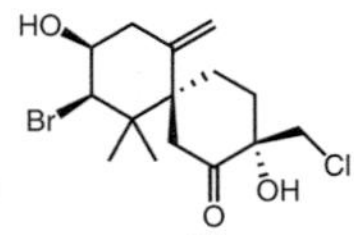

**54**

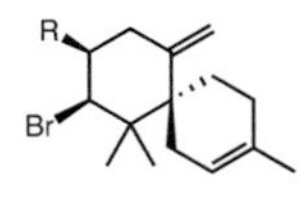

**55** R = H
**56** R = OH
**57** R = OAc

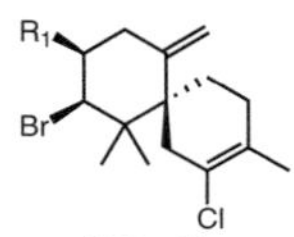

**58** R = H
**59** R = OH
**60** R = OAc

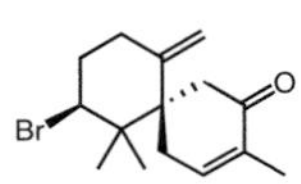

**61**

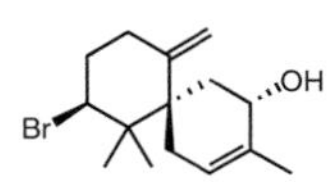

**62**

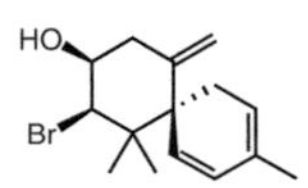

**63**

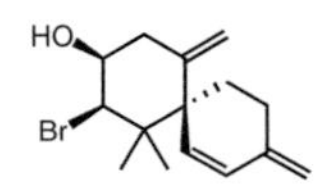

**64**

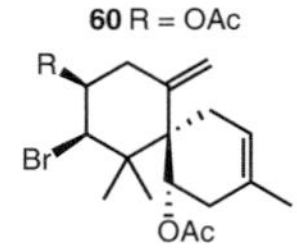

**65** R = OH
**66** R = OAc

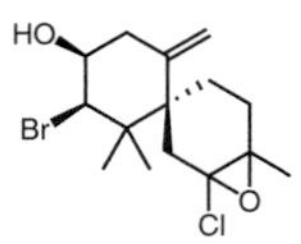

**67**

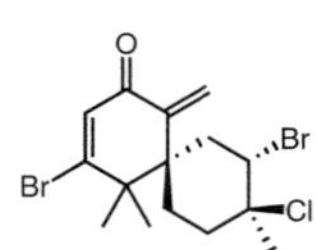

**68**

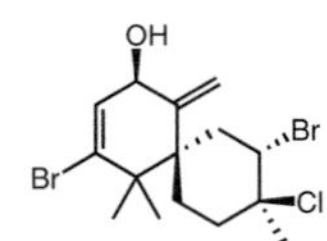

**69**

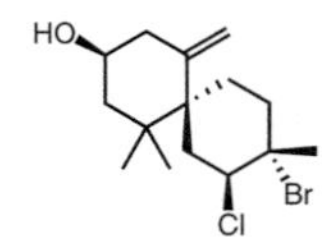

**70**

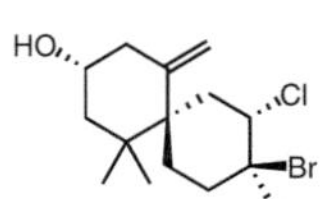

**71**

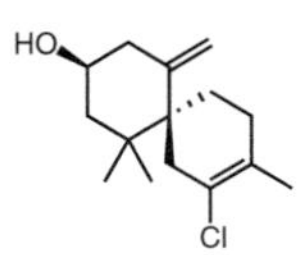

**72**

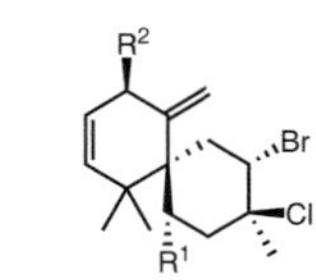

**73** $R^1$ = OH, $R^2$ = H
**74** $R^1$ = H, $R^2$ = OH

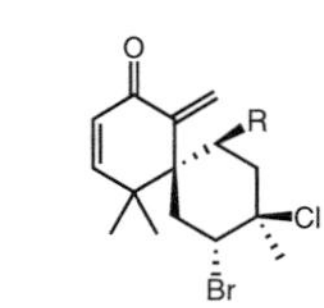

**75** R = H
**76** R = OAc

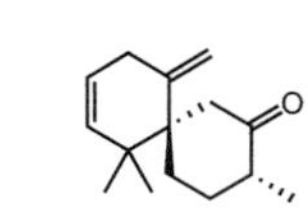

**77**

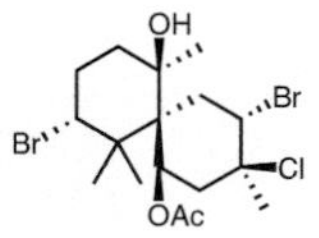

**78**

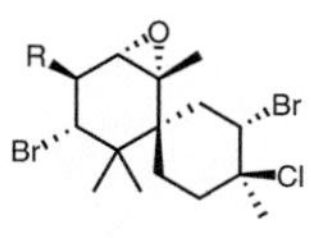

**79** R = H
**80** R = OH

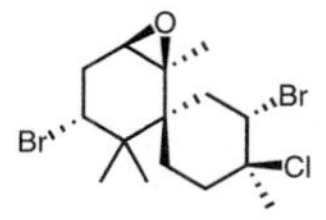

**81**

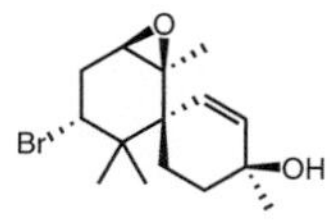

**82**

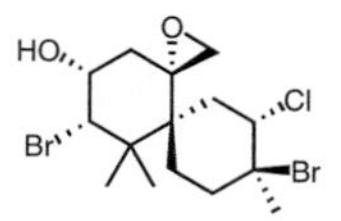

**83**

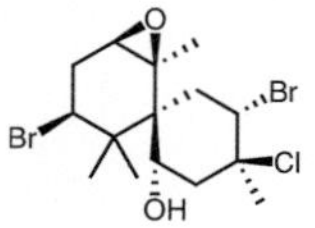

**84**

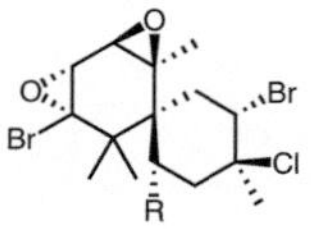

**85** R = H
**86** R = OH

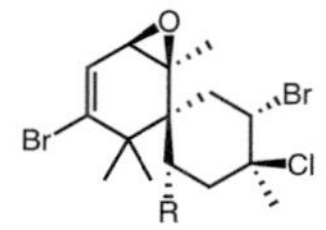

**87** R = H
**88** R = OH
**89** R = OAc

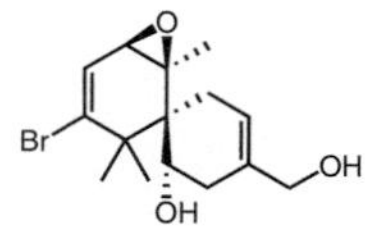

**90**

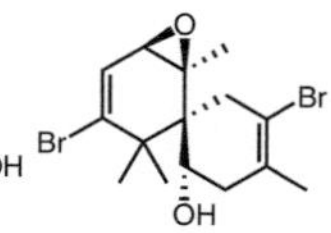

**91**

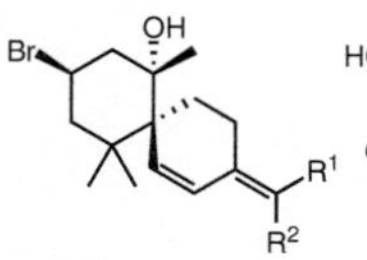

**92** $R^1$ = Br, $R^2$ = H
**93** $R^1$ = H, $R^2$ = Br

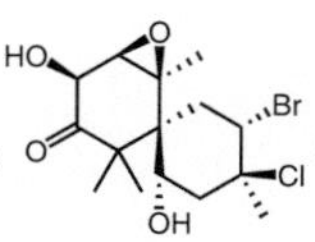

**94**

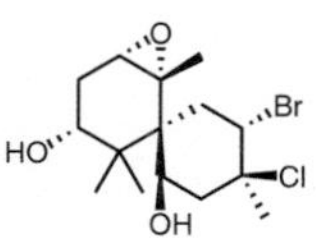

**95**

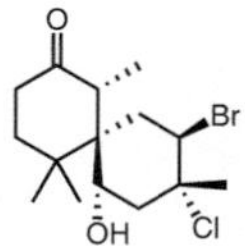

**96**

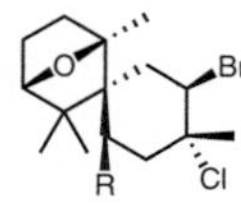

**97** R = H
**98** R = OAc

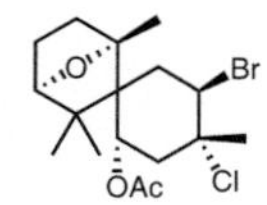

**99**

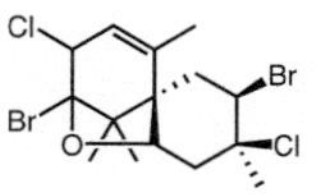

**100**

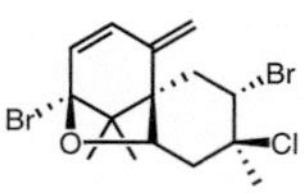

**101**

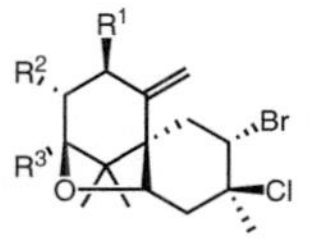

**102** $R^1$ = $R^2$ = OH, $R^3$ = Br
**103** $R^1$ = $R^2$ = $R^3$ = H

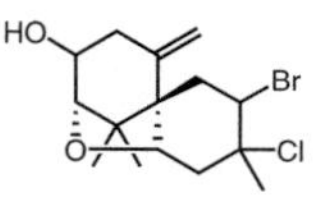

**104**

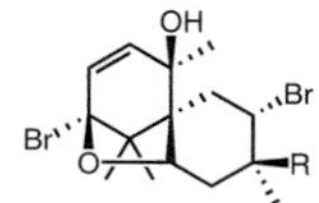

**105** R = Cl
**106** R = OH

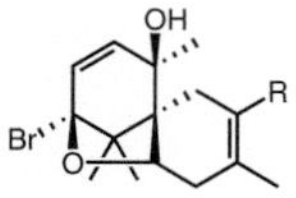

**107** R = H
**108** R = Br

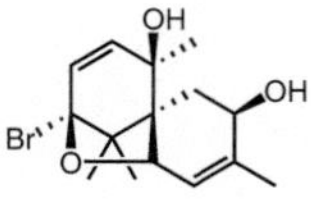

**109**

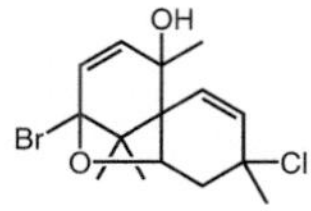

**110**

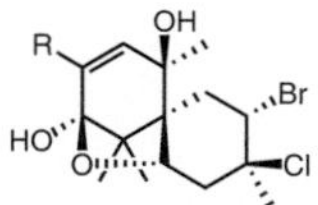

**111** R = Cl
**112** R = Br

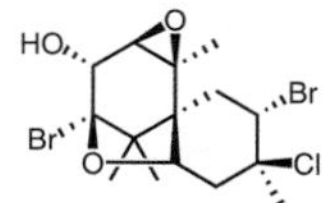

**113**

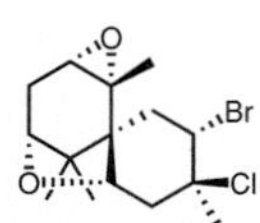

**114**

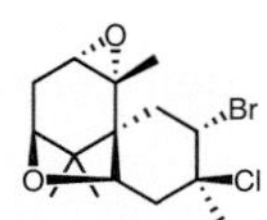

**115**

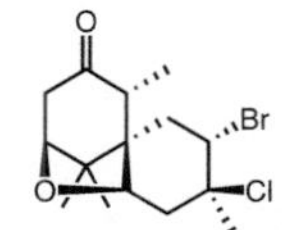

**116**

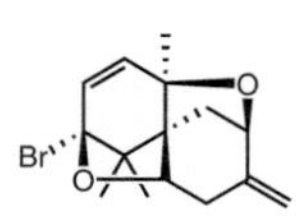

**117**

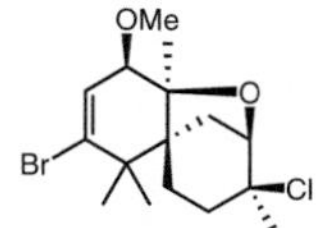

**118**

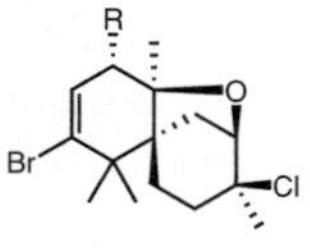

**119** R = OMe
**120** R = Cl

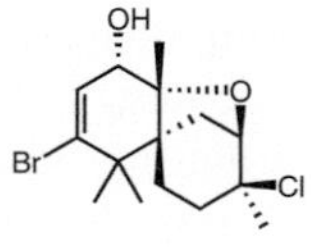

**121**

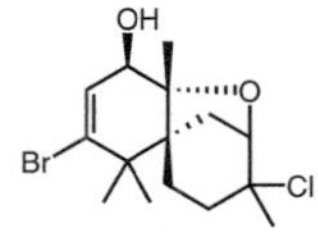

**122**

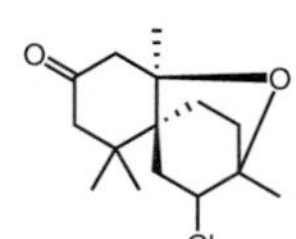

**123**

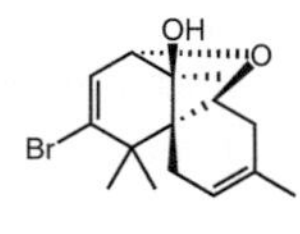

**124**

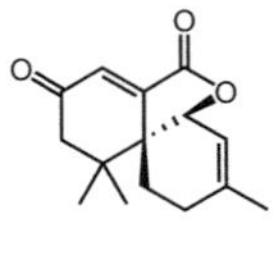
125

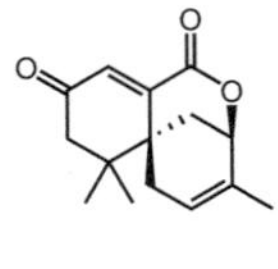
126

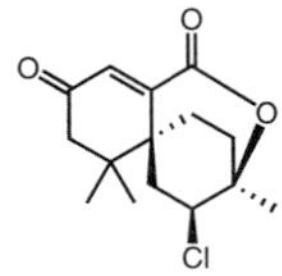
127

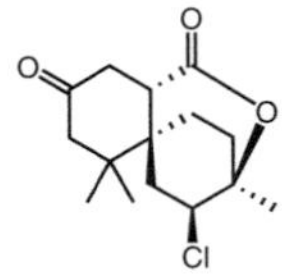
128

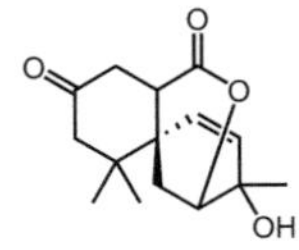
129

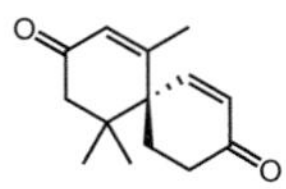
130

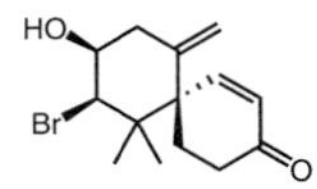
131

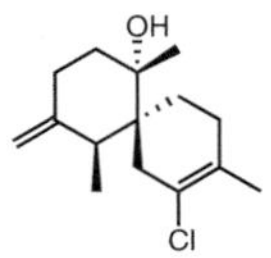
132

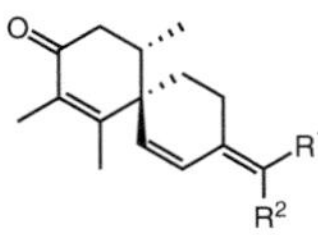
**133** $R^1$ = Br, $R^2$ = H
**134** $R^1$ = H, $R^2$ = Br

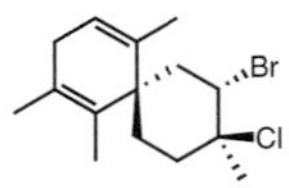
135

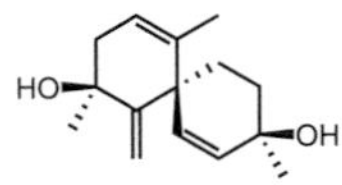
136

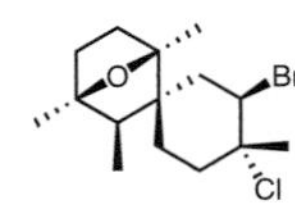
137

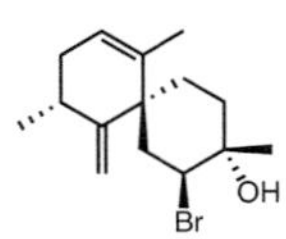
138

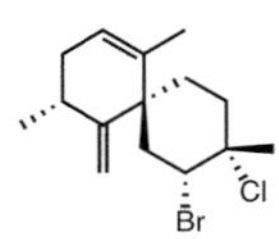
139

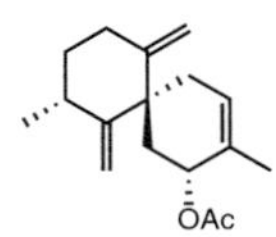
140

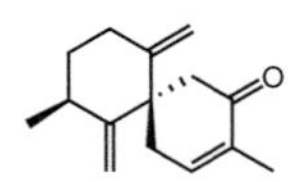
141

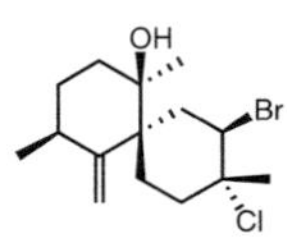
142

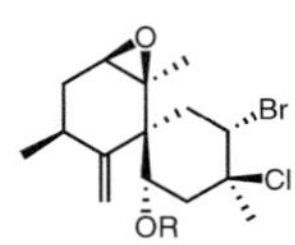
**143** R = H
**144** R = Ac

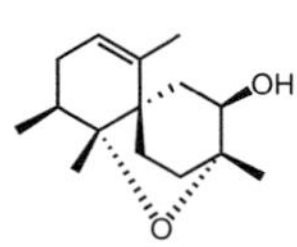
145

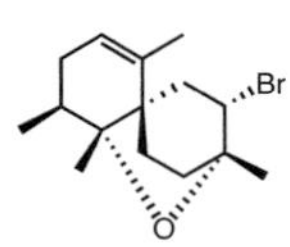
146

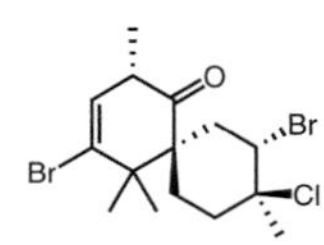
147

### 3.1.2 Lauranes, Cyclolauranes, and Related Sesquiterpenes

Taken together, lauranes, cyclolauranes, and related metabolites consist the second most often encountered class of sesquiterpenes characterized from *Laurencia* spp. and gastropods feeding on them, amounting to 99 compounds in total (**148–246**). Lauranes (**148–224**) and cyclolauranes (**225–237**) possess a 1,2,3-trisubstituted cyclopentane moiety and a 1,2-dimethylbicyclo[3.1.0]hexane system, respectively, connected through a single bond to a six-membered ring, usually aromatic. Frequently, C-10 and/or C-12 are brominated, whereas C-3 is oxygenated. In addition, several of these sesquiterpenes contain 2,7-epoxy (**194–212**), 3,7-epoxy (**213–222**), 4,7-epoxy (**223**), or 5,7-epoxy (**224**) rings. Among them, 16,16-di-chloro-homolaurane (**193**), isolated from *Aplysia dactylomela*, is the only known $C_{16}$ derivative of this carbocycle [277]. Attention should be paid to the case of the isomeric metabolites *iso*-laurenisol (**174**), isolated initially from *Laurencia* sp. cf. *L. gracilis* [103], and isolaurenisol (**176**), obtained initially from

*Laurencia distichophylla* [264], to avoid confusion due to the rather similar common names given. In addition to monomeric cyclolauranes, three dimeric derivatives (**238–240**) have been reported to date from various *Laurencia* species. Metabolites **241–245** feature a rearranged laurane skeleton arising from a methyl migration from C-1 to C-2, while laurentristich-4-ol (**246**), isolated from a Chinese *Laurencia tristicha* [232, 315], possesses an irregular isoprenoid carbocycle, proposed biogenetically to result from the rearrangement of cyclolaurane skeleton. Total synthesis of the originally proposed structure led to its revision [316].

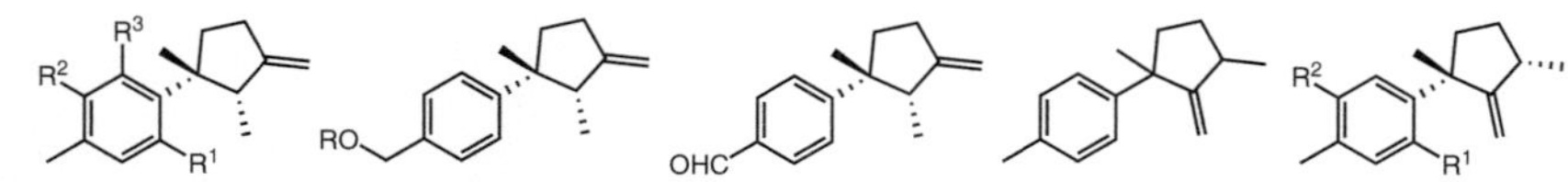

**148** $R^1 = R^2 = R^3 = H$
**149** $R^1 = OH, R^2 = R^3 = H$
**150** $R^1 = OAc, R^2 = R^3 = H$
**151** $R^1 = OH, R^2 = Br, R^3 = H$
**152** $R^1 = OAc, R^2 = Br, R^3 = H$
**153** $R^1 = OH, R^2 = R^3 = Br$
**154** $R^1 = OH, R^2 = Br, R^3 = I$

**155** R = H
**156** R = Ac

**157**

**158**

**159** $R^1 = OH, R^2 = H$
**160** $R^1 = OAc, R^2 = H$
**161** $R^1 = OH, R^2 = Br$
**162** $R^1 = OAc, R^2 = Br$

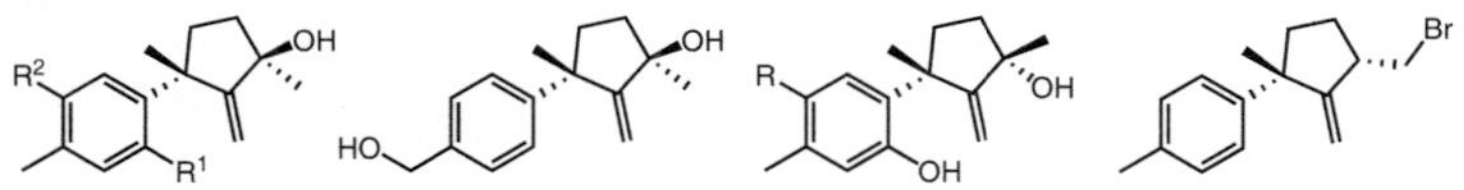

**163** $R^1 = R^2 = H$
**164** $R^1 = H, R^2 = OH$
**165** $R^1 = OH, R^2 = H$
**166** $R^1 = OH, R^2 = Br$

**167**

**168** R = H
**169** R = Br

**170**

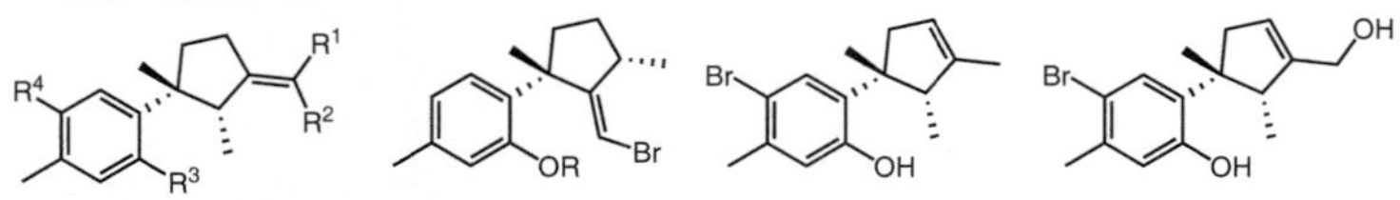

**171** $R^1 = R^4 = H, R^2 = Br, R^3 = OH$
**172** $R^1 = R^4 = H, R^2 = Br, R^3 = OAc$
**173** $R^1 = H, R^2 = R^4 = Br, R^3 = OH$
**174** $R^1 = Br, R^2 = R^4 = H, R^3 = OH$
**175** $R^1 = R^4 = Br, R^2 = H, R^3 = OH$

**176** R = H
**177** R = Ac

**178**

**179**

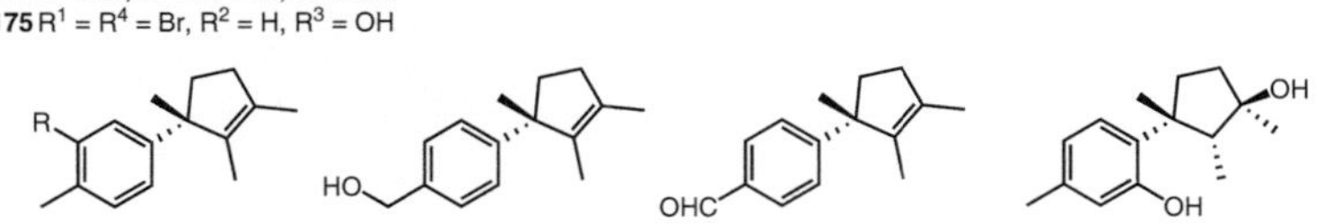

**180** R = H
**181** R = OH

**182**

**183**

**184**

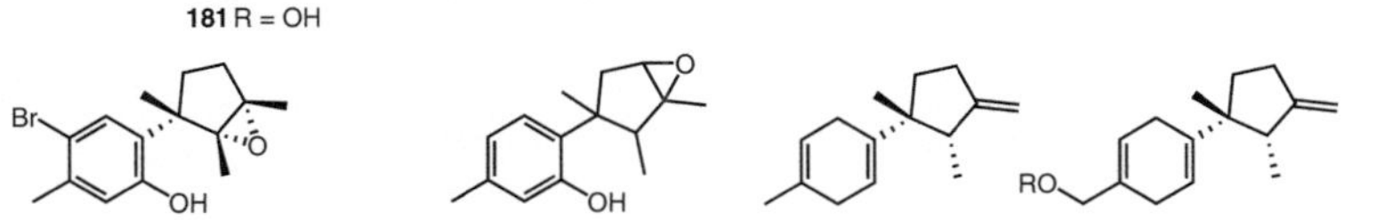

**185**

**186**

**187**

**188** R = H
**189** R = Ac

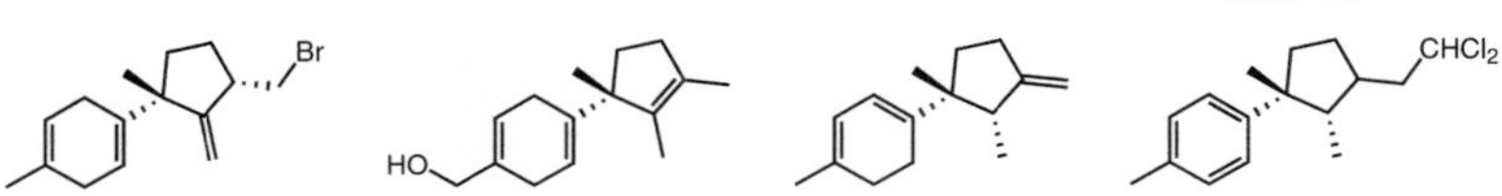

**190**

**191**

**192**

**193**

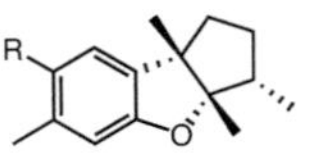

**194** R = H
**195** R = Br

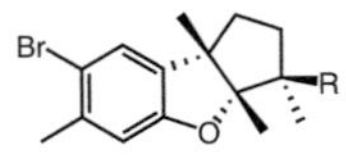

**196** R = OH
**197** R = OOH

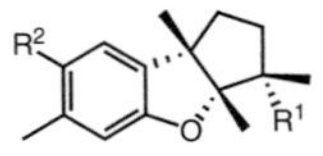

**198** R[1] = OH, R[2] = H
**199** R[1] = OH, R[2] = Br
**200** R[1] = OOH, R[2] = Br

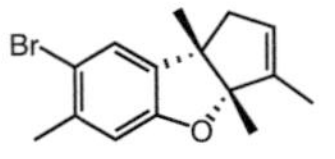

**201**

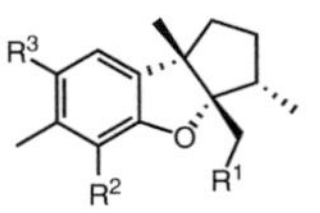

**202** R[1] = OH, R[2] = R[3] = H
**203** R[1] = Br, R[2] = R[3] = H
**204** R[1] = OH, R[2] = H, R[3] = Br
**205** R[1] = OAc, R[2] = H, R[3] = Br
**206** R[1] = OOH, R[2] = H, R[3] = Br
**207** R[1] = R[3] = Br, R[2] = H
**208** R[1] = R[2] = R[3] = Br

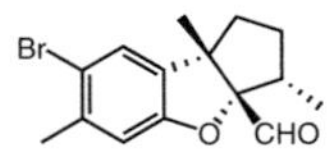

**209**

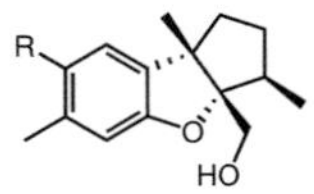

**210** R = H
**211** R = Br

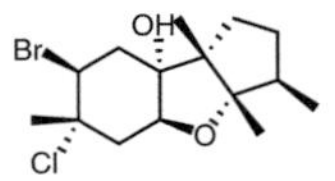

**212**

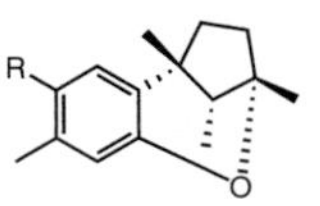

**213** R = H
**214** R = Br

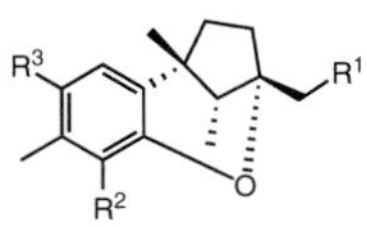

**215** R[1] = Br, R[2] = R[3] = H
**216** R[1] = R[3] = Br, R[2] = H
**217** R[1] = R[2] = R[3] = Br
**218** R[1] = OH, R[2] = H, R[3] = Br
**219** R[1] = OH, R[2] = R[3] = Br
**220** R[1] = I, R[2] = H, R[3] = Br

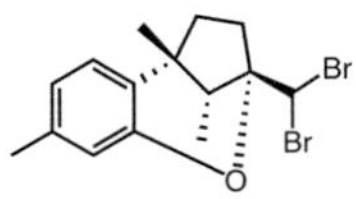

**221**

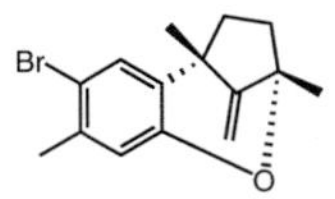

**222**

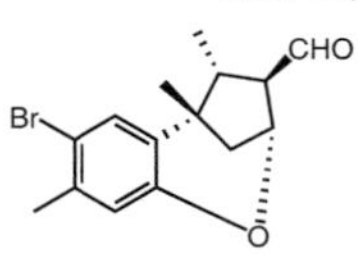

**223**

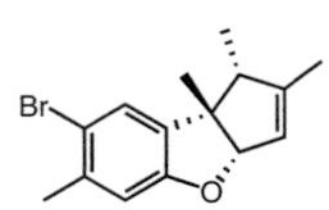

**224**

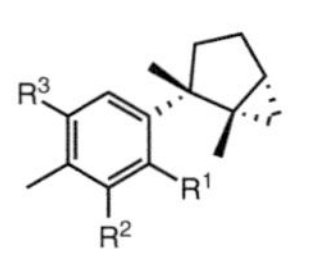

**225** R[1] = R[2] = R[3] = H
**226** R[1] = OH, R[2] = R[3] = H
**227** R[1] = OAc, R[2] = R[3] = H
**228** R[1] = R[3] = H, R[2] = OH
**229** R[1] = R[2] = H, R[3] = OH
**230** R[1] = OH, R[2] = Br, R[3] = H
**231** R[1] = OH, R[2] = H, R[3] = Br
**232** R[1] = OAc, R[2] = H, R[3] = Br
**233** R[1] = Br, R[2] = H, R[3] = OH
**234** R[1] = Br, R[2] = H, R[3] = OAc
**235** R[1] = OH, R[2] = R[3] = Br
**236** R[1] = OH, R[2] = I, R[3] = Br

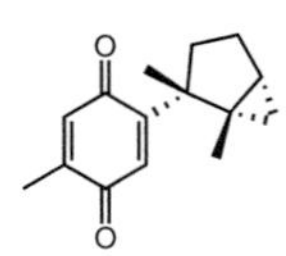

**237**

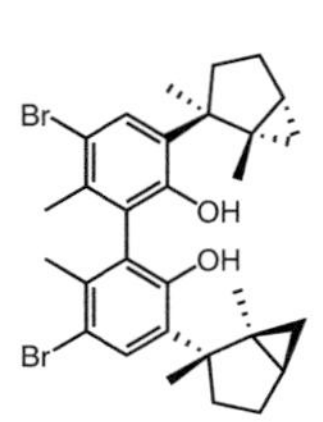

**238**

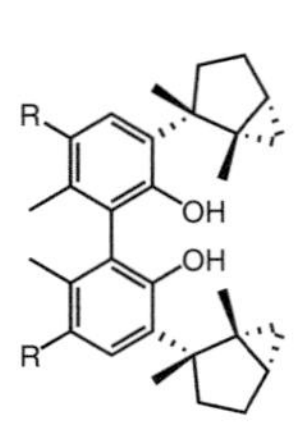

**239** R = H
**240** R = Br

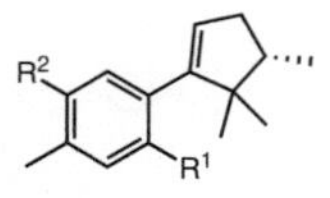

**241** $R^1 = R^2 = H$
**242** $R^1 = OH, R^2 = Br$

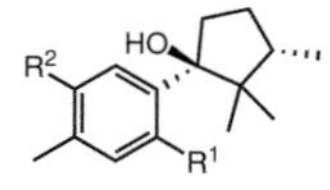

**243** $R^1 = R^2 = H$
**244** $R^1 = OH, R^2 = Br$

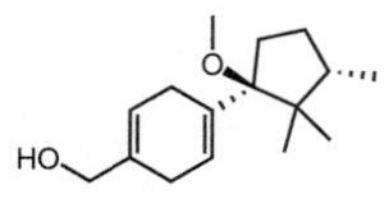

**245**

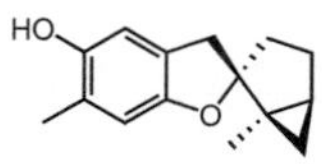

**246**

### 3.1.3 Snyderanes and Related Sesquiterpenes

Snyderanes and related metabolites constitute the third largest class of sesquiterpenes isolated from species of the genus *Laurencia* or the sea hares grazing on them, including up to the present 57 members (**247–303**). They are monocarbocyclic sesquiterpenes arising from the formation of a six-membered ring between carbons C-6 and C-11 of farnesane. Many snyderanes contain a cyclohexane ring with an acyclic side chain (**247–265**), while some of them feature 2,7-epoxy (**266–281**), 3,7-epoxy (**282**), or 4,7-epoxy (**283–289**) rings in their structures, thus forming 6,7-, 6,6-, or 6,5-fused systems. Most often, they are characterized by bromine incorporation at C-10, whereas less frequently they are brominated at C-1. The cyclohexane ring is commonly unsaturated featuring a $\Delta^7$ double bond, as in α-snyderol (**247**), a $\Delta^{7,14}$ double bond, as in β-snyderol (**249**), a $\Delta^6$ double bond, as in 8-hydroxy-γ-snyderol (**263**) or even an additional $\Delta^9$ double bond, as in **259–262**. Bicyclolaurencenol (**290**), obtained from *Laurencia intricata* from Castle Harbour, Bermuda [351], possesses a related bicyclic framework resulting from cyclization between C-8 and C-10 of the snyderane skeleton. Rearranged snyderanes that are derived from a methyl migration from C-11 to C-10, either containing an acyclic side chain (**291–296**), or featuring a 2,7-epoxy (**297** and **298**), or a 4,7-epoxy (**299–303**) ring, have also been isolated from a number of *Laurencia* species. The synthesis of the natural enantiomers of dactyloxenes B (**299**) and C (**300**) from (–)-(*R*)-linalool in seven steps via a common intermediate has been reported [359, 360], while another route based on a stereoselective iodoetherification reaction of a carvone-derived hydroxyalkene unit was also proposed in 2005 [361].

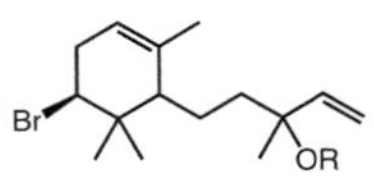

**247** R = H
**248** R = Ac

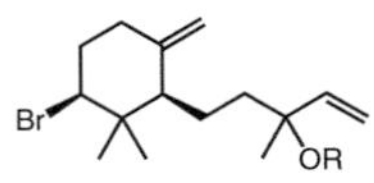

**249** R = H
**250** R = Ac

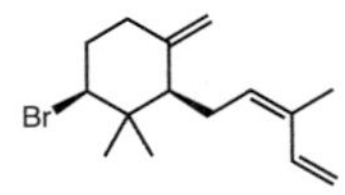

**251**

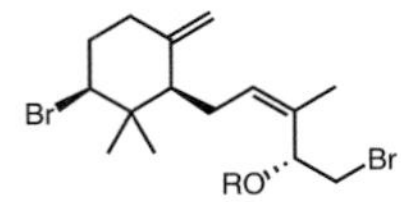

**252** R = H
**253** R = Ac

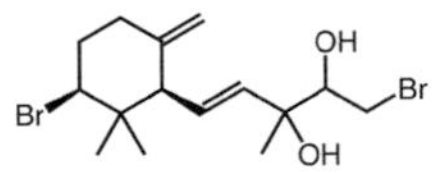

**254**

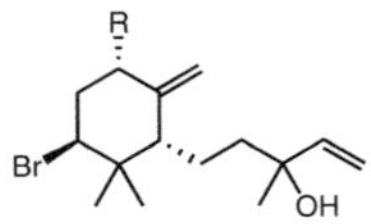

**255** R = H
**256** R = Br

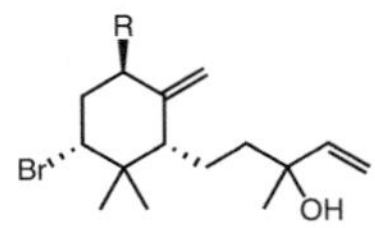

**257** R = OH
**258** R = Br

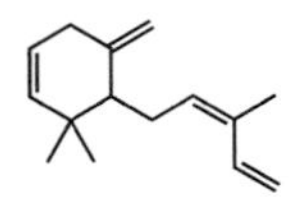

**259**

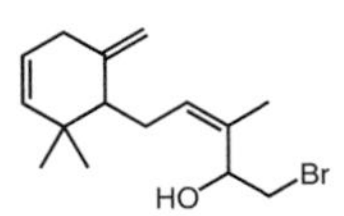

**260**

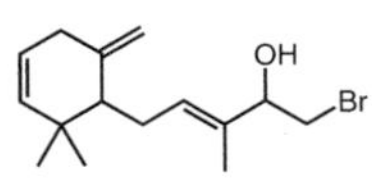

**261**

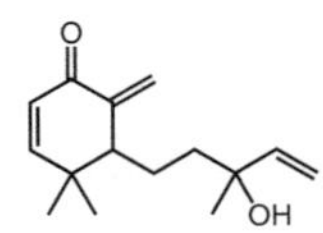

**262**

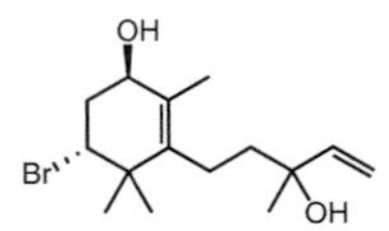

**263**

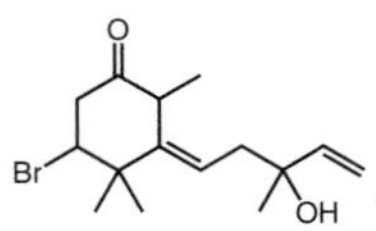

**264**

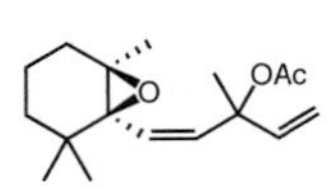

**265**

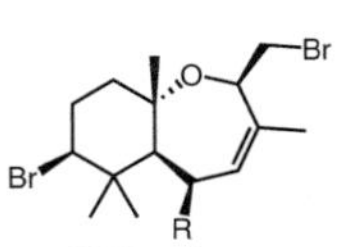

**266** R = H
**267** R = OH
**268** R = OAc

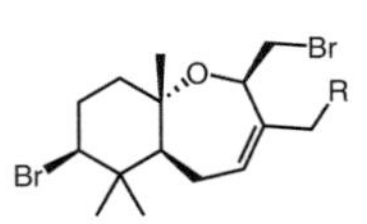

**269** R = OH
**270** R = OAc
**271** R = Br

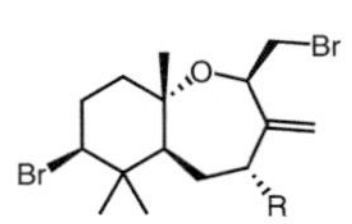

**272** R = OH
**273** R = Br

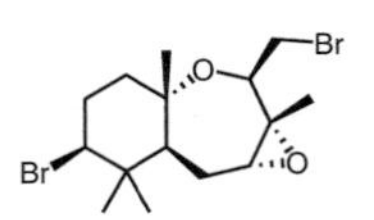

**274**

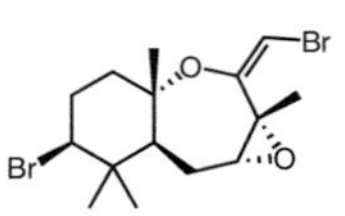

**275**

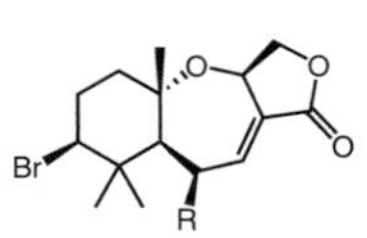

**276** R = H
**277** R = OH

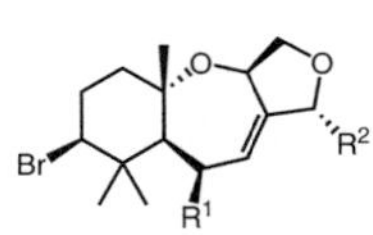

**278** $R^1 = R^2 = H$
**279** $R^1 = H$, $R^2 = OH$
**280** $R^1 = OAc$, $R^2 = H$

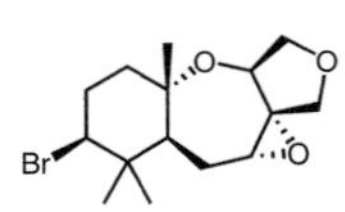

**281**

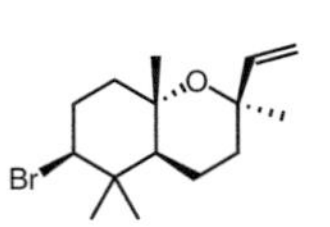

**282**

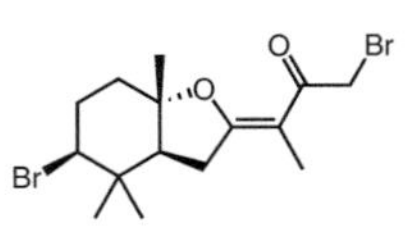

**283**

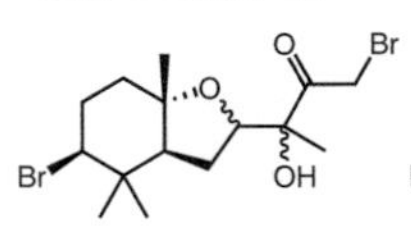

**284** (3*R**,4*S**)
**285** (3*S**,4*R**)

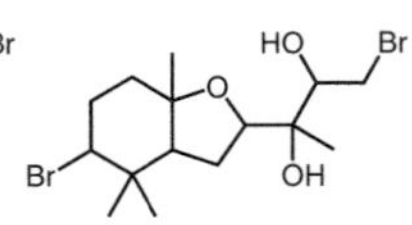

**286**

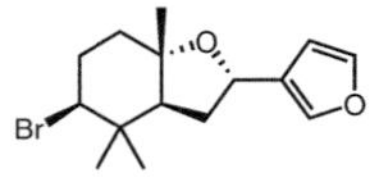

**287**

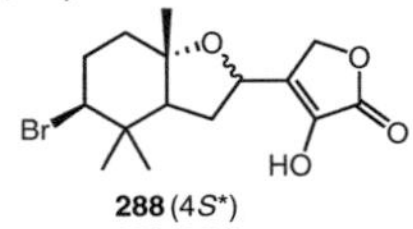

**288** (4*S**)
**289** (4*R**)

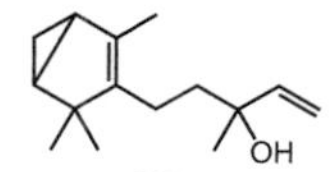

**290**

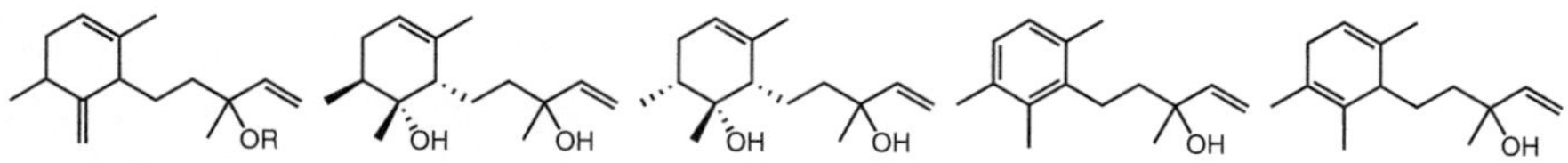

**291** R = H
**292** R = Ac
**293** **294** **295** **296**

**297** **298**

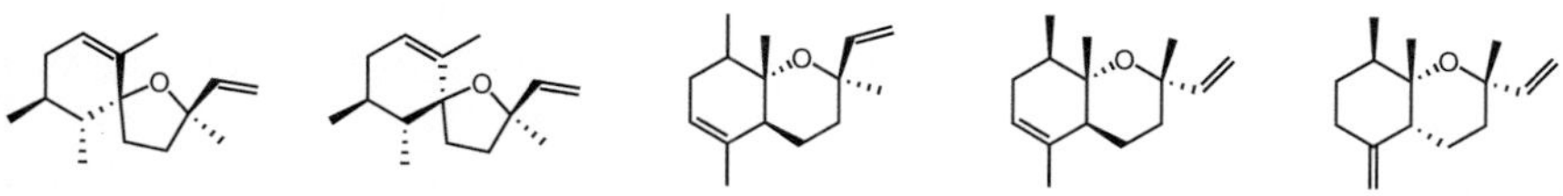

**299** **300** **301** **302** **303**

### 3.1.4 Bisabolanes and Related Sesquiterpenes

Bisabolanes and related metabolites are a fairly large class of sesquiterpenes found mainly as constituents of higher plants. Nonetheless, up to now there are 72 bisabolanes and related sesquiterpenes (**304**–**375**) identified from *Laurencia* species and their grazers. As with snyderanes, bisabolanes are also monocarbocyclic sesquiterpenes arising, however, by the formation of a six-membered ring between carbons C-1 and C-6 of farnesane. Often members of this class are monocyclic (**304**–**320**), but there are also representatives possessing 6,9-epoxy (**321**–**326**), 7,10-epoxy (**327** and **328**), or 7,11-epoxy (**329**–**350**) ring systems, the latter of which are commonly known as caespitanes. Usually, bisabolanes are both brominated and chlorinated at C-2 and C-3, often with a 2,3-*trans*-halide arrangement. It is worth noting that the structure of caespitol (**337**), originally assigned as that of a chamigrane sesquiterpene on the basis of analysis of its spectroscopic data [375], was later revised to that of a bisabolane based on the results of the X-ray crystallographic analysis of isocaespitol (**346**), a sesquiterpene co-occurring in the same sample of *Laurencia caespitosa*, as well as by chemical correlation between the two isomers [376]. In 2003 and 2006, de Carvalho and co-workers reported the isolation of aldingenins A–D from specimens of *Laurencia aldingensis* collected at Castelhanos, Brazil [373, 378], attributing to them bisabolane structures containing multiple oxacyclic rings. The synthesis of the proposed structure of aldingenin B revealed that it was incorrect due to discrepancies observed between the spectroscopic data of the natural and synthetic products [385, 386]. Subsequently, the synthesis of the proposed structure of aldingenin C by Takahashi et al. in 2014 proved that it also merited revision, while re-examination of the $^{13}$C NMR data of the natural product suggested that the correct structure should be that of caespitol (**337**) [374]. In a similar manner, it was suggested that the correct structure of aldingenin D is that of (5*S*)-5-acetoxy-caespitol (**342**) [374], isolated in the interim from a Brazilian *Laurencia catarinensis* collected in Ilha do Arvoredo [370]. In 2015, Mukhina et al. presented computational evidence based primarily on the computed proton spin-spin coupling constants that the correct structures of aldigenins A and B should be those depicted in structures **340** and **350**, respectively [379]. Compounds **351–357** are bromochlorinated $C_{12}$ terpenoids, regarded as degraded bisabolanes [387]. In 1992, de Nys and co-workers reported the isolation of metabolites **358–360** from *Laurencia majuscula* collected from Geoffrey Bay, Magnetic Island, Australia [153]. These compounds feature a bicyclic framework that could result from a further cyclization between C-7 and C-12 of the bisabolane skeleton. Similarly, metabolites **361–371**, isolated from several *Laurencia* species, possess a bicyclic system that could arise from a further cyclization between C-11 and C-14 of the bisabolane skeleton. The structure of **361**, including absolute stereochemistry, was established on the basis of X-ray crystallographic analysis [389]. With the exception of **361** and **364–366**, all of these compounds were obtained as inseparable diastereomeric mixtures. Specimens of *L. caespitosa* collected at Punta del Hidalgo, Tenerife, and near the Island of La Graciosa in the Canary islands, yielded the irregular rearranged bisabolane-related sesquiterpenes, laucapyranoids A-C (**372–374**) [369]. Aplysiadactydiol (**375**), a rearranged bisabolane originally reported as aplysiadiol, but later renamed, since this particular trivial name was already used for a previously published diterpene, was obtained from *A. dactylomela* collected from La Palma Island, Spain [368, 392].

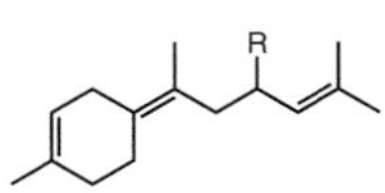


**304** R = H
**305** R = OAc

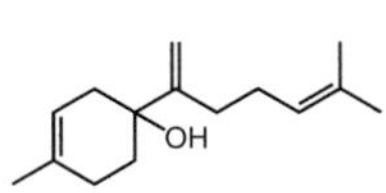


**306**

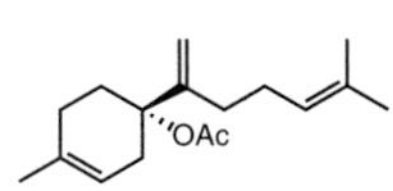


**307**

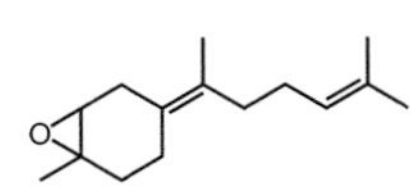


**308**

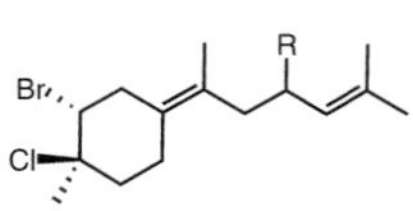


**309** R = H
**310** R = OH

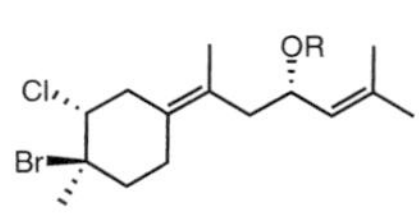


**311** R = H
**312** R = Ac

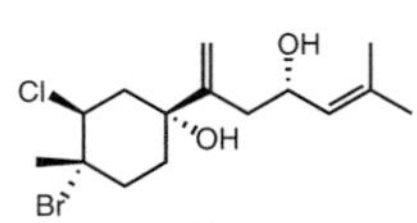


**313**

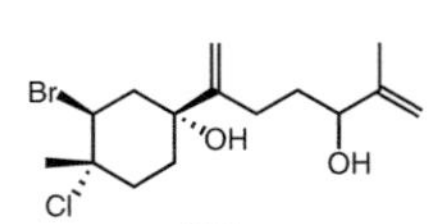


**314**

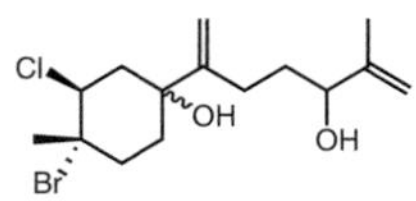


**315** (6*R*)
**316** (6*S*)

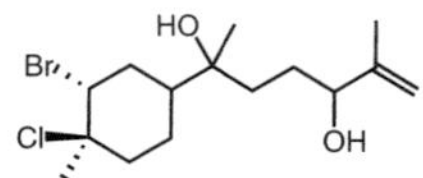


**317**

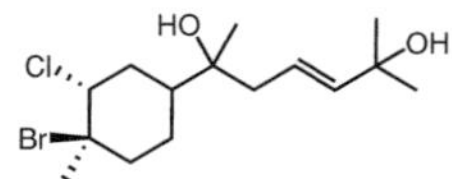


**318**

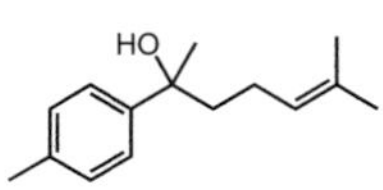


**319**

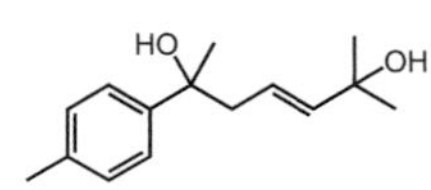


**320**

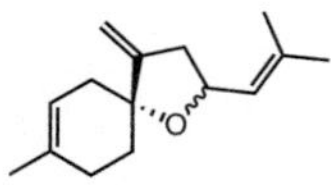

**321** (9*R**)
**322** (9*S**)

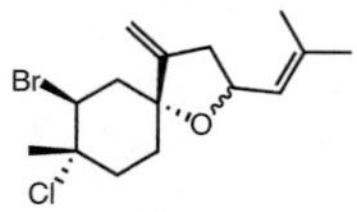

**323** (9*R**)
**324** (9*S**)

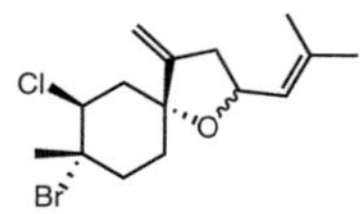

**325**, **326** (epimers at C-9)

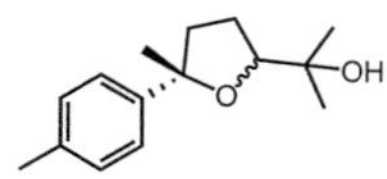

**327** (10*R**)
**328** (10*S**)

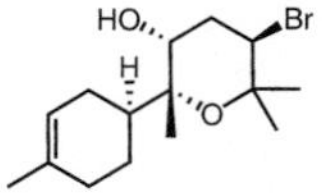

**329**

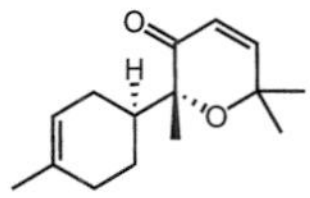

**330**

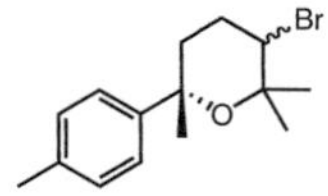

**331** (10*S**)
**332** (10*R**)

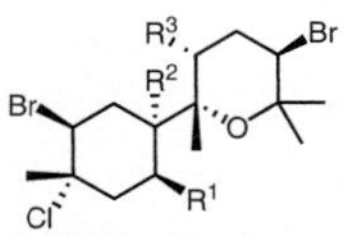

**333** $R^1 = R^2 = R^3$ = H
**334** $R^1$ = OH, $R^2 = R^3$ = H
**335** $R^1$ = OAc, $R^2 = R^3$ = H
**336** $R^1 = R^3$ = H, $R^2$ = OH
**337** $R^1 = R^2$ = H, $R^3$ = OH
**338** $R^1 = R^2$ = H, $R^3$ = OAc
**339** $R^1$ = H, $R^2 = R^3$ = OH
**340** $R^1$ = OH, $R^2$ = H, $R^3$ = OH
**341** $R^1$ = OH, $R^2$ = H, $R^3$ = OAc
**342** $R^1$ = OAc, $R^2$ = H, $R^3$ = OH
**343** $R^1 = R^3$ = OAc, $R^2$ = H
**344** $R^1 = R^2$ = OH, $R^3$ = OAc

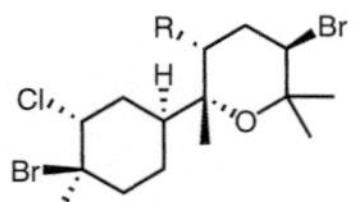

**345** R = H
**346** R = OH
**347** R = OAc

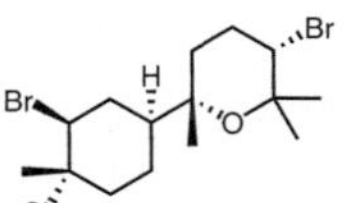

**348**

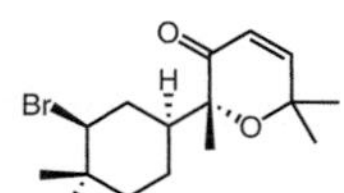

**349**

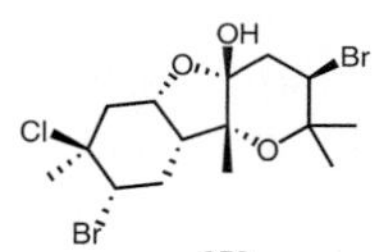

**350**

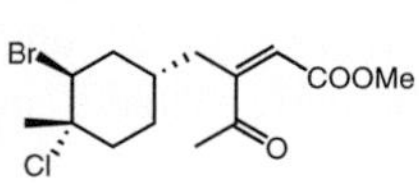

**351**

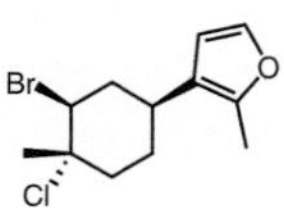

**352**

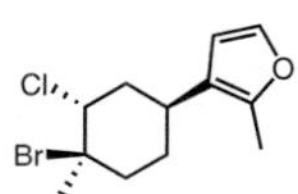

**353**

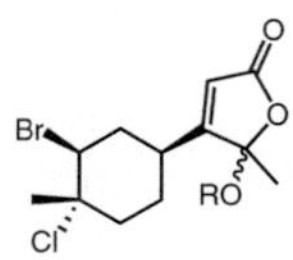

**354** (10*S**), R =H
**355** (10*R**), R = Me
**356** (10*S**), R = Me

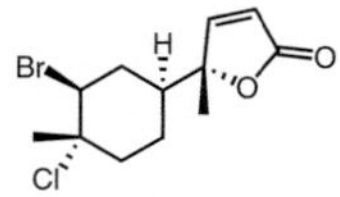

**357**

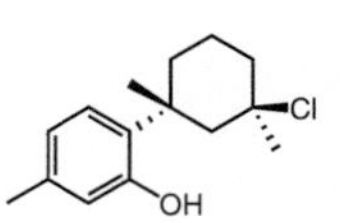

**358**

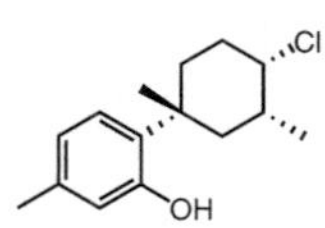

**359**

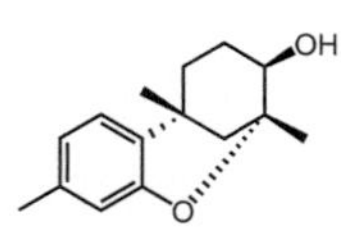

**360**

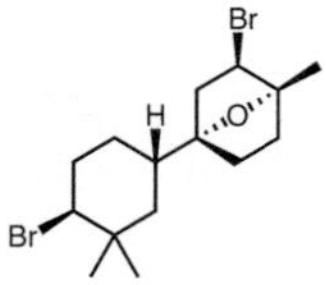

**361**

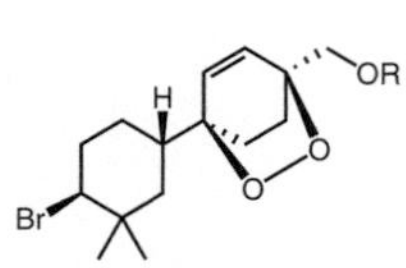

**362** R = H
**363** R = Ac

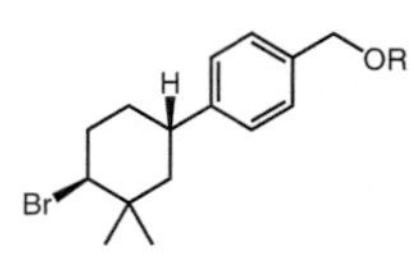

**364** R = H
**365** R = Ac

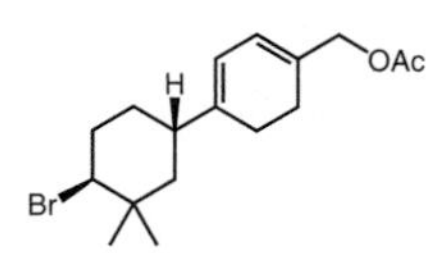

**366**

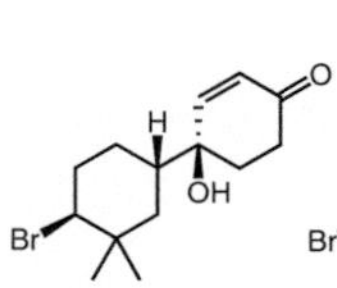

**367**

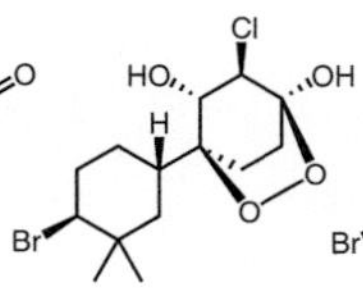

**368**

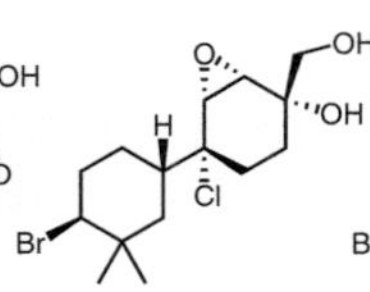

**369**

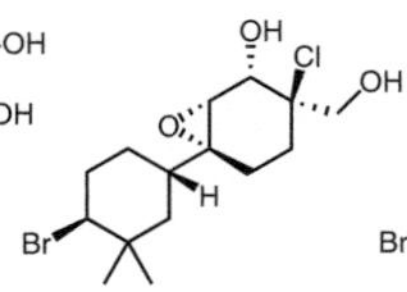

**370**

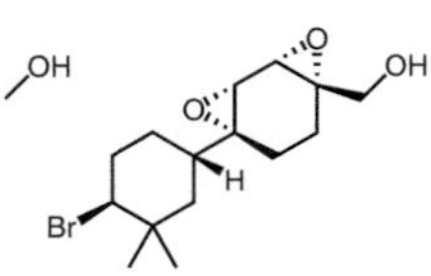

**371**

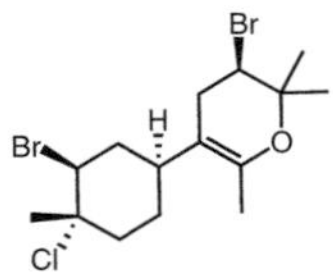

**372**

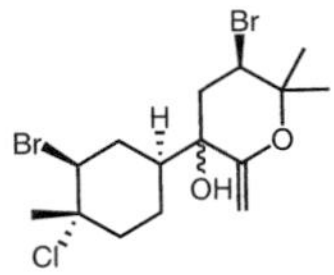

**373** (8*R*)
**374** (8*S*)

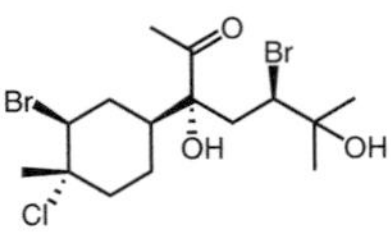

**375**

### 3.1.5 Perforanes, Cycloperforanes, and Related Sesquiterpenes

Perforanes, cycloperforanes, and related metabolites comprise a relatively small group of sesquiterpenes isolated from species of *Laurencia* and mollusks feeding on them, encompassing 22 members (**376–397**). Perforanes (**376–385**) possess a bicyclo[5.4.0]undecane skeleton, while a further cyclization between C-1 and C-9 of the seven-membered ring gives rise to the cycloperforanes (**387–395**) featuring a 4,5,6-fused tricyclic carbocycle. Both carbocycles are postulated to originate biosynthetically from chamigranes [474]. Perforene (**386**), obtained from a *Laurencia perforata* population collected off Corralejos, Fuerteventura, Canary Islands, displayed a perforane-related skeleton resulting from a 1,3-methyl migration [401]. Compounds **396** and **397**, possessing a rearranged cycloperforane carbocycle, were isolated from *Laurencia tenera* [402] and *L. perforata* [399], respectively, both collected near Magnetic Island in Australia.

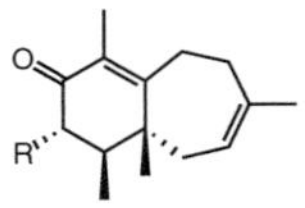

**376** R = H
**377** R = OH
**378** R = Cl

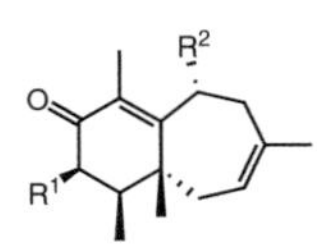

**379** $R^1$ = OH, $R^2$ = H
**380** $R^1$ = OMe, $R^2$ = H
**381** $R^1$ = $R^2$ = OH

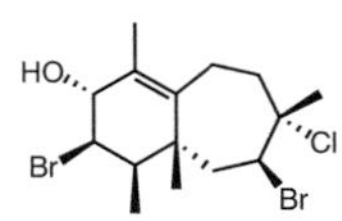

**382**

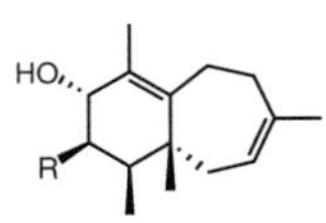

**383** R = H
**384** R = Br

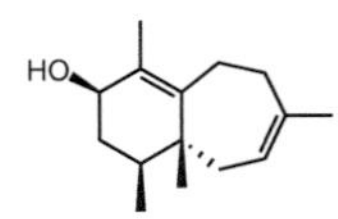

**385**

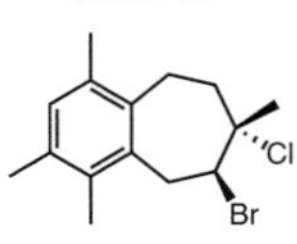

**386**

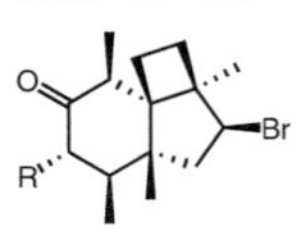

**387** R = H
**388** R = Br

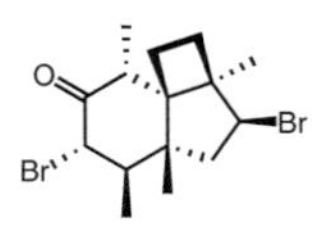

**389**

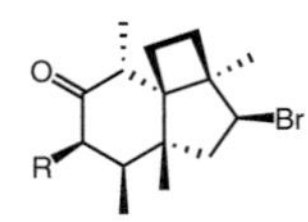

**390** R = H
**391** R = Br
**392** R = OMe

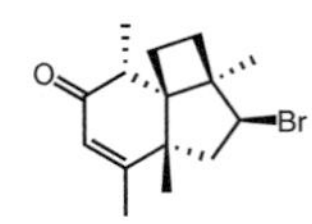

**393**

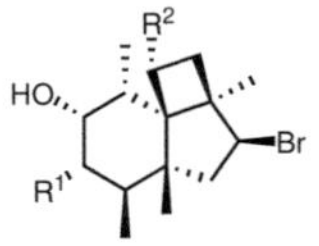

**394** $R^1$ = $R^2$ = H
**395** $R^1$ = Br, $R^2$ = OAc

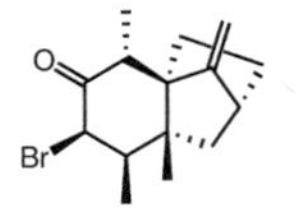

**396**

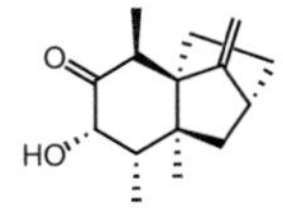

**397**

### 3.1.6 Brasilanes

To date, 17 sesquiterpenes featuring the brasilane skeleton (**398–414**) have been isolated from species of *Laurencia* and the sea hares consuming them. Brasilanes possess a rearranged bicyclo[4.3.0]nonane carbocycle with a *gem*-dimethyl group at C-3, an isopropyl group at C-5, and a fifth methyl group at C-9. The presence of a tetrasubstituted double bond between C-1 and C-6 is rather common, while in three metabolites (**411–413**), isolated from *L. obtusa* collected in Milos Island, Greece [405], a 1,6-epoxy moiety can be observed. A number of these derivatives bear an OH substitution at C-9, whereas in four brasilanes C-8 is halogenated. Brasilenol (**398**) and epibrasilenol (**400**) were isolated isolated from *Aplysia brasiliana* collected in Padre Island, southern Texas in 1978 [406]. However, their dietary origin was later confirmed when Howard and Fenical reported their isolation from a Mediterranean *L. obtusa* population [320]. Subsequently, two total syntheses of brasilenol (**398**) were published by Greene et al., determining its absolute stereochemistry [408, 409].

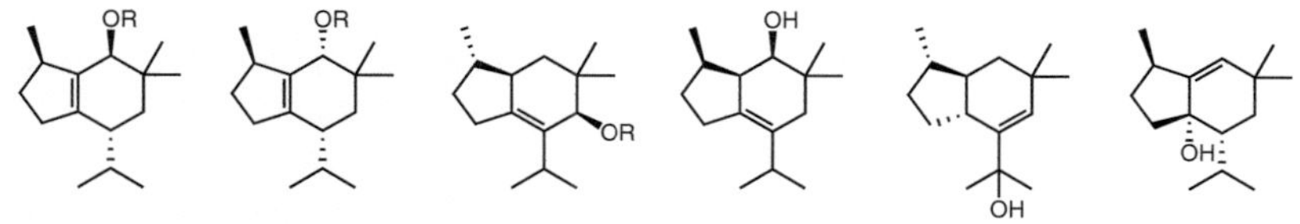

**398** R = H, **399** R = Ac; **400** R = H, **401** R = Ac; **402** R = H, **403** R = Ac; **404**; **405**; **406**

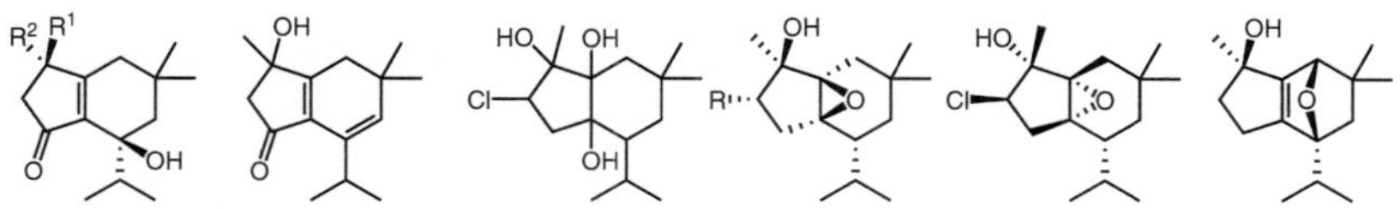

**407** R[1] = OH, R[2] = Me, **408** R[1] = Me, R[2] = OH; **409**; **410**; **411** R = Cl, **412** R = Br; **413**; **414**

### 3.1.7 Cuparanes

Cuparanes are found in higher plants and liverworts and occasionally in marine organisms. Up to the present, 14 members of this class of sesquiterpenes (**415–428**) have been discovered from the genus *Laurencia* and mollusks feeding on these red algae. Cuparanes arise by a further cyclization between carbons C-7 and C-11 of the bisabolane skeleton forming a five-membered ring. In several cases, C-10 is brominated. α-Bromocuparene (**415**) and its epimer at C-10, α-isobromocuparene (**417**), isolated initially from *Laurencia glandulifera* in 1975, were the first cuparanes to be reported from macroalgae [414].

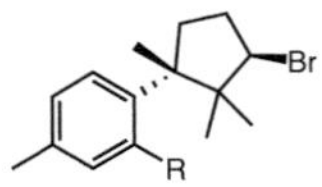

**415** R = H
**416** R = OH

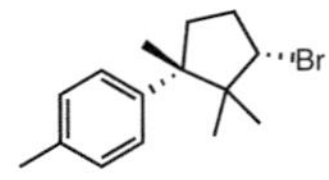

**417**

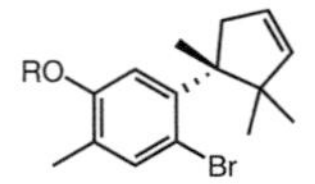

**418** R = H
**419** R = Ac

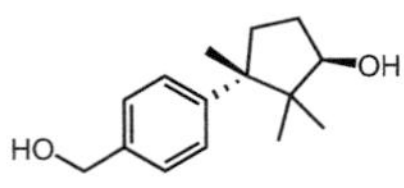

**420**

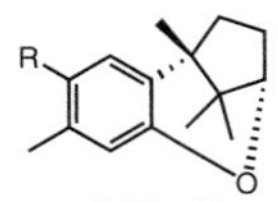

**421** R = H
**422** R = Br

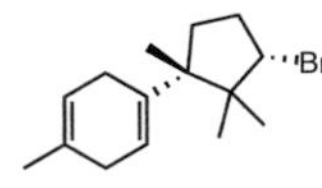

**423**

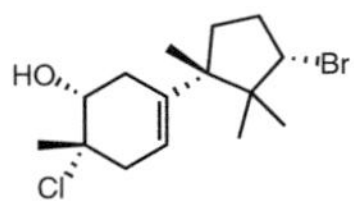

**424**

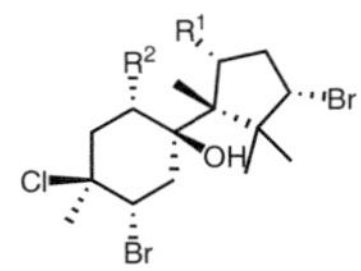

**425** R¹ = R² = H
**426** R¹ = OH, R² = H
**427** R¹ = OH, R² = OAc

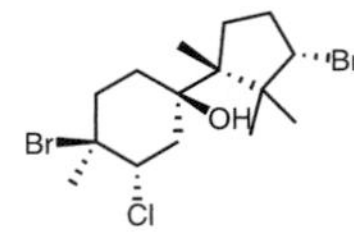

**428**

### 3.1.8 Eudesmanes, Cycloeudesmanes, and Related Sesquiterpenes

Even though eudesmanes, also referred to as selinanes, and cycloeudesmanes are commonly encountered in higher plants, to date only 21 related metabolites (**429**–**449**) have been reported as constituents of *Laurencia* spp. or their grazers. Among them, 17 eudesmanes (**429**–**445**), one 5,7-cycloeudesmane (**446**), and one 6,8-cycloeudesmane (**447**) can be distinguished. Chabrolidione B (**448**) and teuhetenone A (**449**), isolated from *L. obtusa*, are a *seco*-eudesmane and a eudesmane-related $C_{12}$ terpene, respectively [423]. Several of these metabolites feature bromine substitution at C-1.

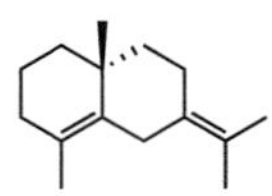

**429**

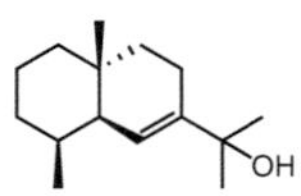

**430**

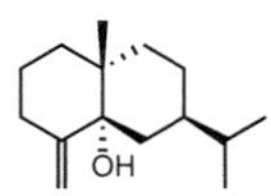

**431**

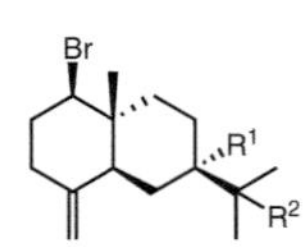

**432** R¹ = H, R² = OH
**433** R¹ = OH, R² = H

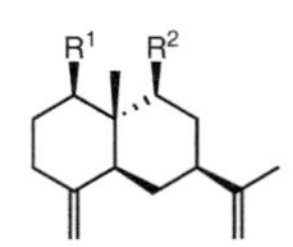

**434** R¹ = Br, R² = H
**435** R¹ = H, R² = Br

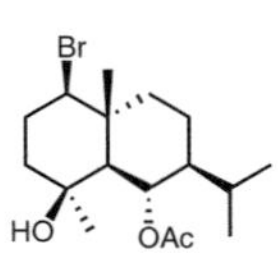

**436**

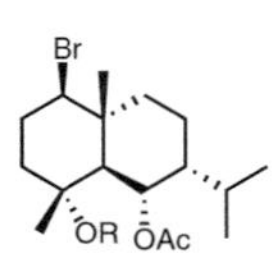

**437** R = H
**438** R = Ac

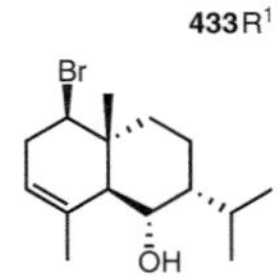

**439**

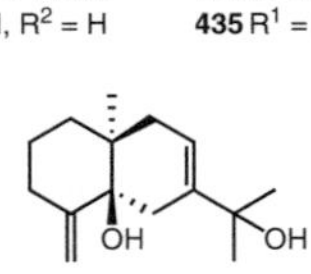

**440**

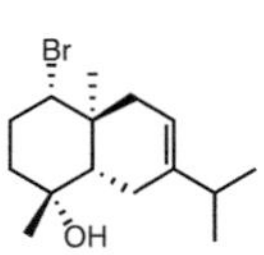

**441**

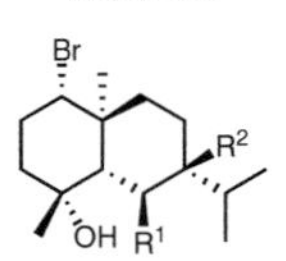

**442** R¹ = H, R² = Cl
**443** R¹ = Cl, R² = H

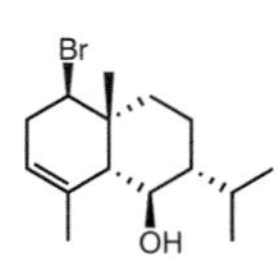

**444**

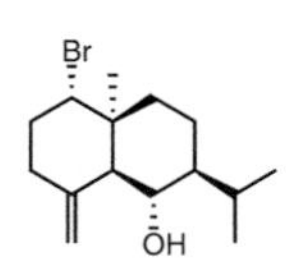

**445**

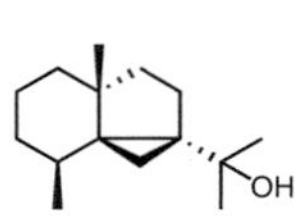

**446**

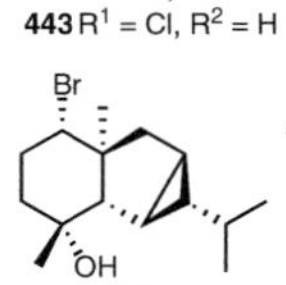

**447**

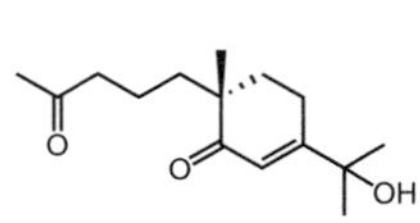

**448**

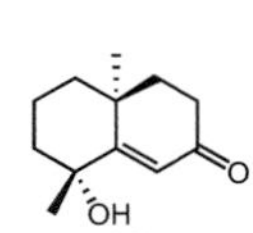

**449**

### 3.1.9 Aristolanes

Aristolanes, sesquiterpenes arising from cyclization of eremophilanes between carbons C-6 and C-11, which in turn are derived from eudesmanes after migration of the methyl group at C-10 to C-5, are rarely detected in red algae of the genus *Laurencia* or in gastropods feeding on them. Nonetheless, up to now 14 aristolanes (**450**–**463**) have been reported from such sources. Aristolan-1$\alpha$-bromo-9$\beta$,10$\beta$-epoxide (**463**), isolated from *Laurencia similis* collected in Hainan Island, People's Republic of China, is the only halogenated derivative characterized so far [431].

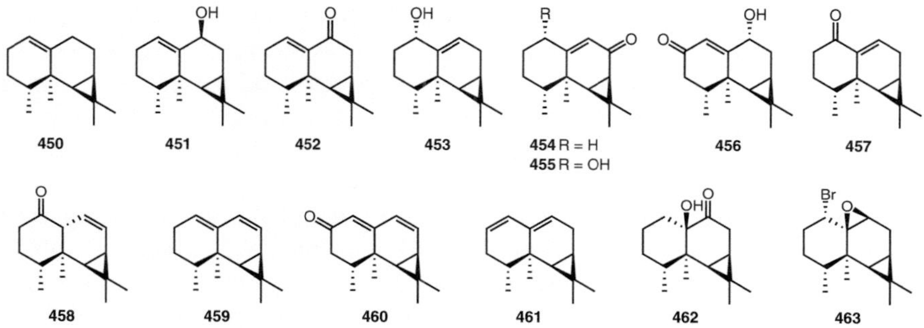


### 3.1.10 Miscellaneous Sesquiterpenes

In addition to the sesquiterpenes belonging to the classes discussed above, there are 49 more sesquiterpenes (**464**–**512**) possessing skeletons that are rarely encountered either in the genus *Laurencia* or from natural sources in general. Among these, derivatives of known skeletons, such as the germacrane **466**, the humulanes **467** and **468**, the cadinanes **475** and **476**, the oplopane **477**, the guaiane **480**, the triquinanes **490**–**492**, the cedrane **494**, the africanane **495**, and the aromadendranes **496**–**498**, have been isolated, and are all non-halogenated with the exception of **494**. The linear bromohydrin **464** and the tetrahydropyran-containing sesquiterpene obtusenol (**465**) were isolated from *Laurencia luzonensis* collected in Kudaka Island, Okinawa [323] and *Laurencia obtusa* collected in Gökçeada Island, Aegean Sea [435], respectively. *Laurencia okamurai* collected near Nanji Island in the East China Sea afforded *seco*-laurokamurone (**469**) featuring a monocarbocyclic framework, possibly formed by oxidative cleavage of the rearranged laurane skeleton [275]. *L. nipponica* from Hokkaido, Japan, yielded the *seco*-chamigranes lauracetals A (**470**), B (**471**), C (**474**), D (**472**), and E (**473**), possessing a carbocycle that could arise from a 10-bromo-chamigrene derivative by oxidative cleavage of the bond between C-1 and C-2, followed by acetal formation [75, 199, 438–440]. Dactylol (**478**) and poitediol (**479**) were reported as constituents of *Aplysia dactylomela* [442] and *L. poitei* [441], respectively, almost simultaneously in 1978 by two different research groups. They feature an uncommon rearranged bicyclo[6.3.0]undecane skeleton, designated as poitane or precapnellane. The total synthesis of racemic poitediol (**479**) was accomplished in 1984 [447] and one year later the conversion of **479** to dactylol (**478**) was achieved by the same group [443]. In 1977, Wratten and Faulkner reported the isolation of oppositol (**481**) and the related epimeric diols **482** and **483**, possessing an 8(7 → 6)-*abeo*-eudesmane skeleton named oppositane, from *Laurencia subopposita* collected at La Jolla, California [248, 448]. The bromoketone **484**, isolated from *Laurencia mariannensis* collected from Geoffrey Bay, Magnetic Island, Australia [198], is the sole representative of this carbocycle containing a bicyclo[4.3.0]nonane core, different, however, from the oplopane skeleton. Pacifigorgiol (**485**) is a highly rearranged sesquiterpene representing the pacifigorgiane skeleton and has been reported as a constituent of several *Laurencia* species [118, 149, 323, 335, 451]. Laurencial (**486**), containing a *spiro*[4.5]decane core, was obtained from *Laurencia nipponica* collected off the Pacific Coast of Hokkaido [452] and its synthesis was achieved utilizing the stereospecific bromination-cyclization of spirocyclic cyclohexanones [453]. Another Japanese *L. nipponica* population afforded spirolaurenone (**487**) [78, 199, 454, 455], which also features a *spiro*[4.5]decane system. Its total synthesis has been accomplished based on the cyclization of geranonitrile [456]. Laurenones A (**488**) and B (**489**), isolated from a different *L. nipponica* population from Japan, were considered as biogenetic transformation products of **487** [457]. The isolation of viridianol (**493**), a sesquiterpene with a 3,6-cyclo-precapnellane skeleton, was reported from *Laurencia viridis* in 1994 by Norte and co-workers [459]. *Laurencia obtusa* from Cayo Coco, Cuba, afforded bromocyclococanol (**499**), possessing the tricyclic skeleton cyclococane, postulated to be biosynthetically related to the bisabolanes [461]. Calenzanol (**500**) and debromoisocalenzanol (**501**), obtained from *Laurencia microcladia* collected from Calenzana Bay, Italy, represent a 3,6,5-fused tricyclic carbocycle named calenzanane [462, 463]. A tricyclic humulane-related sesquiterpene named laurobtusol (**502**) was isolated from a Mediterranean *L. obtusa* [464]. Synthesis of several stereoisomers of **502** provided evidence that the structure was not correctly assigned but failed to propose another structure for laurobtusol [465]. Güimarediol (**503**) and dactylomelatriol (**504**), are the first tricyclic sesquiterpenes featuring the omphalane skeleton to be isolated from the marine environment [132, 135, 466]. Chromatographic separations of a *L. majuscula* extract from La Gomera in the Canary islands resulted in the isolation of gomerones A (**505**), B (**506**), and C (**507**), possessing the tricyclic skeleton designated as gomerane [467]. The total synthesis of racemic gomerone C (**507**) led to the revision of the relative configurations of the asymmetric centers for **506** and **507** [468]. Compounds **508–511**, bearing the uncommon tricyclic skeleton rhodolaurane, have been reported from a number of *Laurencia* species [74, 132, 158, 469], whereas aplydactone (**512**) is a highly strained metabolite featuring a related tetracyclic carbon framework that has been isolated from *A. dactylomela* collected in the northern part of Magadascar [471]. Its structure, including absolute stereochemistry, was determined by single crystal X-ray diffraction analysis [471].

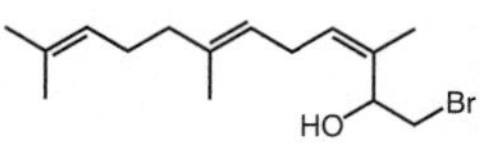
**464**

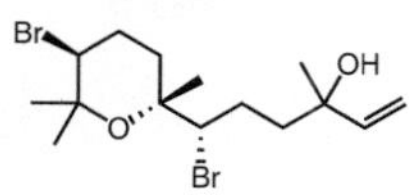
**465**

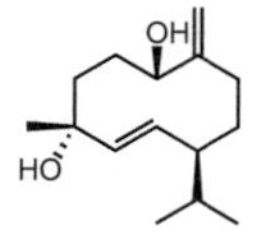
**466**

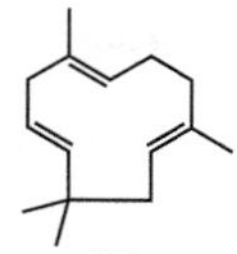
**467**

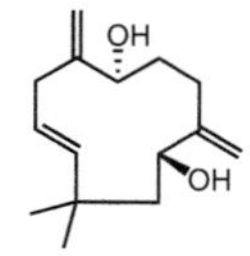
**468**

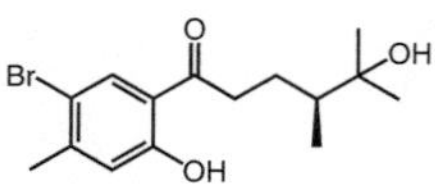
**469**

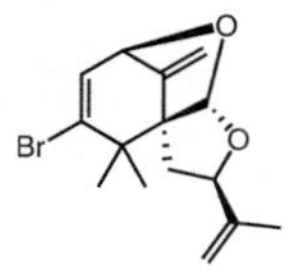
**470**

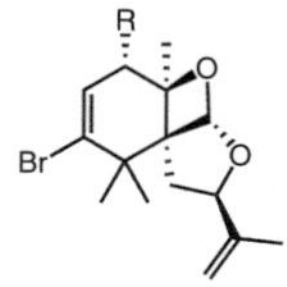
**471** R=OH
**472** R=Br
**473** R=Cl

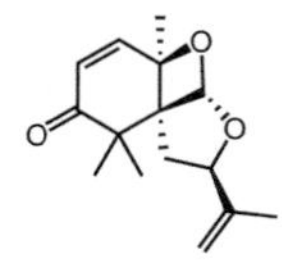
**474**

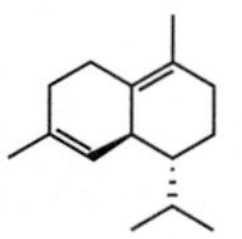
**475**

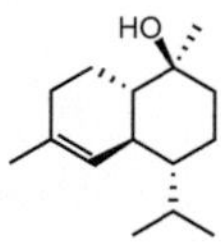
**476**

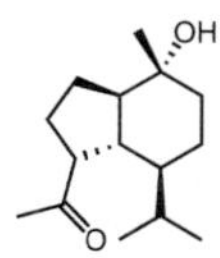
**477**

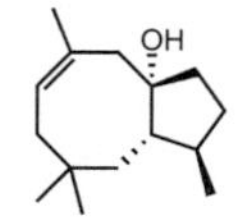
**478**

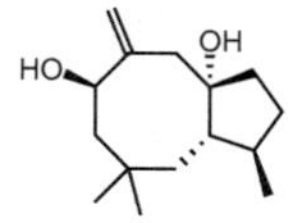
**479**

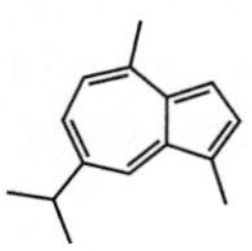
**480**

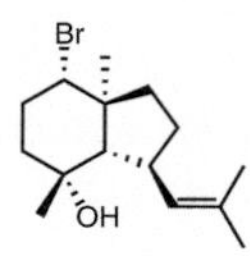
**481**

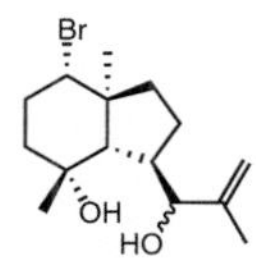
**482** 8*S**
**483** 8*R**

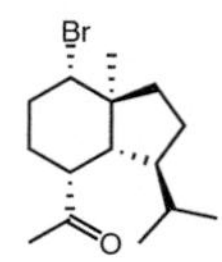
**484**

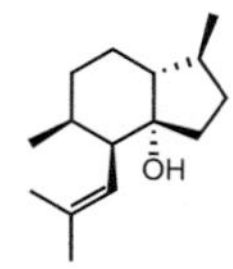
**485**

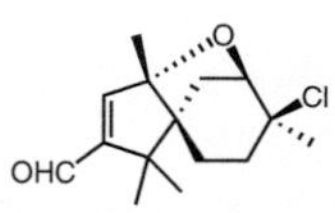
**486**

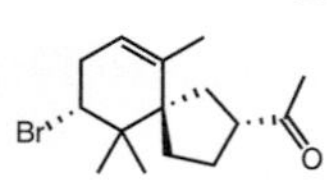
**487**

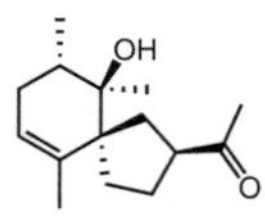
**488**

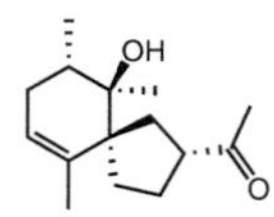
**489**

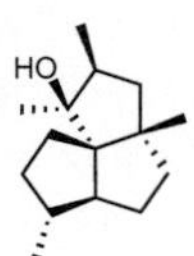
**490**

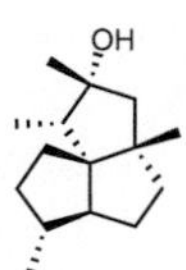
**491**

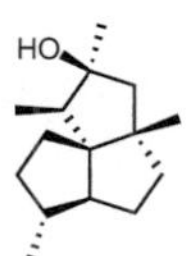
**492**

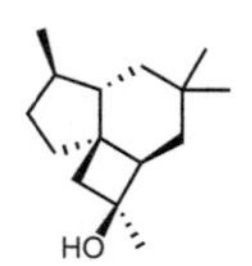
**493**

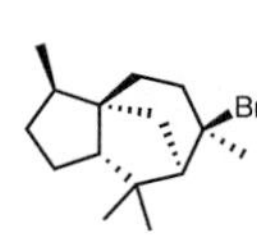
**494**

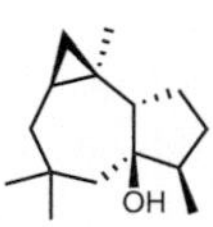
**495**

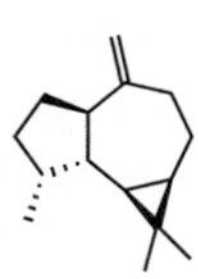
**496**

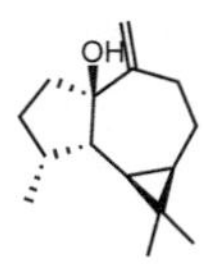
**497**

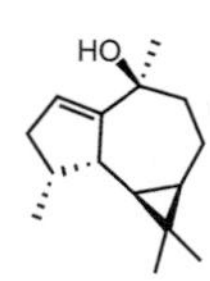
**498**

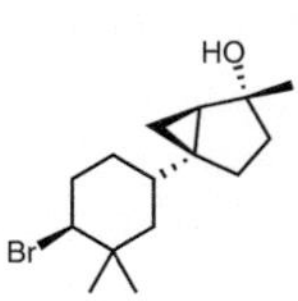
**499**

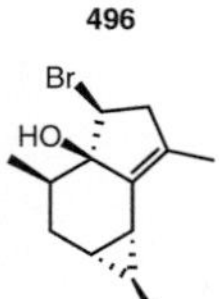
**500**

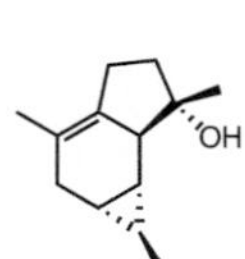
**501**

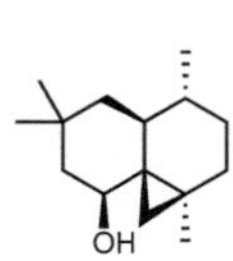
**502**

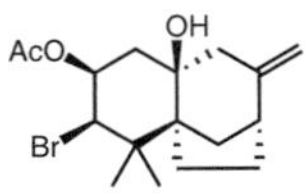
**503**

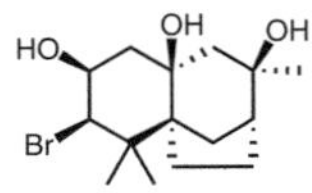
**504**

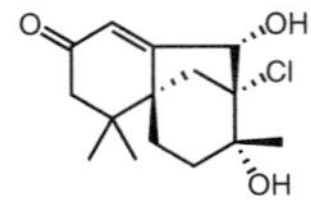
**505**

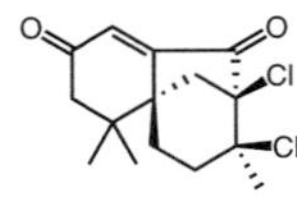
**506**

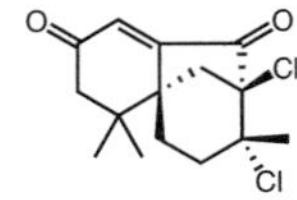
**507**

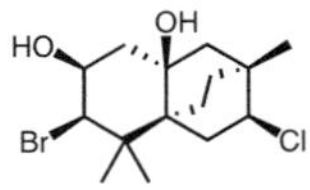
**508**

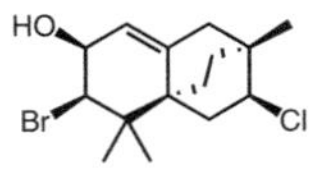
**509**

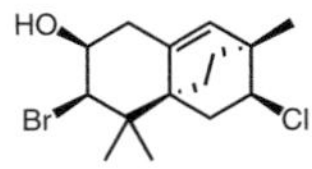
**510**

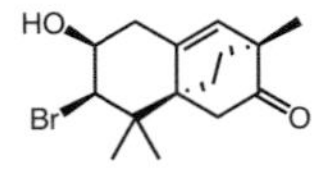
**511**

**512**

## *3.2 Diterpenes*

Diterpenes are $C_{20}$ compounds consisting of four isoprene units that are derived from geranylgeranyl pyrophosphate. In contrast to the high number and variety of sesquiterpenes from species of the genus *Laurencia*, as well as mollusks grazing on them, diterpenes constitute a small group and up to now 133 diterpenes and related metabolites (**513–645**, Table 2), featuring 25 different skeletons (see Figs. 6 and 7), have been reported.

### 3.2.1 Pargueranes, Isopargueranes, and Related Diterpenes

Pargueranes, isopargueranes, and related metabolites (**513–545**), featuring a modified pimarane skeleton, comprise the major group of diterpenes isolated from *Laurencia* and sea hares feeding on these red algae, encompassing 33 derivatives. Pargueranes (**513–538**) contain a cyclopropane ring between C-3 and C-4 of the pimarane skeleton, while the rearranged tetracyclic skeleton of isopargueranes (**539–543**) contains a cyclobutane ring incorporating C-19. The first representatives of this group of metabolites, namely, parguerol (**514**) and its 16-acetate (**517**), deoxyparguerol (**525**), isoparguerol (**540**), and its 16-acetate (**541**), were reported in 1982 by Schmitz et al. from *A. dactylomela* collected in Puerto Rico [156, 270]. The algal biosynthetic origin of these compounds was confirmed when deoxyparguerol 16-acetate (**527**) was isolated from *L. obtusa* collected at Kimmeridge Bay, Dorset, United Kingdom [482]. Two *seco*-analogues of pargueranes and isopargueranes featuring tricyclic carbocycles containing either a seven- (**544**) or an eight-membered (**545**) ring A, respectively, have been isolated. It is worth noting that both rare skeletons were named neopargueranes [249, 483]. In the majority of these diterpenes C-15 is brominated, while C-16 is oxygenated. Frequently, C-2, C-7 and C-19 are also oxygenated. Furthermore, most of these derivatives, with the exception of **533** and **538**, possess a $\Delta^{9,11}$ double bond.

**Table 2** Diterpenes from *Laurencia* spp.

| No. | Name | Source of isolation/synthesis |
|---|---|---|
| **513** | Deacetylparguerol | *L. saitoi* [475[a]], *A. pulmonica* [476], *A. punctata* [249] |
| **514** | Parguerol | *L. saitoi* [477], *A. dactylomela* [156[a], 178, 270], *A. kurodai* [478], *A. pulmonica* [476], *A. punctata* [249] |
| **515** | Parguerol 19-acetate | *L. saitoi* [477[a], 480] |
| **516** | Parguerol 7-acetate | *L. saitoi* [477] |
| **517** | Parguerol 16-acetate | *L. saitoi* [477], *A. dactylomela* [156[a], 178] |
| **518** | Parguerol 7,19-diacetate | *L. saitoi* [480] |
| **519** | Parguerol 16,19-diacetate | *L. saitoi* [475[a], 477, 480] |
| **520** | Parguerol 7,16-diacetate | *L. filiformis* [421[a]], *L. saitoi* [477] |
| **521** | Parguerol triacetate | *L. filiformis* [421], *L. saitoi* [475, 477, 479[a], 480] |
| **522** | 15-Bromo-parguer-9(11)-en-16-ol | *L. nipponica* [97[a]], *L. saitoi* [480] |
| **523** | 2-Deacetoxy-deoxyparguerol | *L. saitoi* [475[a], 477] |
| **524** | 15-Bromo-2,7,16-trihydroxy-parguer-9(11)-ene | *L. saitoi* [475, 479[a,c,d]], *A. pulmonica* [476] |
| **525** | Deoxyparguerol | *L. saitoi* [475, 477], *A. dactylomela* [156[a,d], 178, 270] |
| **526** | Deoxyparguerol 7-acetate | *L. filiformis* [421[a,d]] |
| **527** | Deoxyparguerol 16-acetate | *L. filiformis* [421], *L. lageniformis* [481], *L. obtusa* [482[a,d]], *L. saitoi* [477, 480], *A. fasciata* [326] |
| **528** | Deoxyparguerol 7,16-deacetate | *L. filiformis* [421[a], 479[d]], *A. kurodai* [478] |
| **529** | – | *L. filiformis* [421] |
| **530** | 15-Bromoparguer-9(11)-ene-2,16-diol 2-acetate | *L. filiformis* [421] |
| **531** | 16-Acetoxy-15-bromo-7-hydroxy-parguer-9(11)-ene | *A. fasciata* [326] |
| **532** | – | *L. saitoi* [475] |
| **533** | 15-Bromo-parguer-7-en-16-ol | *L. nipponica* [97[a]], *L. saitoi* [480] |
| **534** | 15-Bromo-isoparguer-9(11)-en-16-ol | *L. filiformis* [421] |
| **535** | – | *L. saitoi* [477] |
| **536** | – | *L. saitoi* [477] |
| **537** | – | *L. saitoi* [477] |
| **538** | Preparguerene | *L. filiformis* [421[a]], *L. saitoi* [477] |
| **539** | Deacetylisoparguerol | *L. saitoi* [475[a]], *A. pulmonica* [476] |
| **540** | Isoparguerol | *L. saitoi* [477], *A. dactylomela* [156[a], 270], *A. kurodai* [478], *A. pulmonica* [476] |
| **541** | Isoparguerol 16-acetate | *A. dactylomela* [156[a], 270] |
| **542** | Isoparguerol 7,16-diacetate | *L. saitoi* [475[a], 477] |
| **543** | – | *L. saitoi* [477] |
| **544** | – | *L. saitoi* [475, 483[a]] |

(continued)

**Table 2** (continued)

| No. | Name | Source of isolation/synthesis |
|---|---|---|
| **545** | Neoparguerodione | *A. punctata* [249] |
| **546** | Aplysin-20 | *L. venusta* [484], *A. kurodai* [485, 486[a,c,d]]; Synthesis [487[f], 488[f]] |
| **547** | *epi*-Aplysin-20 | *A. kurodai* [489[d]] |
| **548** | Venustanol | *L. venusta* [484] |
| **549** | Concinndiol | *L. concinna* [490[c,d]], *L. snyderae* [354] |
| **550** | Isoconcinndiol | *L. snyderae* var. *guadalupensis* [491[a]], *A. dactylomela* [492[b,c,d]], *A. punctata* [249]; Synthesis [488[b,f], 493[b,f]] |
| ***ent*-550** | *ent*-Isoconcinndiol | *A. kurodai* [489[d]] |
| **551** | Pinnatol A | *L. pinnata* [494[d]] |
| **552** | Pinnatol B | *L. pinnata* [494[d]] |
| **553** | Isopinnatol B | *A. dactylomela* [50] |
| **554** | Pinnatol C | *L. pinnata* [494[d]] |
| **555** | Pinnatol D | *L. pinnata* [494]; Synthesis [493[f]] |
| **556** | – | *Laurencia* sp. [105] |
| **557** | Isoaplysin-20 | *L. perforata* [495], *L. snyderae* var. *guadalupensis* [491], *A. kurodai* [496[a]]; Synthesis [497[b,e], 498[b,c,f], 499[b,c,f]] |
| **558** | – | *L. perforata* [495[d]] |
| **559** | Paniculatol | *L. paniculata* [500[c,d]] |
| **560** | – | *Laurencia* sp. [106[c,d]] |
| **561** | – | *Laurencia* sp. [106] |
| **562** | – | *L. obtusa* [501] |
| **563** | – | *L. obtusa* [501] |
| **564** | 3-Bromobarekoxide | *L. luzonensis* [502[c,d]] |
| **565** | Laukarlaol | *L. karlae* [333]; Synthesis [503[b,f]] |
| **566** | Neoconcinndiol hydroperoxide | *L. snyderae* [354[c]] |
| **567** | 3,15-Dibromo-7,16-dihydroxy-isopimar-9(11)-ene | *L. perforata* [495] |
| **568** | 3,15-Dibromo-7,12,16-trihy-droxy-isopimar-9(11)-ene | *L. perforata* [495] |
| **569** | Irieol | *Laurencia* cf. *irieii* [504[a]], *L. pinnata* [505] |
| **570** | Iriediol | *Laurencia* cf. *irieii* [504[b,d], 506[a,c]] |
| **571** | Irieol A | *Laurencia* cf. *irieii* [504[d], 506[a,c]] |
| **572** | Irieol C | *Laurencia* cf. *irieii* [504[a,d], 507], *L. mariannensis* [74] |
| **573** | Pinnaterpene A | *L. pinnata* [508[c,d]] |
| **574** | Irieol E | *Laurencia* cf. *irieii* [504[a], 507] |
| **575** | Irieol B | *Laurencia* cf. *irieii* [504] |
| **576** | Pinnaterpene B | *L. pinnata* [508[d]] |
| **577** | Irieol F | *Laurencia* cf. *irieii* [504] |

(continued)

**Table 2** (continued)

| No. | Name | Source of isolation/synthesis |
|---|---|---|
| **578** | Irieol G | *Laurencia* cf. *irieii* [504] |
| **579** | Irieol D | *Laurencia* cf. *irieii* [504[a], 507] |
| **580** | 11-Deacetylpinnaterpene C | *L. decumbens* [160[a], 509[a]], *L. nangii* [94] |
| **581** | Pinnaterpene C | *L. decumbens* [160], *L. mariannensis* [74], *L. pinnata* [494, 508[a,d]] |
| **582** | Pinnaterpene D | *L. pinnata* [494[a], 508[d]] |
| **583** | Angasiol | *L. nangii* [94], *A. angasi* [510[a,c,d]] |
| **584** | Angasiol acetate | *A. juliana* [511[c,d]] |
| **585** | 10-Acetoxy-angasiol | *Laurencia* sp. [187] |
| **586** | Prepinnaterpene | *L. pinnata* [505[a]]; Synthesis [449[f]] |
| **587** | Neoirieone | *Laurencia* cf. *irieii* [507[a,c,d]], *L. nangii* [94] |
| **588** | Neoirietriol | *L. yonaguniensis* [512[c,d]] |
| **589** | Neoirietetraol | *L. nangii* [94], *L. yonaguniensis* [513[a], 514[c,d]] |
| **590** | Luzodiol | *L. luzonensis* [324[a,d]], *A. fasciata* [326], *A. depilans* [515] |
| **591** | Laurendecumtriol | *L. decumbens* [160[a], 509[a]] |
| **592** | – | *A. depilans* [515] |
| **593** | – | *A. depilans* [515] |
| **594** | – | *A. depilans* [515] |
| **595** | – | *A. depilans* [515] |
| **596** | – | *A. depilans* [515] |
| **597** | – | *A. depilans* [515] |
| **598** | – | *A. depilans* [515] |
| **599** | – | *A. depilans* [515] |
| **600** | Dactylomelol | *Laurencia* sp. [516], *A. dactylomela* [50, 517[a,c,d]] |
| **601** | Punctatene | *Laurencia* sp. [516[d]] |
| **602** | Punctatene acetate | *A. punctata* [249] |
| **603** | (*E*)-Dactylohydroperoxide A | *Laurencia* sp. [516] |
| **604** | (*Z*)-Dactylohydroperoxide A | *Laurencia* sp. [516] |
| **605** | Dactylo-3,14-diol | *Laurencia* sp. [516] |
| **606** | Dactylohydroperoxide B | *Laurencia* sp. [516] |
| **607** | Dactylohydroperoxide C | *Laurencia* sp. [516] |
| **608** | Obtusadiol | *L. microcladia* [518], *L. obtusa* [320[a]] |
| **609** | 14-Bromo-obtus-1-ene-3,11-diol | *A. dactylomela* [519[c,d]] |
| **610** | Rogioldiol A | *L. microcladia* [518[d]] |
| **611** | Rogioldiol B | *L. microcladia* [520] |
| **612** | Rogioldiol C | *L. microcladia* [520] |
| **613** | Laurenditerpenol | *L. intricata* [521]; Synthesis [522[f], 523[f], 524[d,g], 525[e], 526[g]] |
| **614** | Rogiolal | *L. microcladia* [518[c,d]] |
| **615** | Isorogiolal | *L. microcladia* [518[d]] |
| **616** | Prevezol A | *L. obtusa* [527] |

(continued)

**Table 2** (continued)

| No. | Name | Source of isolation/synthesis |
|---|---|---|
| **617** | Prevezol B | *L. obtusa* [527, 528[b]]; Synthesis [529[b,g]] |
| **618** | Prevezol C | *L. obtusa* [528]; Synthesis [530[b,g], 531[e]] |
| **619** | Prevezol D | *L. obtusa* [528] |
| **620** | Prevezol E | *L. obtusa* [528] |
| **621** | Neorogioldiol | *L. microcladia* [532[a,d]], *L. obtusa* [528] |
| **622** | Neorogioldiol B | *L. obtusa* [528] |
| **623** | Neorogioltriol | *L. glandulifera* [533] |
| **624** | Viridiol A | *L. viridis* [534] |
| **625** | Viridiol B | *L. viridis* [534] |
| **626** | Laurencianol | *L. obtusa* [535[c,d]] |
| **627** | Punctatol | *A. punctata* [249] |
| **628** | Glandulaurencianol A | *L. glandulifera* [536] |
| **629** | Glandulaurencianol B | *L. glandulifera* [536[a]], *A. punctata* [536[a]] |
| **630** | Glandulaurencianol C | *A. punctata* [536] |
| **631** | Dactylopyranoid | *A. dactylomela* [50] |
| **632** | Kahukuene B | *L. majuscula* [537] |
| **633** | 10-Hydroxy-kahukuene B | *L. mariannensis* [74[a]], *L. nangii* [94], *L. obtusa* [271] |
| **634** | Kahukuene A | *L. majuscula* [537] |
| **635** | Dactyloditerpenol acetate | *A. dactylomela* [178[d]] |
| **636** | Rogioldiol D | *L. microcladia* [532] |
| **637** | *O*-11,15-Cyclo-14-bromo-14,15-dihydrorogiol-3,11-diol | *L. glandulifera* [538], *L. microcladia* [532[a]], *L. obtusa* [528] |
| **638** | Aplysiadiol | *L. caduciramulosa* [195], *L. japonensis* [93], *L. nangii* [310], *Laurencia* sp. [187], *A. kurodai* [539[a]]; Synthesis [540] |
| **639** | Aplysiadiol methyl ether | *A. kurodai* [539] |
| **640** | Anhydroaplysiadiol | *L. japonensis* [93] |
| **641** | (*E*)-Phytol | *L. implicata* [91], *L. nipponica* [96, 97], *L. tristicha* [290, 541], *C. papillosus* [342] |
| **642** | (*E*)-Phytol acetate | *L. nipponica* [96] |
| **643** | – | *L. pinnatifida* [219] |
| **644** | 2,3-Epoxy-phytyl acetate | *L. composita* [214] |
| **645** | (–)-Geranyl-linalool | *L. microcladia* [532] |

[a]The isolation and structure elucidation of this compound was reported for the first time
[b]The structure of this compound has been revised
[c]The structure of this compound or of a derivative has been confirmed by X-ray crystallographic analysis
[d]The absolute stereochemistry of this compound has been established
[e]Partial synthesis
[f]Stereoselective total synthesis
[g]Enantioselective total synthesis
[h]Formal synthesis

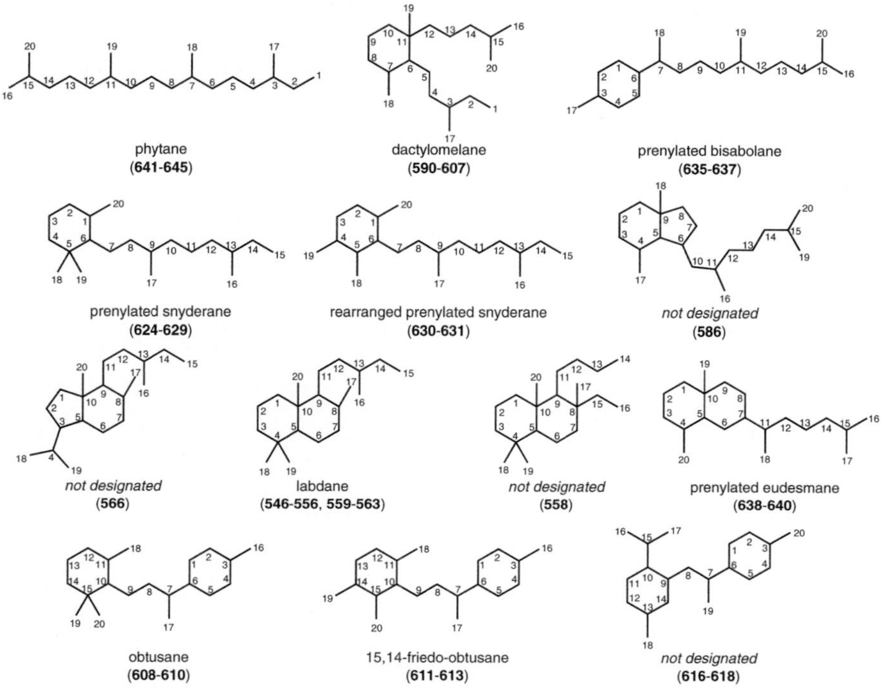


**Fig. 6** Acyclic, monocyclic, and bicyclic carbon skeletons of diterpenes isolated from *Laurencia* species

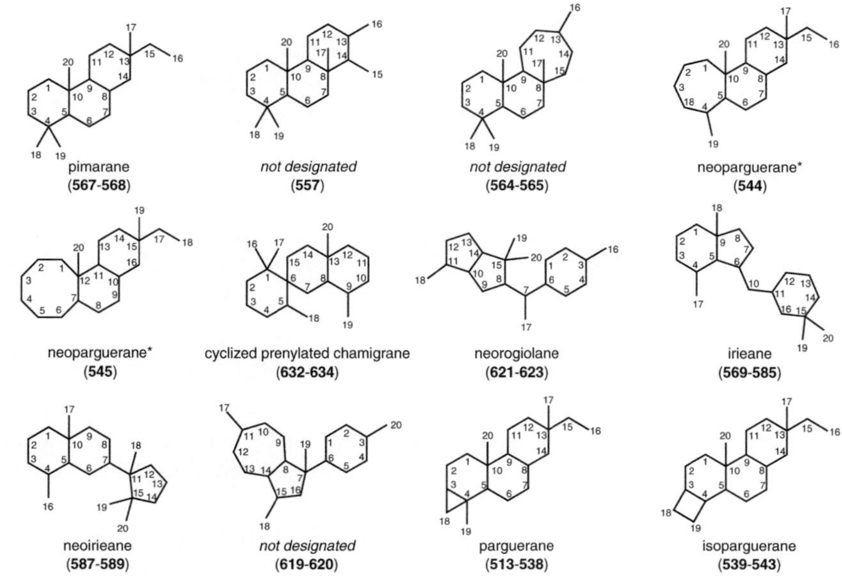


**Fig. 7** Tricyclic and tetracyclic carbon skeletons of diterpenes isolated from *Laurencia* species

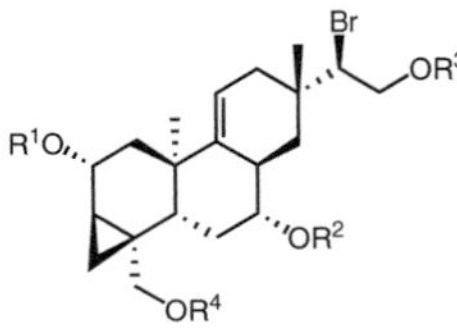

**513** $R^1 = R^2 = R^3 = R^4 = H$
**514** $R^1 = Ac$, $R^2 = R^3 = R^4 = H$
**515** $R^1 = R^4 = Ac$, $R^2 = R^3 = H$
**516** $R^1 = R^2 = Ac$, $R^3 = R^4 = H$
**517** $R^1 = R^3 = Ac$, $R^2 = R^4 = H$
**518** $R^1 = R^2 = R^4 = Ac$, $R^3 = H$
**519** $R^1 = R^3 = R^4 = Ac$, $R^2 = H$
**520** $R^1 = R^2 = R^3 = Ac$, $R^4 = H$
**521** $R^1 = R^2 = R^3 = R^4 = Ac$

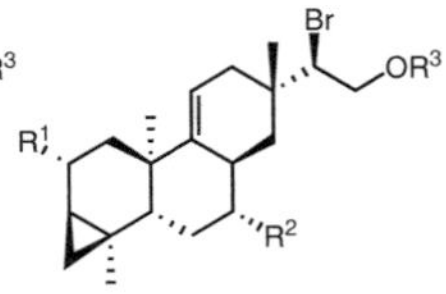

**522** $R^1 = R^2 = R^3 = H$
**523** $R^1 = R^3 = H$, $R^2 = OH$
**524** $R^1 = R^2 = OH$, $R^3 = H$
**525** $R^1 = OAc$, $R^2 = OH$, $R^3 = H$
**526** $R^1 = R^2 = OAc$, $R^3 = H$
**527** $R^1 = OAc$, $R^2 = OH$, $R^3 = Ac$
**528** $R^1 = R^2 = OAc$, $R^3 = Ac$

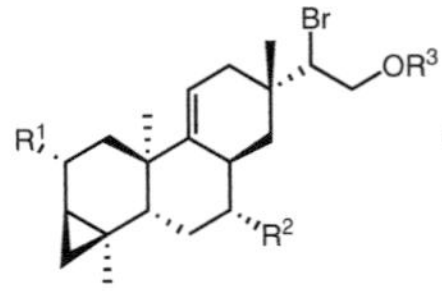

**529** $R^1 = R^2 = R^3 = H$
**530** $R^1 = OAc$, $R^2 = R^3 = H$
**531** $R^1 = H$, $R^2 = OH$, $R^3 = Ac$

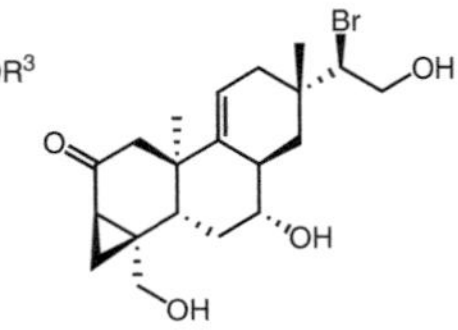

**532**

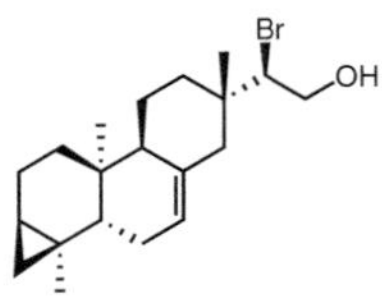

**533**

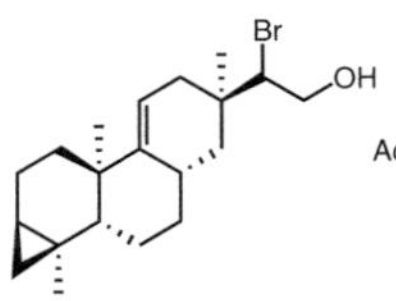

**534**

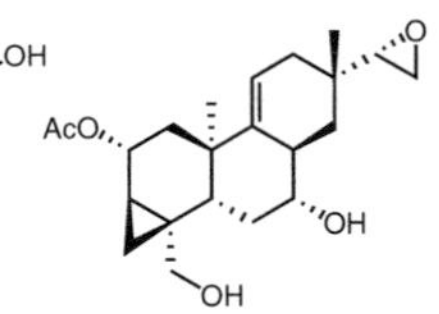

**535**

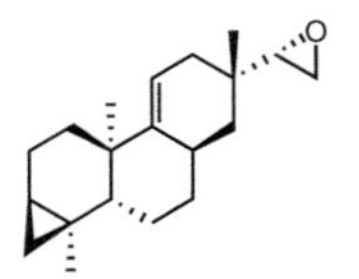

**536**

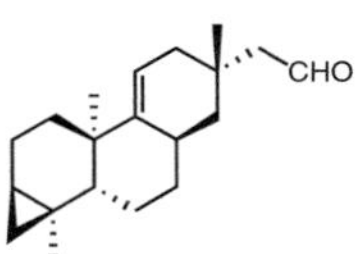

**537**

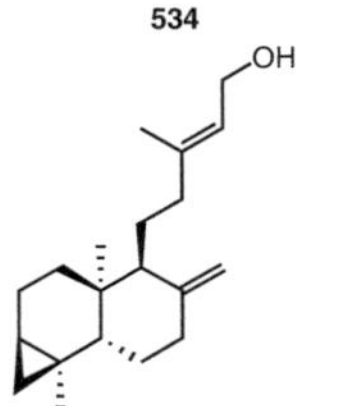

**538**

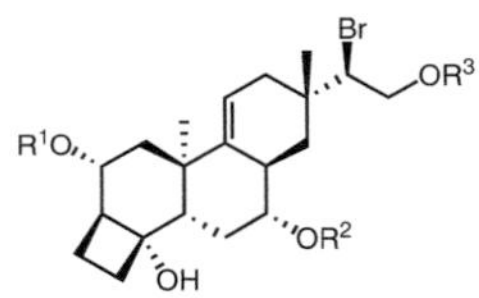

**539** $R^1 = R^2 = R^3 = H$
**540** $R^1 = Ac$, $R^2 = R^3 = H$
**541** $R^1 = R^3 = Ac$, $R^2 = H$
**542** $R^1 = R^2 = R^3 = Ac$

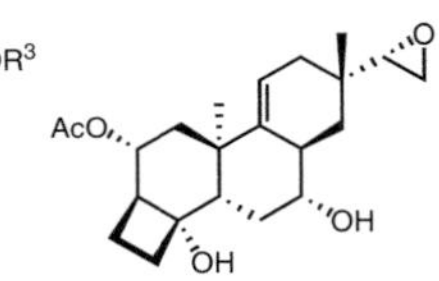

**543**

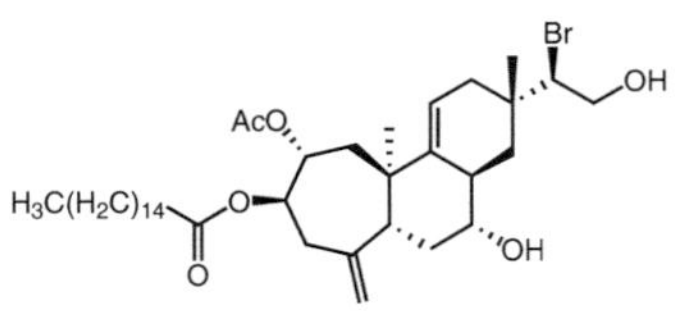

**544**

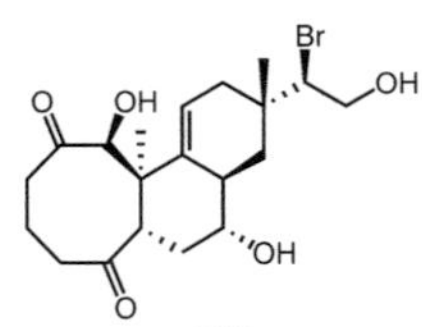

**545**

### 3.2.2 Labdanes, Pimaranes, and Related Diterpenes

Labdanes, pimaranes, and related metabolites (**546**–**568**) constitute the second largest group of diterpenes isolated from species of the genus *Laurencia* and their grazers, including 23 compounds so far. Each of them is brominated at C-3, with the exception of laukarlaol (**565**) and neoconcinndiol hydroperoxide (**566**). Furthermore, many of them have OH substitutions at C-8, C-9, or C-13. Aplysin-20 (**546**) and isoaplysin-20 (**557**) were the first members of this group to be isolated from the sea hare *Aplysia kurodai* in 1971 by Yamamura and Hirata [485, 486] and in 1977 by Yamamura and Terada [496], respectively.

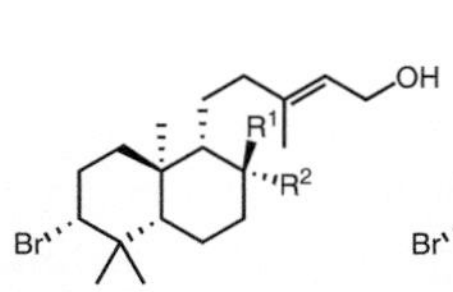
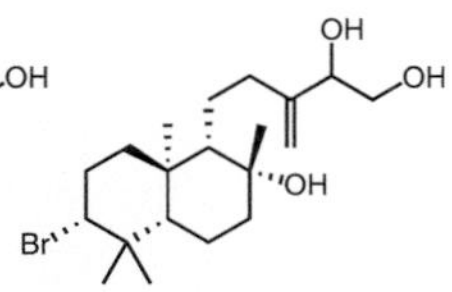
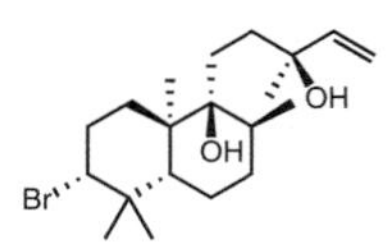
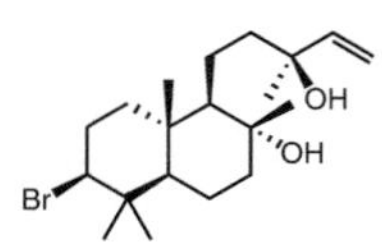

**546** $R^1$ = Me, $R^2$ = OH
**547** $R^1$ = OH, $R^2$ = Me
**548**
**549**
**550**

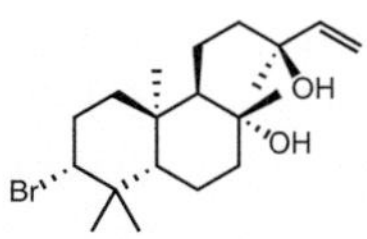
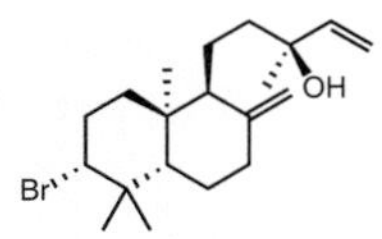
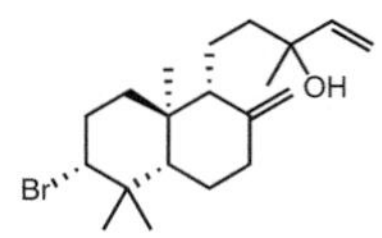
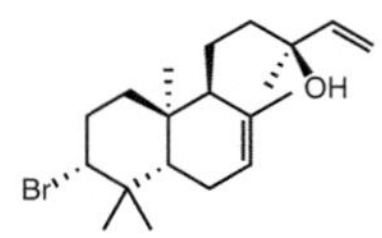

**551** **552** **553** **554**

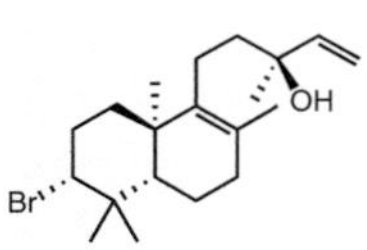
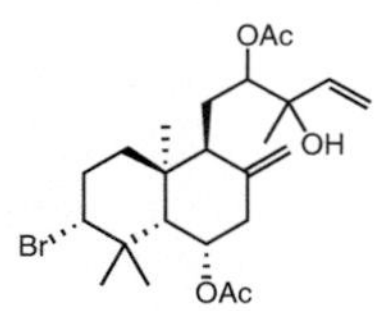
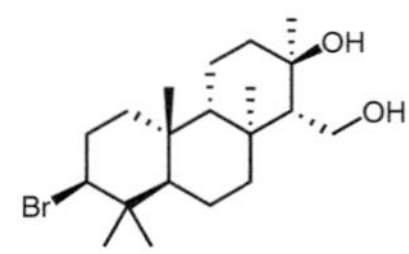
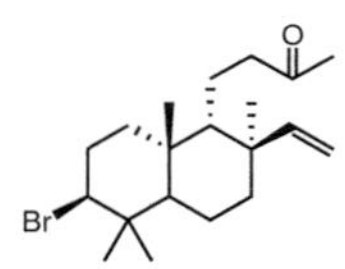

**555** **556** **557** **558**

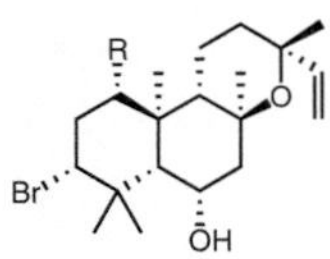
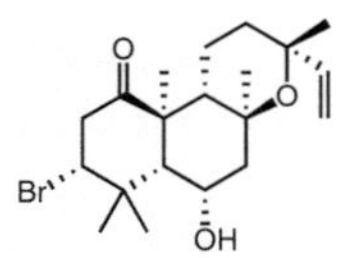
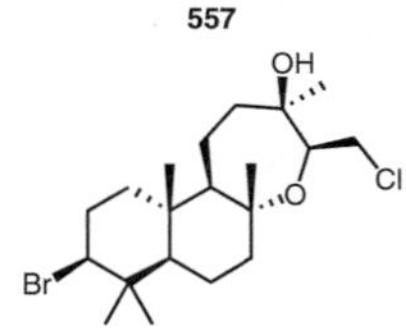
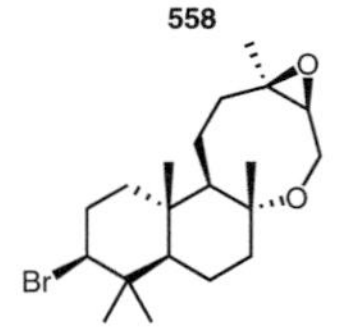

**559** R = H
**560** R = OAc
**561**
**562**
**563**

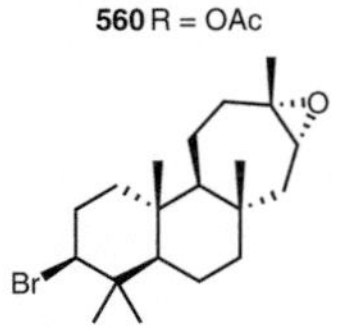
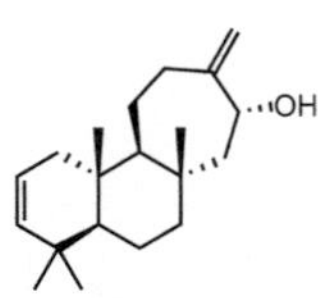
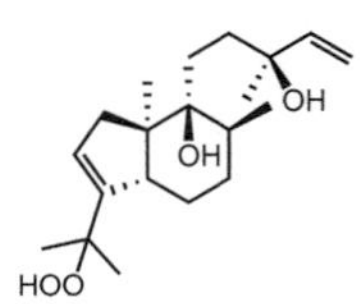
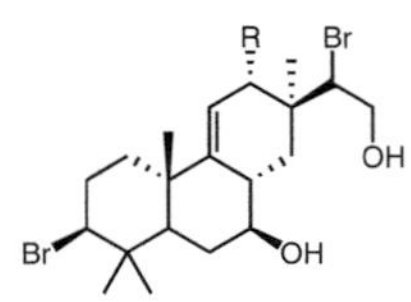

**564** **565** **566**
**567** R = H
**568** R = OH

The initially proposed structures were revised after the total syntheses of the natural products [488, 497–499]. The rearranged labdane **558** was obtained from *Laurencia perforata*, along with the pimarane derivatives **567** and **568** [495]. Paniculatol (**559**) and the related metabolites **560** and **561** possess a tetrahydropyran ring [106, 500], while metabolites **562** and **563**, isolated from *Laurencia obtusa* collected from Mitikas Bay in the Ionian Sea, contain oxepane and oxocane moieties, respectively [501]. *Laurencia snyderiae* from Santa Catalina Island afforded neoconcinndiol hydroperoxide (**566**), featuring a modified ring A, the structure of which was established on the basis of single-crystal X-ray diffraction analysis [354].

### 3.2.3 Irieanes and Neoirieanes

Taken together, irieanes and neoirieanes consist the third most often encountered group of diterpenes characterized from *Laurencia* spp. and gastropods consuming them, amounting to 21 compounds in total (**569–589**). The irieane and neoirieane skeletons are unique in Nature and so far have only been found in *Laurencia* species and sea hares feeding on them. In 1975 and 1978, Fenical and co-workers reported irieol (**569**) and iriediol (**570**), as well as irieols A (**571**), B (**575**), C (**572**), D (**579**), E (**574**), F (**577**), and G (**578**), from *Laurencia* cf. *irieii* collected near Puerto Peñasco, Mexico [504, 506]. These compounds possess a regular tricarbocyclic framework that closely resembles the sesquiterpene skeleton oppositane. The structure of irieol A (**571**) was determined on the basis of X-ray crystallographic analysis [506]. In 1982, reinvestigation of *L.* cf. *irieii* by the same group led to the isolation of neoirieone (**587**), a diterpene possessing a different but closely related to irieane skeleton [507]. The structure of neoirieone (**587**) was elucidated by X-ray crystallographic analysis of the reduction product of **587** [507]. All irieane and neoirieane derivatives are dibrominated at C-1 and C-14, with the exception of prepinnaterpene (**586**) that bears only one bromine atom at C-1. Furthermore, apart from compounds **580–585**, all others possess an α-oriented OH substitution at C-4.

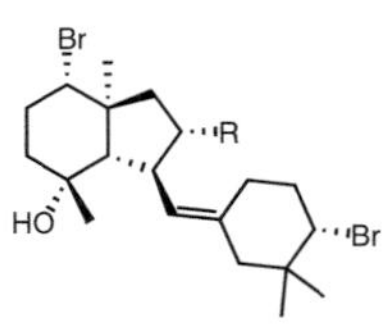

**569** R = H
**570** R = OH

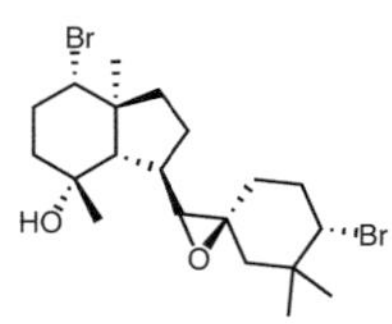

**571**

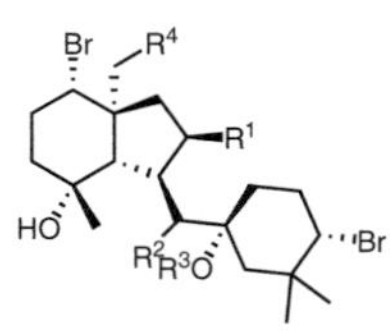

**572** R$^1$ = R$^2$ = R$^3$ = R$^4$ = H
**573** R$^1$ = R$^2$ = R$^4$ = H, R$^3$ = Ac
**574** R$^1$ = R$^3$ = R$^4$ = H, R$^2$ = OH
**575** R$^1$ = OH, R$^2$ = R$^3$ = R$^4$ = H
**576** R$^1$ = R$^2$ = H, R$^3$ = Ac, R$^4$ = OH
**577** R$^1$ = R$^2$ = OH, R$^3$ = R$^4$ = H
**578** R$^1$ = OH, R$^2$ = OAc, R$^3$ = R$^4$ = H
**579** R$^1$ = OAc, R$^2$ = OH, R$^3$ = R$^4$ = H

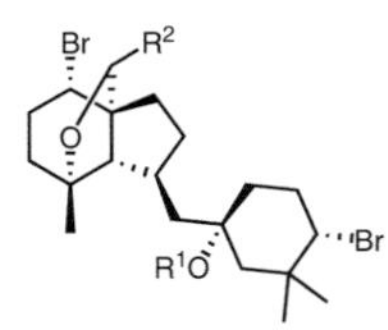

**580** R$^1$ = H, R$^2$ = OH
**581** R$^1$ = Ac, R$^2$ = OH
**582** R$^1$ = Ac, R$^2$ = H

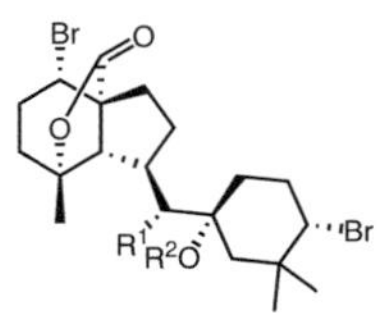

**583** R$^1$ = R$^2$ = H
**584** R$^1$ = H, R$^2$ = Ac
**585** R$^1$ = OAc, R$^2$ = H

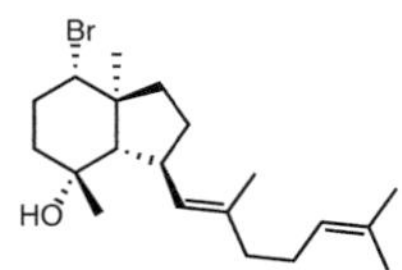

**586**

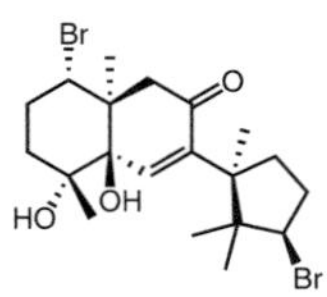

**587**

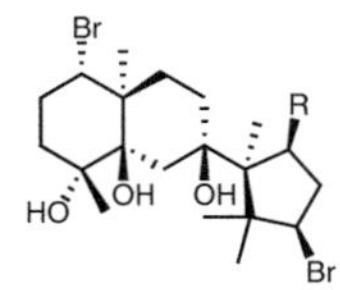

**588** R = H
**589** R = OH

### 3.2.4 Dactylomelanes

To date, 18 diterpenes featuring the dactylomelane skeleton (**590–607**) have been isolated from species of *Laurencia* and the sea hares consuming them. Dactylomelanes possess a monocarbocyclic framework arising by cyclization between carbons C-6 and C-11 of the phytane skeleton. Dactylomelol (**600**), the first representative of this class, was reported initially as a constituent of *A. dactylomela* collected from Tenerife, Canary islands, and its structure, including the absolute stereochemistry, was determined by single-crystal X-ray diffraction analysis [517]. Compounds **590–599** bear an α-oriented Br substitution at C-10, whereas the remaining substances (**600–607**) possess a 7,10-epoxy ring. In all cases, the side chain attached to C-6 is the same, with the exception of **602**, in which the hydroxy group at C-3 is acetylated.

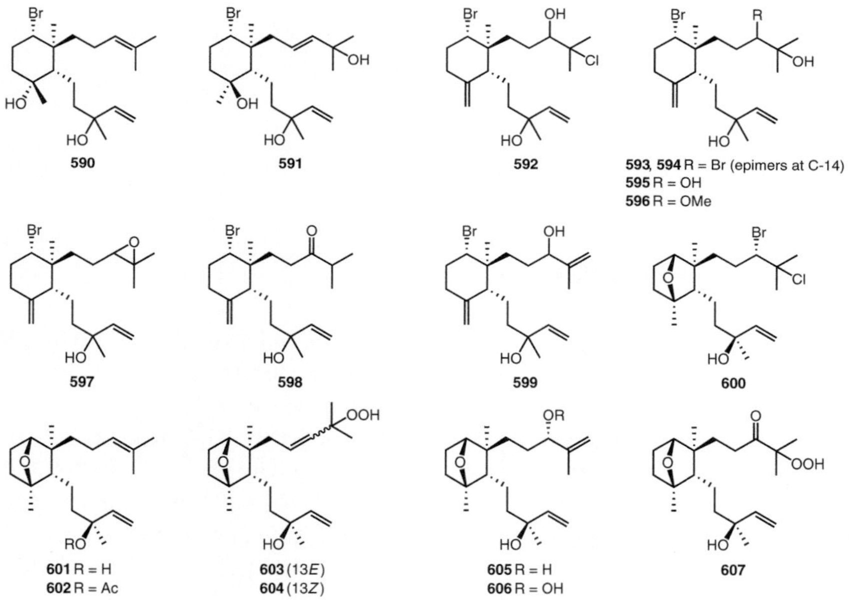


**590** **591** **592** **593**, **594** R = Br (epimers at C-14)
**595** R = OH
**596** R = OMe

**597** **598** **599** **600**

**601** R = H
**602** R = Ac
**603** (13*E*)
**604** (13*Z*)
**605** R = H
**606** R = OH
**607**

### 3.2.5 Obtusanes, 15,14-Friedoobtusanes, and Related Diterpenes

Obtusanes, 15,14-friedoobtusanes, and related metabolites comprise a small group of diterpenes isolated from species of *Laurencia* and their grazers, encompassing eight members (**608–615**). Three diterpenes featuring the bicyclic obtusane skeleton (**608–610**) and three 15,14-friedoobtusanes (**611–613**), resulting from a 1,2-methyl migration to C-14, have been reported so far. Obtusadiol (**608**), the first diterpene featuring the bicyclic obtusane skeleton, was originally isolated from

*L. obtusa* collected from the Greek shores near Athens [320] and later from *L. microcladia* from Il Rogiolo, Italy [518]. The latter population afforded also two related degraded $C_{11}$ derivatives, namely, rogiolal (**614**) and isorogiolal (**615**) [518]. Most of the metabolites belonging to this group are characterized by a *cis*-1,2-bromohydrin functionality, with the exception of compound **609** and laurenditerpenol (**613**).

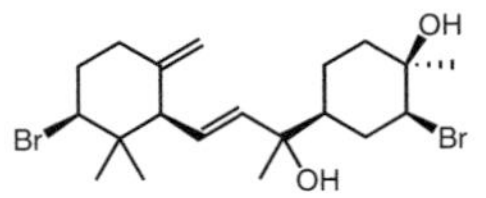


**608**

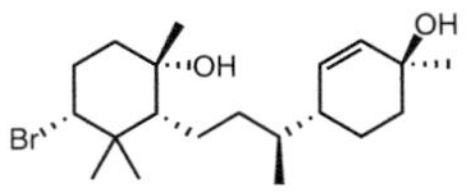


**609**

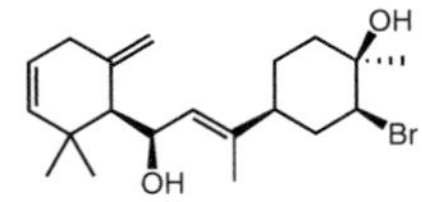


**610**

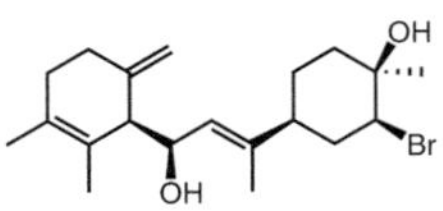


**611**

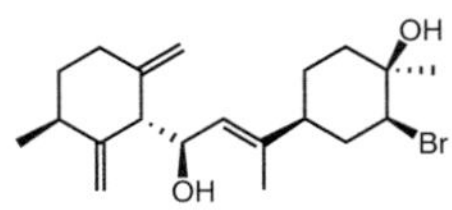


**612**

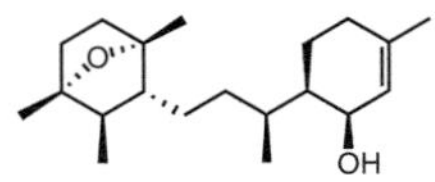


**613**

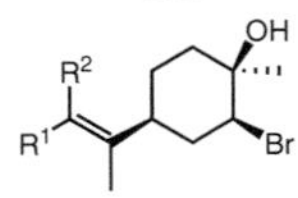


**614** R[1] = CHO, R[2] = H
**615** R[1] = H, R[2] = CHO

### 3.2.6 Prevezanes and Neorogiolanes

Diterpenes belonging to the group of the prevezanes and neorogiolanes (**616–623**) feature three rare carbon frameworks that so far are found exclusively in species of *Laurencia*. To date, eight diterpenes featuring these skeletons have been reported. In 2001, Mihopoulos et al. reported the isolation of prevezols A (**616**) and B (**617**) from a *L. obtusa* population collected from Preveza, Greece [527]. In 2003, the same group published the structures of prevezols C (**618**), D (**619**), and E (**620**), obtained from another population of *L. obtusa* collected again from Preveza, while prevezol B (**617**) was re-isolated and its structure revised [528]. Prevezols A (**616**), B (**617**) and C (**618**) incorporate a 1,2-dicyclohexylethane core into their structure, while prevezols D (**619**) and E (**620**) contain a 1-cyclohexyldecahydroazulene ring system. Comparison of the spectroscopic data of the natural products **617** and **618** with those of the products obtained through total syntheses of the proposed structures for prevezols B [529] and C [530], respectively, exhibited slight discrepancies, albeit without suggesting alternative structures for the natural products. *L. microcladia* collected from Il Rogiolo, Italy, yielded neorogioldiol (**621**) [532], whereas its epimer at C-11, neorogioldiol B (**622**), was isolated from *L. obtusa* collected from Preveza [528]. The related metabolite neorogioltriol (**623**) was isolated from *L. glandulifera* collected in Kefalonia Island, Greece [533]. Compounds **621–623** possess a 2-(cyclohexylmethyl)octahydropentalene system. Without exception, all of these metabolites feature a *cis*-1,2-bromohydrin functionality.

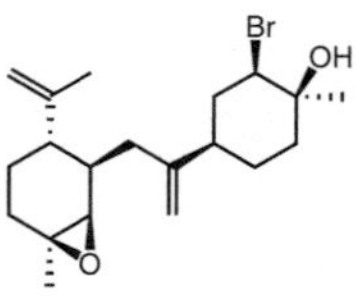


**616**

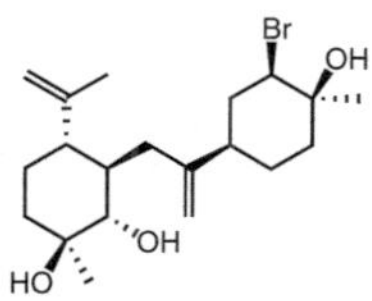


**617**

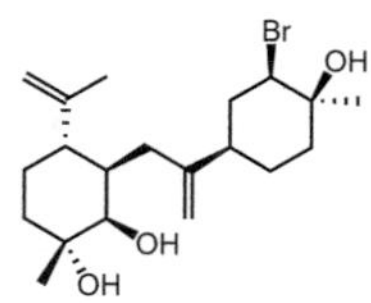


**618**

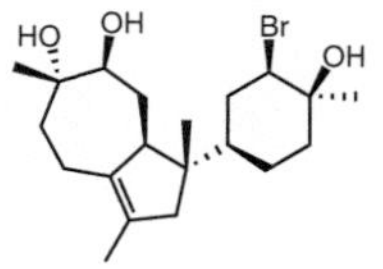


**619**

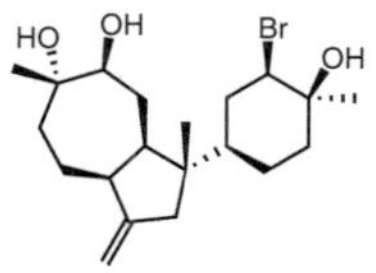


**620**

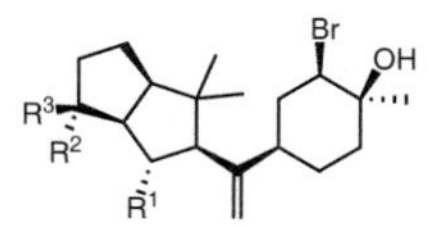


**621** $R^1$ = Br, $R^2$ = OH, $R^3$ = Me
**622** $R^1$ = Br, $R^2$ = Me, $R^3$ = OH
**623** $R^1$ = $R^2$ = OH, $R^3$ = Me

### 3.2.7 Miscellaneous Diterpenes

In addition to the diterpenes belonging to the classes discussed above, there are 22 more diterpenes (**624**–**645**) that are either linear or possess prenylated sesquiterpene skeletons reported so far from species of the genus *Laurencia* or mollusks feeding on them. Compounds **624**–**629** feature a prenylated snyderane (or 10,15-cyclophytane) framework, whereas metabolites **630** and **631** possess a prenylated rearranged snyderane carbocycle. Among these, derivatives **626**–**631** represent an etherification between carbons C-9 and C-13. Laurencianol (**626**) and punctatol (**627**) are dibrominated, bearing bromine atoms at both C-4 and C-10, glandulaurencianols A (**628**) and B (**629**) have only one Br substitution at C-4, whereas dactylopyranoid (**631**) has a Br substitution at C-10. Kahukuenes A (**634**) and B (**632**), isolated from a Hawaiian *L. majuscula* [537] collection, as well as compound **633** are the only examples of diterpenes possessing a prenylated chamigrane backbone that has been further cyclized to yield a tricarbocyclic skeleton. Dactyloditerpenol acetate (**635**), obtained from *A. dactylomela* collected off Mona Island, Puerto Rico [178], as well as rogioldiol D (**636**) and *O*-11,15-cyclo-14-bromo-14,15-dihydrorogiol-3,11-diol (**637**), isolated from *L. microcladia* from Il Rogiolo [532], feature a prenylated bisabolane skeleton. The latter compound exhibits an etherification between carbons C-11 and C-15, while both **636** and **637** display a *cis*-1,2-bromohydrin moiety in their structures. In 1990, Yamada and co-workers reported the isolation of aplysiadiol (**638**) and its methyl ether derivative **639**, diterpenes possessing a prenylated eudesmane skeleton, from *A. kurodai* [539], whereas anhydroaplysiadiol (**640**) was later isolated together with **638** from *Laurencia japonensis* [93]. Furthermore, five linear diterpenes (**641**–**645**) have been isolated from a number of *Laurencia* species.

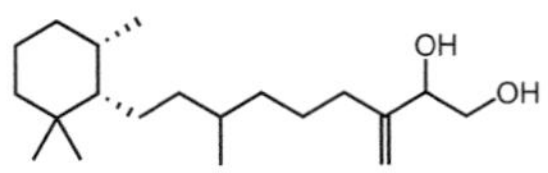


624

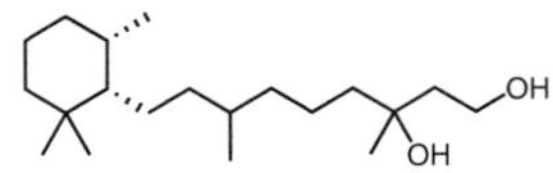


625

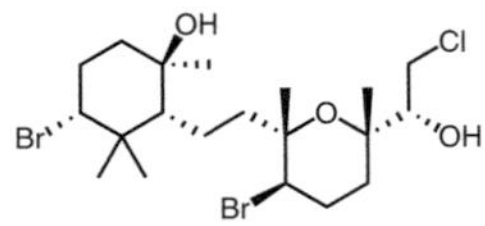


626

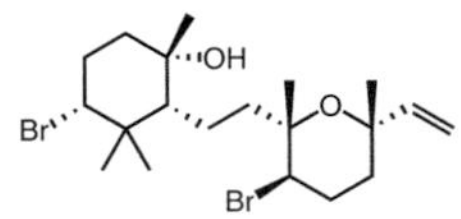


627

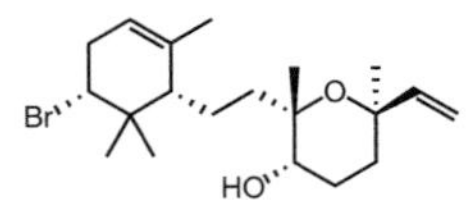


628

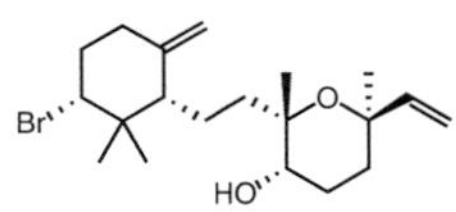


629

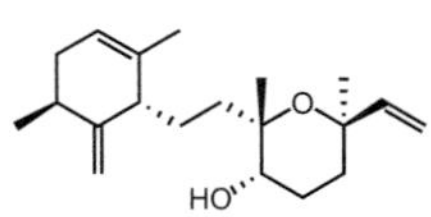


630

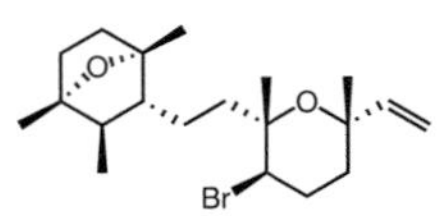


631

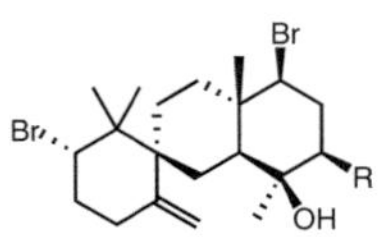


632 R = H
633 R = OH

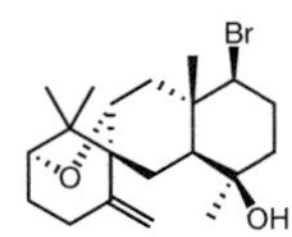


634

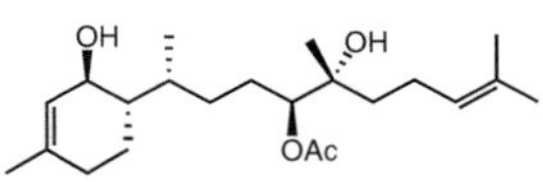


635

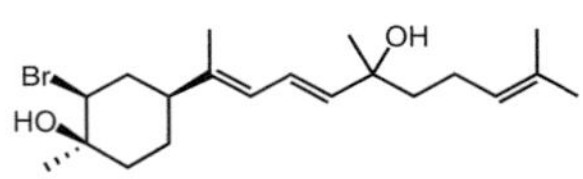


636

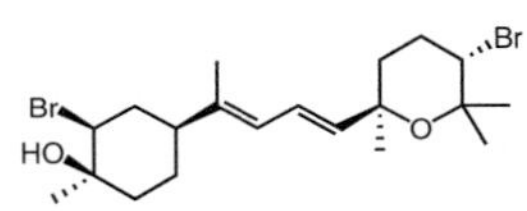


637

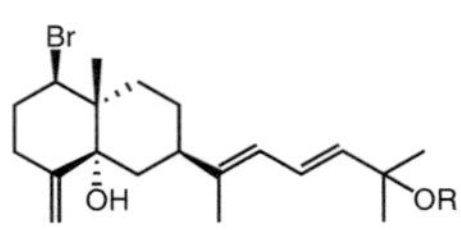


638 R = H
639 R = Me

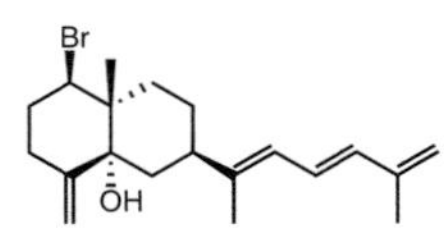


640

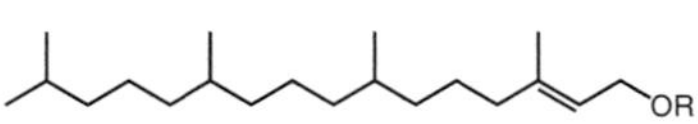


641 R = H
642 R = Ac

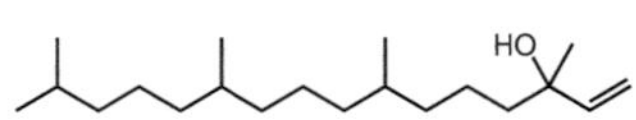


643

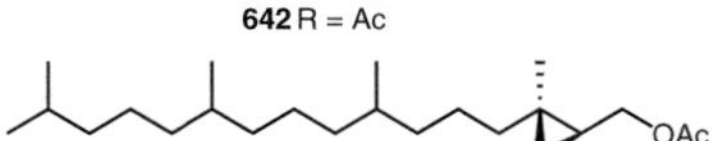


644

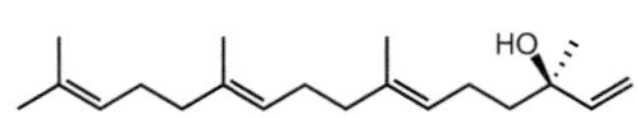


645

## 3.3 *Triterpenes*

Triterpenes isolated from species of the genus *Laurencia*, as well as mollusks grazing on them, encompass to date 59 non-carbocyclic polyethers and related metabolites (**646–704**, Table 3) that display a diversity of ring sizes and functionalization. From a biogenetic point of view, triterpenes may arise from a common precursor,

**Table 3** Triterpenes from *Laurencia* spp.

| No. | Name | Source of Isolation/Synthesis |
|---|---|---|
| **646** | Thyrsiferol | *L. catarinensis* [542], *L. mariannensis* [543], *L. saitoi* [480, 544], *L. tryrsifera* [545[a,c], 546], *L. venusta* [547], *L. viridis* [548], *A. dactylomela* [549]; Synthesis [550[e], 551[e], 552[d,g], 553[f]] |
| **647** | Thyrsiferyl 23-acetate | *L. saitoi* [480, 544[a], 546], *L. venusta* [547]; Synthesis [552[d,g], 553[f]] |
| **648** | Aplysiol A (21$\alpha$-Hydroxy-thyrsiferol) | *L. mariannensis* [543], *A. dactylomela* [549[a,d]] |
| **649** | 15(28)-Anhydrothyrsiferol (Dehydrothyrsiferol) | *L. pinnatifida* [272[a,d]], *L. viridis* [554, 555] |
| **650** | Dehydrothyrsiferyl 23-acetate | *L. obtusa* [556] |
| **651** | 15(28)-Anhydrothyrsiferyl diacetate | *L. saitoi* [557[d]] |
| **652** | 16-Hydroxy-dehydrothyrsiferol | *L. viridis* [554] |
| **653** | 16-*epi*-Hydroxy-dehydrothyrsiferol | *L. viridis* [558] |
| **654** | 3-*epi*-Dehydrothyrsiferol | *L. viridis* [559] |
| **655** | 10-*epi*-Dehydrothyrsiferol | *L. viridis* [560] |
| **656** | 15,16-Anhydrothyrsiferol | *L. omaezakiana* [561[d]] |
| **657** | 15-Anhydrothyrsiferyl diacetate | *L. saitoi* [557[d]] |
| **658** | 10-*epi*-15,16-Dehydrothyrsiferol | *L. viridis* [554] |
| **659** | Venustatriol | *L. venusta* [547[c,d]], *A. dactylomela* [549]; Synthesis [552[g], 562[g]] |
| **660** | Dehydrovenustatriol | *L. viridis* [554] |
| **661** | 22-Hydroxy-15(28)-dehydro-venustatriol | *L. viridis* [563] |
| **662** | 15,16-Dehydrovenustatriol | *L. viridis* [554] |
| **663** | 15,16-Epoxy-thyrsiferol A | *L. viridis* [564] |
| **664** | 15,16-Epoxy-thyrsiferol B | *L. viridis* [564] |
| **665** | Magireol A | *L. saitoi* [557[d]] |
| **666** | Magireol B | *L. saitoi* [557] |
| **667** | Magireol C | *L. saitoi* [557] |
| **668** | Aplysiol B (Laurenmariannol) | *L. mariannensis* [543], *A. dactylomela* [549[a,d], 565[b]] |
| **669** | Saiyacenol A | *L. viridis* [555] |
| **670** | Saiyacenol B | *L. viridis* [555] |
| **671** | 28-Hydroxy-saiyacenol B | *L. viridis* [564] |
| **672** | Callicladol | *L. calliclada* [566] |
| **673** | Isodehydrothyrsiferol | *L. viridis* [560] |
| **674** | Iubol | *L. viridis* [563] |
| **675** | Aplysqualenol A | *A. dactylomela* [567] |
| **676** | Aplysqualenol B | *A. dactylomela* [567] |
| **677** | Thyrsenol A | *L. viridis* [568] |

(continued)

**Table 3** (continued)

| No. | Name | Source of Isolation/Synthesis |
|---|---|---|
| **678** | Thyrsenol B | *L. viridis* [568] |
| **679** | 15-Dehydroxythyrsenol A | *L. viridis* [569] |
| **680** | Spirodehydrovenustatriol | *L. viridis* [570] |
| **681** | Martiriol | *L. viridis* [558] |
| **682** | 1,2-Dehydropseudodehydro-thyrsiferol | *L. viridis* [563] |
| **683** | – | *L. viridis* [559] |
| **684** | – | *L. viridis* [559] |
| **685** | Pseudodehydrothyrsiferol | *L. viridis* [558]; Synthesis [571[d,g]] |
| **686** | Lactodehydrothyrsiferol | *L. viridis* [559]; Synthesis [572] |
| **687** | 14-Keto-dehydrothyrsiferol | *L. viridis* [570] |
| **688** | Clavidol | *L. viridis* [559] |
| **689** | Prethyrsenol A | *L. viridis* [569] |
| **690** | 13-Hydroxy-prethyrsenol A | *L. viridis* [569] |
| **691** | Saiyacenol C | *L. viridis* [564] |
| **692** | Predehydrovenustatriol acetate | *L. viridis* [554] |
| **693** | Secodehydrothyrsiferol | *L. viridis* [563] |
| **694** | Adejen A | *L. viridis* [570] |
| **695** | Adejen B | *L. viridis* [570] |
| **696** | Aurilol | *D. auricularia* [573]; Synthesis [574[d,g]] |
| **697** | Enshuol | *L. omaezakiana* [575]; Synthesis [576[d,g]] |
| **698** | Dioxepandehydrothyrsiferol | *L. viridis* [558] |
| **699** | Teurilene | *L. saitoi* [544[a,c]]; Synthesis [577–582] |
| **700** | Intricatetraol | *L. intricata* [583[a]], *L. nangii* [94], *Laurencia* sp. [584]; Synthesis [585[d,g]] |
| **701** | Squalene | *Laurencia* spp. [586] |
| **702** | (+)-(10*R*,11*R*)-Squalene-10,11-epoxide | *L. okamurai* [587[a,d,g], 588] |
| **703** | Auriculol | *D. auricularia* [589[d,g]] |
| **704** | Omaezakianol | *L. omaezakiana* [561]; Synthesis [590[d,g]] |

[a]The isolation and structure elucidation of this compound was reported for the first time
[b]The structure of this compound has been revised
[c]The structure of this compound or of a derivative has been confirmed by X-ray crystallographic analysis
[d]The absolute stereochemistry of this compound has been established
[e]Partial synthesis
[f]Stereoselective total synthesis
[g]Enantioselective total synthesis

(+)-(10*R*,11*R*)-squalene-10,11-epoxide (**702**), which could evolve to the hypothetical common intermediate (6*S*,7*S*,10*R*,11*R*,14*R*,15*R*,18*S*,19*S*)-squalene tetraepoxide [591]. From this intermediate, three main routes can be proposed and as such three different subgroups can be distinguished, namely: (1) those possessing a

2,7-dioxabicyclo[4.4.0]decane ring system, such as thyrsiferol (**646**), venustatriol (**659**) and their congeners; (2) those possessing a 2,8-dioxabicyclo[5.4.0]undecane ring system, such as enshuol (**697**); and (3) those possessing a symmetry element, such as teurilene (**699**). A fourth group would include miscellaneous triterpenes, including linear analogues.

### 3.3.1 Triterpenes Possessing a 2,7-Dioxabicyclo[4.4.0]decane Ring System and Related Metabolites

Since the isolation of thyrsiferol (**646**) from *Laurencia thyrsifera* collected in Seal Reef-Kaikoura, New Zealand by Blunt et al. in 1978 [545], a number of related compounds have been reported. In particular, 50 triterpenes and related metabolites possessing a 2,7-dioxabicyclo[4.4.0]decane moiety (**646–695**) have been isolated from *Laurencia* species and the sea hares consuming them so far. In most cases (**646–679** and **681**), the 2,7-dioxabicyclo[4.4.0]decane core is connected through a single bond to a tetrahydropyran ring in the left side, whereas in the right side it is connected to a side chain. The latter differs from C-15 to C-23 presenting variations in the number, size and position of the cyclic ether rings, varying degrees of oxidation, and different configurations of the chiral centers. In the case of spirodehydrovenustatriol (**680**), the 2,7-dioxabicyclo[4.4.0]decane core has been modified to a 1,6-dioxaspiro[4.5]decane ring system. With the exception of 3-epidehydrothyrsiferol (**654**) that bears an $\alpha$-oriented Br substitution at C-3, and martiriol (**681**), with a $\beta$-oriented OH substitution at C-3, all the other derivatives (**646–653** and **655–679**) feature a $\beta$-oriented Br substitution at C-3. In compounds **682–686**, the 2,7-dioxabicyclo[4.4.0]decane core is instead connected through a single bond to a tetrahydrofuran ring on the left side, while in metabolites **692** and **693** the basic ring system is connected to a linear side chain on the left side. Furthermore, triterpenes **687–691** lack the 10,14-ether ring, but also bear a β-oriented Br substitution at C-3. Adejens A (**694**) and B (**695**) are two unusual $C_{17}$ terpenoids obtained from *L. viridis* and their biogenetic origin has been proposed on the basis of an oxidative degradation of the squalene skeleton [570]. It is worth noting that the spectroscopic data of 21$\alpha$-hydroxy-thyrsiferol and laurenmariannol, reported from a Chinese *L. mariannensis* sample in 2008 as new natural products [543], were found to be the same as those of aplysiols A (**648**) and B (**668**), respectively, isolated by Manzo et al. in 2007 from the mantle of the sea hare *A. dactylomela*, collected from the South China Sea [549]. The relative configurations of the asymmetric centers of aplysiol B (**668**) were later revised [555, 565].

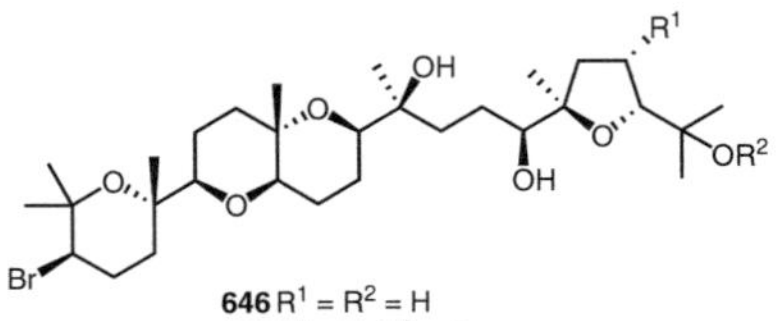

**646** R¹ = R² = H
**647** R¹ = H, R² = Ac
**648** R¹ = OH, R² = H

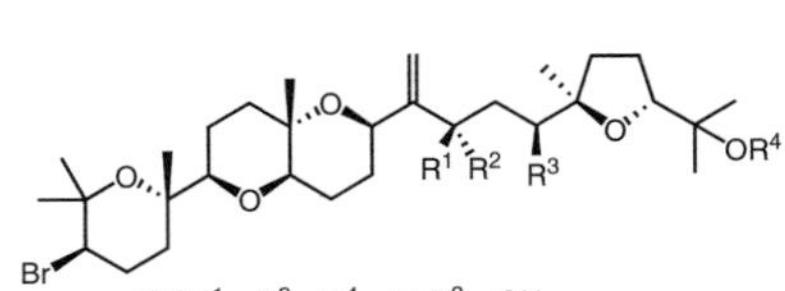

**649** R¹ = R² = R⁴ = H, R³ = OH
**650** R¹ = R² = H, R³ = OH, R⁴ = Ac
**651** R¹ = R² = H, R³ = OAc, R⁴ = Ac
**652** R¹ = R⁴ = H, R² = R³ = OH
**653** R¹ = R³ = OH, R² = R⁴ = H

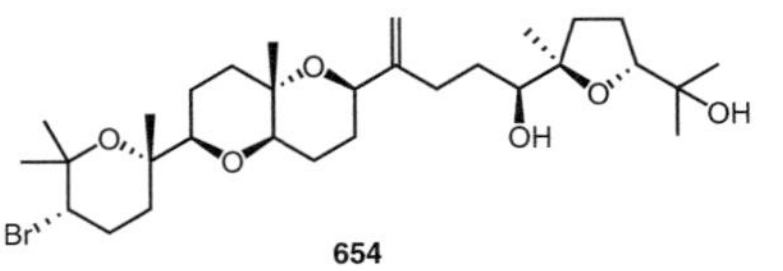

**654**

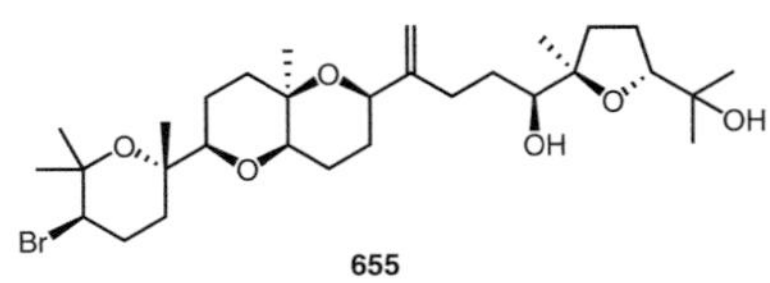

**655**

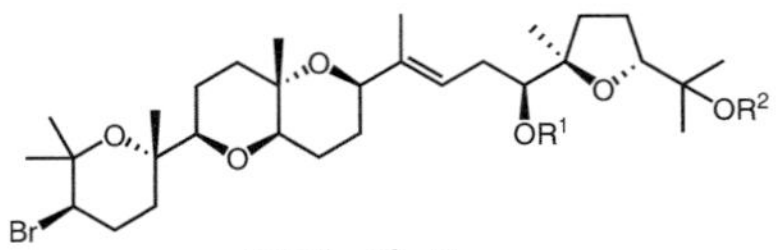

**656** R¹ = R² = H
**657** R¹ = R² = Ac

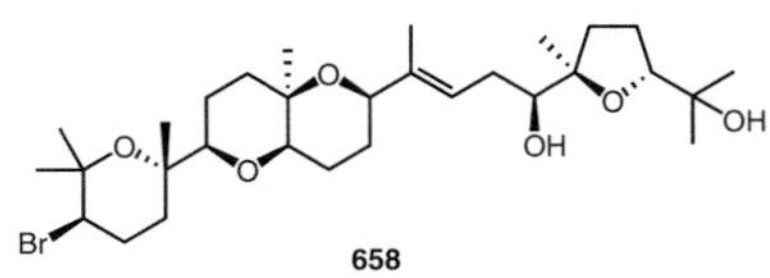

**658**

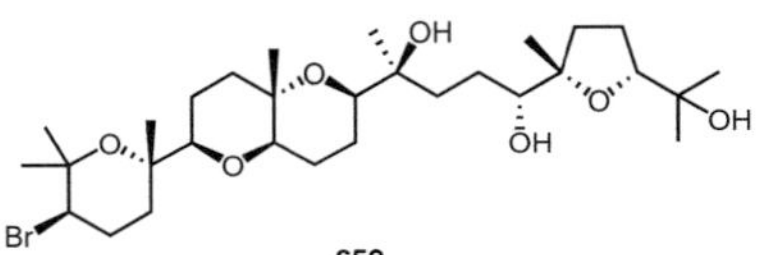

**659**

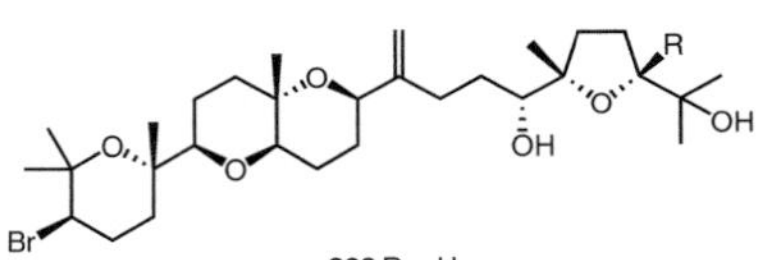

**660** R = H
**661** R = OH

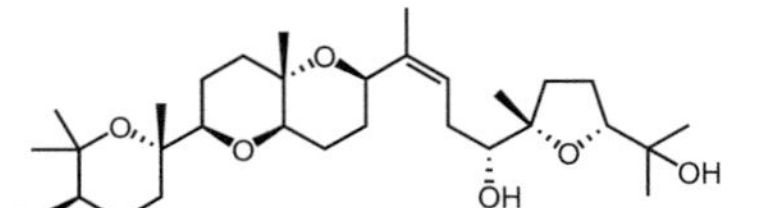

**662**

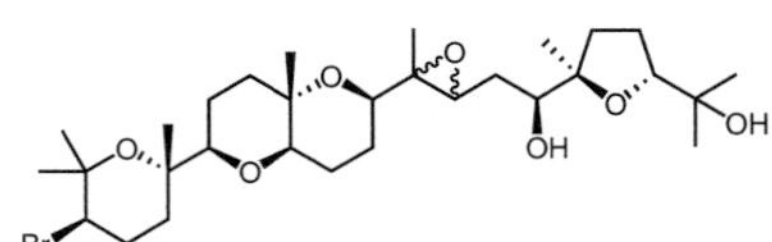

**663** (15*S**,16*R**)
**664** (15*R**,16*S**)

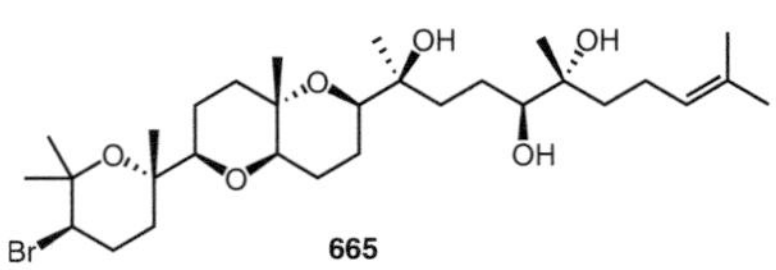

**665**

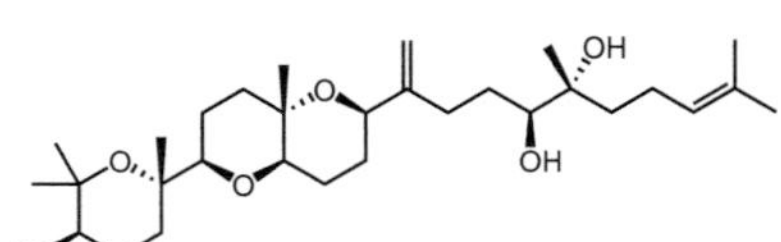

**666**

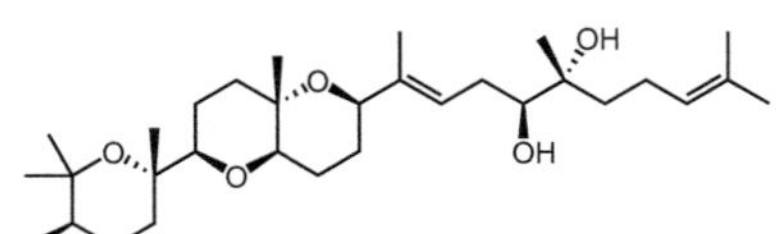

**667**

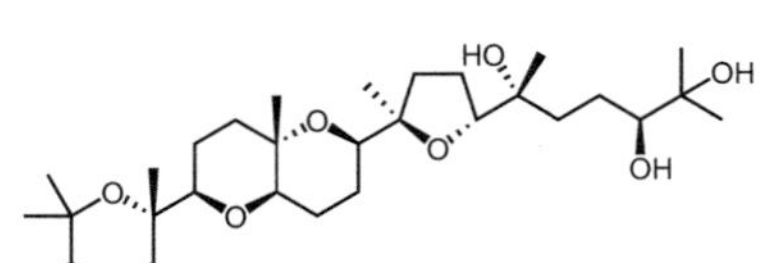

**668**

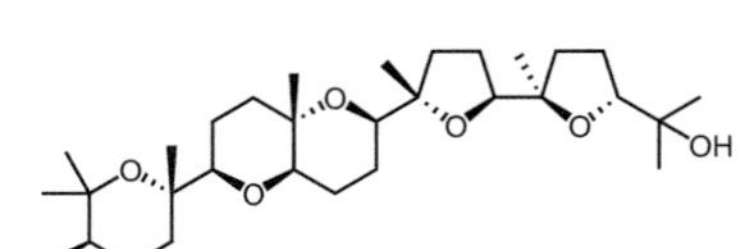

**669**

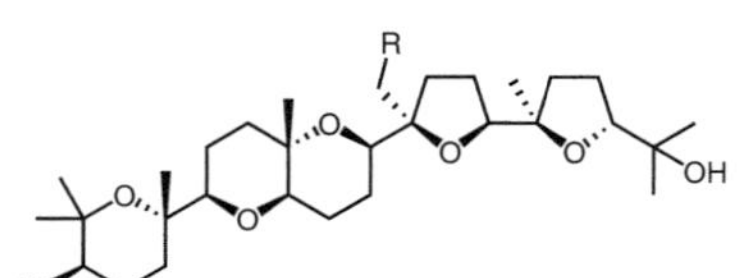

**670** R = H
**671** R = OH

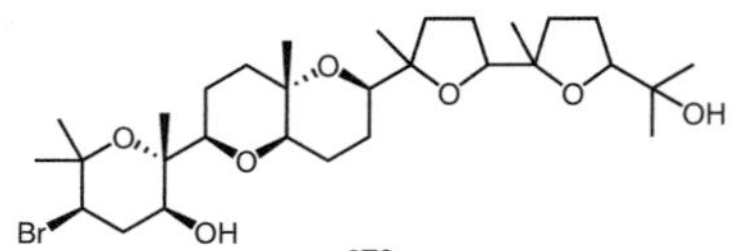

**672**

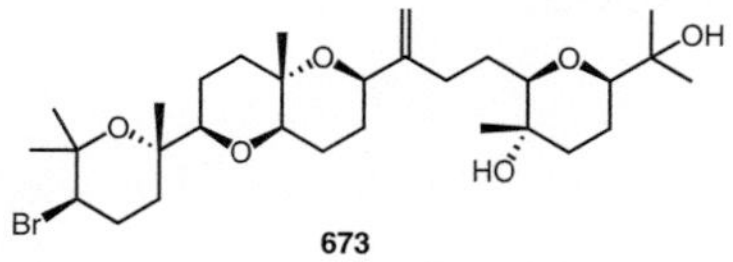

**673**

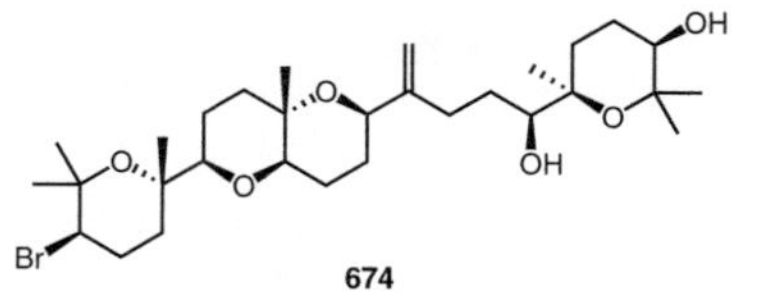

**674**

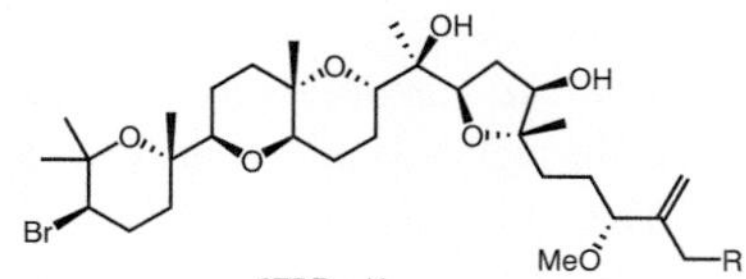

**675** R = H
**676** R = OH

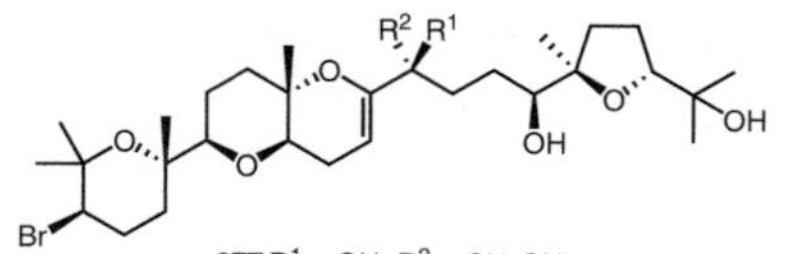

**677** $R^1$ = OH, $R^2$ = $CH_2OH$
**678** $R^1$ = $CH_2OH$, $R^2$ = OH
**679** $R^1$ = H, $R^2$ = $CH_2OH$

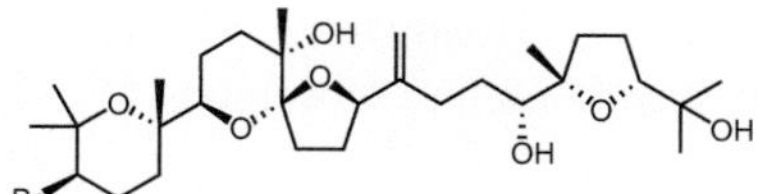

**680**

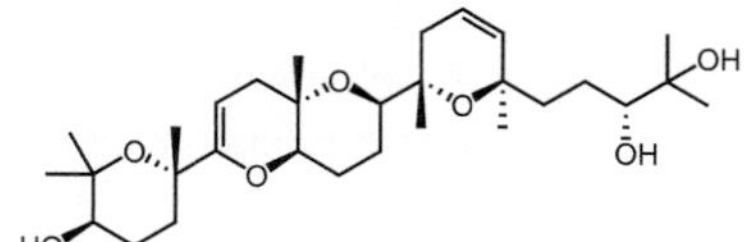

**681**

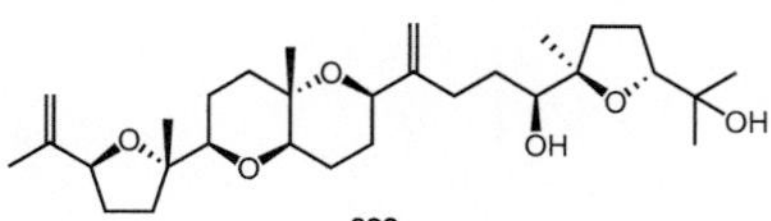

**682**

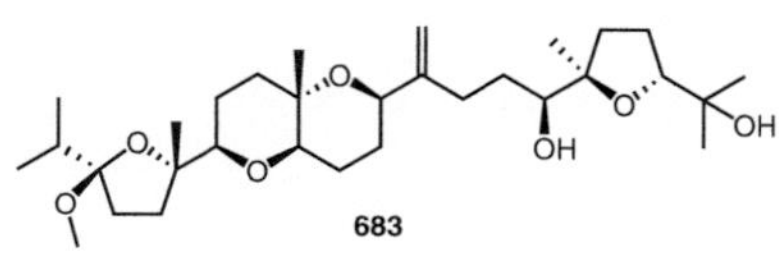

**683**

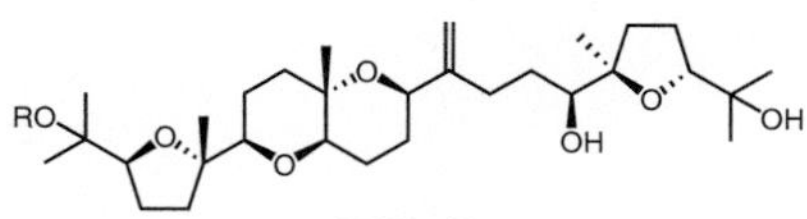

**684** R = Me
**685** R = H

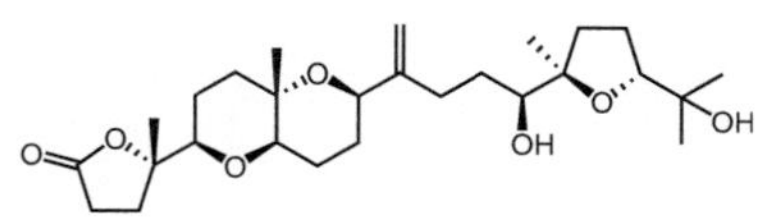

**686**

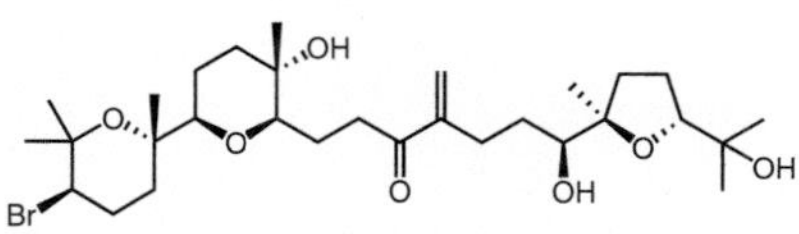

**687**

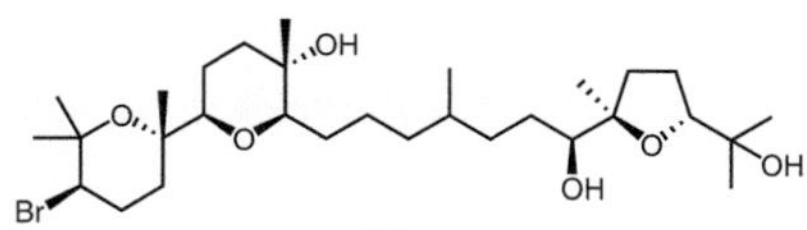

**688**

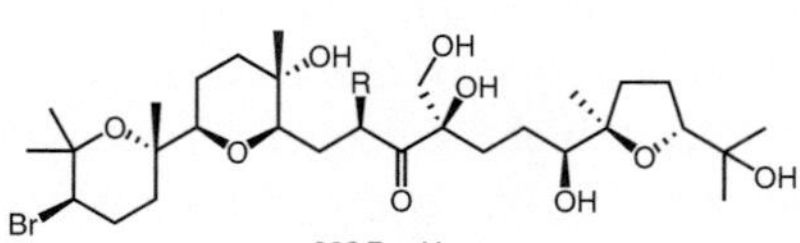

**689** R = H
**690** R = OH

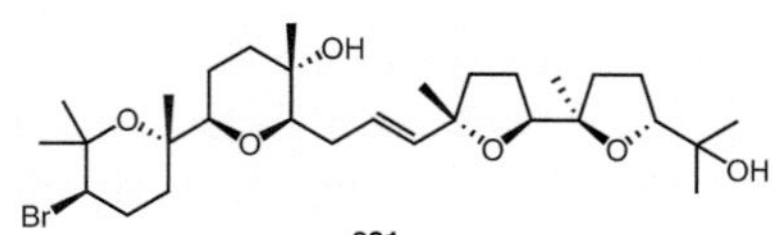

**691**

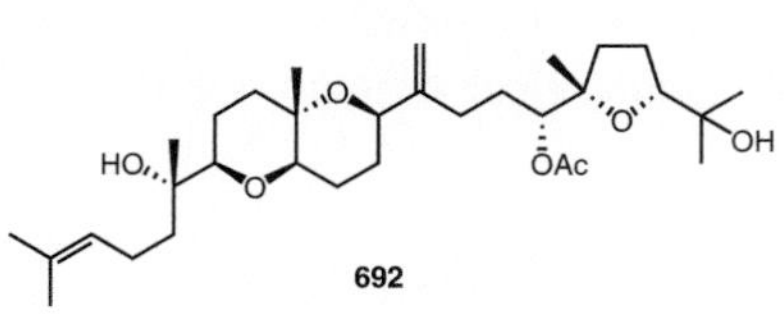

**692**

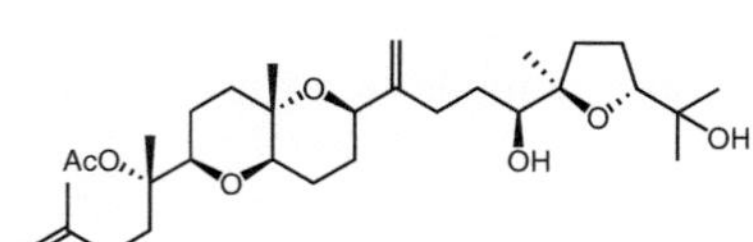

**693**

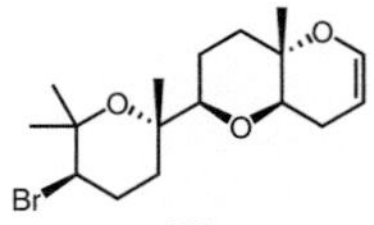

**694**

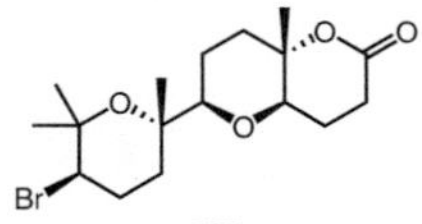

**695**

### 3.3.2 Triterpenes Possessing a 2,8-Dioxabicyclo[5.4.0]undecane Ring System

Up to the present, only three triterpenes possessing a 2,8-dioxabicyclo[5.4.0]undecane unit (**696–698**) have been identified in red algae of the genus *Laurencia* or in gastropods feeding on them. In aurilol (**696**), isolated from the sea hare *Dolabella auricularia* [574], and enshuol (**697**), obtained from *Laurencia omaezakiana* [575], the 2,8-dioxabicyclo[5.4.0]undecane unit is connected through a single bond to two and three isolated tetrahydrofuran rings, respectively. The absolute stereochemistry of **697** was established by total synthesis [575, 576]. Dioxepandehydrothyrsiferol (**698**), isolated from *L. viridis* collected in Macaronesia, Canary islands [558], features a second oxepane ring that is fused to the 2,8-dioxabicyclo[5.4.0]undecane core to yield a 7,7,6-fused ring system, which further connects to a tetrahydropyran ring through single bonds. All three triterpenes share a $\beta$-oriented bromine substitution at C-3 and a hydroxy group at C-23.

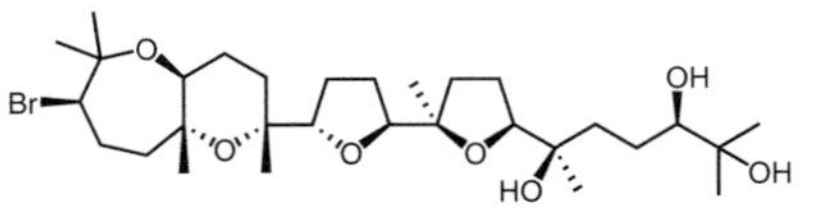


**696**

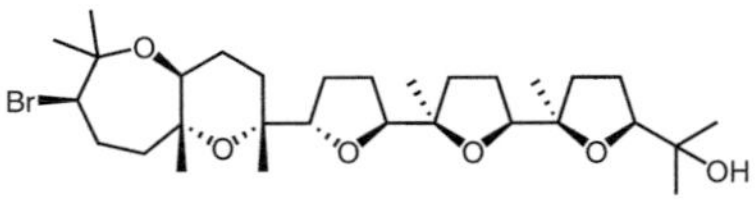


**697**

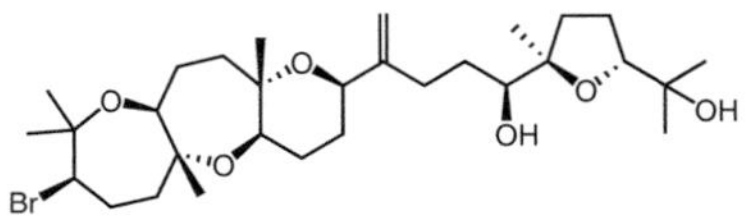


**698**

### 3.3.3 Triterpenes Possessing Symmetry Elements

Only two triterpenes possessing symmetric elements (**699** and **700**) have been isolated so far from *Laurencia* species or their grazers. In particular, the non-halogenated teurilene (**699**), isolated from *Laurencia saitoi* collected from Teuri Island, Hokkaido, Japan, in spite of having eight stereogenic centers and three tetrahydrofuran rings in its structure, has a mirror symmetry ($C_s$) (*meso* form), thus making it achiral [544]. The bromochlorinated intricatetraol (**700**), initially obtained from *Laurencia intricata* collected at different sites near Hokkaido, Japan, is the only example of a triterpene with a $C_2$ symmetrical structure [583].

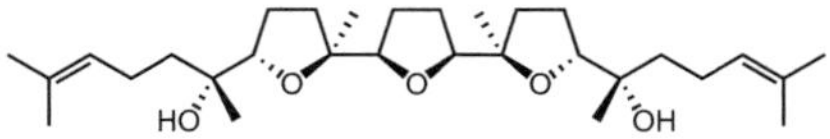


**699**

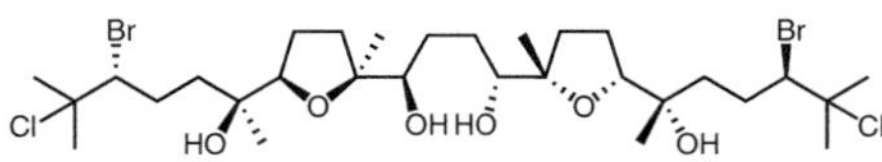


**700**

### 3.3.4 Miscellaneous Triterpenes

There are four triterpenes (**701–704**) reported so far from *Laurencia* species or mollusks feeding on them that cannot be classified into the three subgroups discussed above. Squalene (**701**) and (+)-(10*R*,11*R*)-squalene-10,11-epoxide (**702**) are considered precursors in the biogenesis of all other triterpenes. The structure of auriculol (**703**), a linear non-halogenated triterpene isolated from a Japanese *D. auricularia* specimen, was established by analysis of its spectroscopic data and from organic synthesis [589]. *L. omaezakiana* collected from Omaezaki, Shizuoka Prefecture, Japan, yielded omaezakianol (**704**) consisting of four tetrahydrofuran rings and its total synthesis was achieved from squalene via a biomimetic epoxide-opening cascade reaction in six steps [561, 590].

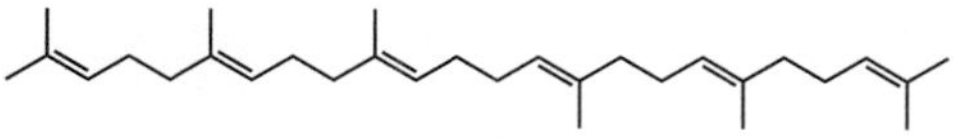

**701**

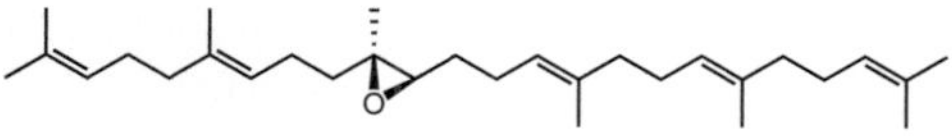


**702**

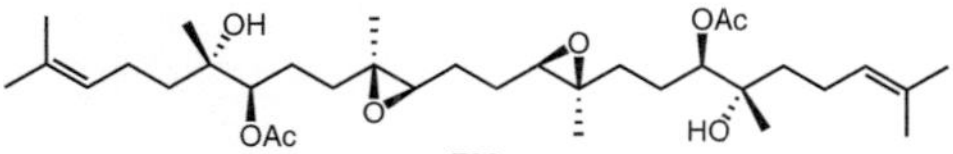


**703**

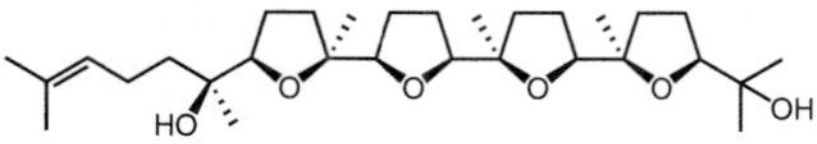


**704**

## 3.4 $C_{15}$ Acetogenins

Red algae of the genus *Laurencia* are a unique source of non-terpenoid $C_{15}$ acetogenins, polyketides arising in all probability from a $C_{16}$ fatty acid. Up to now chemical investigations on species of the genus *Laurencia*, as well as mollusks grazing on them have afforded 244 acetogenins and related metabolites (**705–948**, Table 4). In general, these are brominated, quite frequently bromochlorinated or dibrominated, while it is not uncommon to contain three or more halogen atoms, and only seldom do they lack halogen atoms. These acetogenins are usually characterized by a conjugated (*E*)- or (*Z*)-enyne or bromoallene terminus, and, with the exception of a few linear acetogenins, most of them contain at least one cyclic ether of variable ring size. The latter can be further classified into groups according to the size of the largest ether ring present, namely, acetogenins containing five-, six-, seven-, eight-, nine-, ten-, or twelve-membered rings.

**Table 4** $C_{15}$ Acetogenins from *Laurencia* spp.

| No. | Name | Source of Isolation/Synthesis |
|---|---|---|
| **705** | Graciosallene | *L. obtusa* [592[a,d], 593] |
| **706** | Graciosin | *L. obtusa* [318[a], 592[b,c,d], 593] |
| **707** | Itomanallene B | *L. intricata* [422[a]], *L. nangii* [94, 594] |
| **708** | Dihydroitomanallene B | *L. nangii* [594] |
| **709** | Omaezallene | *Laurencia* sp. [584[d,g]] |
| **710** | – | *Laurencia* sp. [584[d,g]] |
| **711** | – | *Laurencia* sp. [584] |
| **712** | (*E*)-Deacetylkumausyne | *L. nipponica* [595[d]]; Synthesis [596[g]] |
| **713** | (*Z*)-Deacetylkumausyne | *L. nipponica* [595[d]] |
| **714** | (*E*)-Kamausyne | *L. nipponica* [111, 595[a]]; Synthesis [597[f], 598[d,g], 599[e], 600[g], 601[h], 602[h], 603[h], 604[f]] |
| **715** | (*Z*)-Kamausyne | *L. nipponica* [595[d]] |
| **716** | 12-Acetoxy-marilzafurenyne | *L. marilzae* [605] |
| **717** | Marilzafurollene A | *L. marilzae* [605] |
| **718** | Marilzafurollene B | *L. marilzae* [605] |
| **719** | Marilzafurollene C | *L. marilzae* [605] |
| **720** | Marilzafurollene D | *L. marilzae* [605] |
| **721** | Laureoxolane | *L. nipponica* [606] |
| **722** | – | *L. glandulifera* [607] |
| **723** | – | *L. glandulifera* [607] |
| **724** | – | *L. glandulifera* [607] |
| **725** | – | *L. glandulifera* [607] |
| **726** | – | *L. glandulifera* [607] |
| **727** | Bisezakyne A | *Laurencia* sp. [608] |
| **728** | Laureepoxide | *L. nipponica* [418, 606, 609[a,d]] |
| **729** | Hurgadenyne | *L. obtusa* [145] |
| **730** | – | *L. obtusa* [593[d]] |
| **731** | – | *L. obtusa* [610] |
| **732** | Notoryne | *L. nipponica* [138[d]] |
| **733** | Laurendecumbenyne B | *L. decumbens* [611[a], 612[b]]; Synthesis [613[g]] |
| **734** | (3*E*)-Elatenyne | *L. majuscula* [614]; Synthesis [613[g]] |
| **735** | Elatenyne | *L. decumbens* [611], *L. elata* [196[a], 221, 615, 617, 618[b]]; Synthesis [613[g], 616[b]] |
| **736** | Laurefurenyne B | *Laurencia* sp. [619]; Synthesis [620, 621[b]] |
| **737** | Laurefurenyne A | *Laurencia* sp. [619]; Synthesis [620, 621[b,d,g]] |
| **738** | – | *L. majuscula* [416, 615]; Synthesis [616[b]] |
| **739** | Kumausallene | *L. nipponica* [111, 622[a,c,d]]; Synthesis [623[f], 624[g,h], 625[g], 626[g], 627[g,h]] |
| **740** | Aplysiallene | *L. okamurai* [192[a]], *Laurencia* sp. [105, 584], *A. kurodai* [308, 628]; Synthesis [629[b,d]] |
| **741** | Laurenmariallene | *L. mariannensis* [74] |
| **742** | Okamurallene | *L. okamurai* [69, 630[a], 631[b], 632[d]] |
| **743** | Deoxyokamurallene | *L. okamurai* [631[b], 633[a]] |
| **744** | Isookamurallene | *L. okamurai* [631[b], 633[a]] |

(continued)

**Table 4** (continued)

| No. | Name | Source of Isolation/Synthesis |
|---|---|---|
| **745** | – | *L. okamurai* [631[d]] |
| **746** | Panacene | *A. brasiliana* [634]; Synthesis [635[f], 636[b,f], 637[f]] |
| **747** | Obtusin | *L. obtusa* [318, 389, 593, 638[a,c,d]] |
| **748** | Neoobtusin | *L. obtusa* [639] |
| **749** | Chinzallene | *Laurencia* sp. [105] |
| **750** | Laurenenyne A | *L. caduciramulosa* [195], *L. japonensis* [92[a,d], 93]; Synthesis [640[e]] |
| **751** | Laurenenyne B | *L. caduciramulosa* [195], *L. japonensis* [92[a,d], 93]; Synthesis [640[e]] |
| **752** | Laurenidificin | *L. nidifica* [641] |
| **753** | (*E*)-Ocellenyne | *A. oculifera* [642] |
| **754** | (*Z*)-Ocellenyne | *A. oculifera* [642] |
| **755** | – | *L. majuscula* [416] |
| **756** | – | *L. majuscula* [416] |
| **757** | – | *L. majuscula* [416] |
| **758** | (*Z*)-Maneonene A | *L. nidifica* [643[a], 644], *L. obtusa* [645]; Synthesis [646[b,f]] |
| **759** | (12*E*)-Lembyne A | *L. mariannensis* [62[a]], *Laurencia* sp. [118], *A. dactylomela* [118] |
| **760** | (*Z*)-Maneonene C | *L. nidifica* [644, 647[a]], *A. dactylomela* [182] |
| **761** | (*Z*)-Maneonene E | *L. obtusa* [645[a]], *L. papillosa* [101] |
| **762** | (*Z*)-Maneonene B | *L. nidifica* [643[a], 644]; Synthesis [646[b,f]] |
| **763** | Lembyne A | *Laurencia* sp. [118, 165[a]], *A. dactylomela* [118] |
| **764** | (*Z*)-Maneonene D | *L. obtusa* [645] |
| **765** | (*E*)-Maneonene B | *L. nidifica* [643[a], 644], *L. papillosa* [101]; Synthesis [648[b,f]] |
| **766** | (*E*)-Maneonene C | *L. obtusa* [645] |
| **767** | Isomaneonene A | *L. nidifica* [644, 647[a]] |
| **768** | Isomaneonene B | *L. nidifica* [644, 647[a]] |
| **769** | Lembyne B | *Laurencia* sp. [165] |
| **770** | Isodactylyne | *A. dactylomela* [649]; Synthesis [650[d,g], 651[g]] |
| **771** | Dactylyne | *L. lageniformis* [310], *Laurencia* sp. [270, 608], *A. dactylomela* [652[a,c,d]]; Synthesis [650[g], 651[g]] |
| **772** | Scanlonenyne | *L. obtusa* [349]; Synthesis [653[d,g]] |
| **773** | Srilankenyne | *A. oculifera* [654] |
| **774** | Bisezakyne B | *Laurencia* sp. [608] |
| **775** | – | *L. obtusa* [593[d]] |
| **776** | – | *L. obtusa* [411], *L. paniculata* [610[c,d]] |
| **777** | – | *L. paniculata* [610] |
| **778** | Japonenyne A | *L. japonensis* [655] |
| **779** | Japonenyne B | *L. japonensis* [655] |
| **780** | Japonenyne C | *L. japonensis* [655] |
| **781** | Laurobtusin | *L. obtusa* [656] |
| **782** | (3*E*)-Dactylomelyne | *L. obtusa* [413], *A. dactylomela* [372[a]]; Synthesis [657[e], 658[f], 659[f]] |

(continued)

**Table 4** (continued)

| No. | Name | Source of Isolation/Synthesis |
|---|---|---|
| **783** | (3*Z*)-Dactylomelyne | *A. dactylomela* [372]; Synthesis [657[e], 658[f], 659[f]] |
| **784** | Isolaurepinnacin | *L. pinnata* [660]; Synthesis [661[e], 662[d,g], 663[g], 664[g,h], 665[g,h], 666[g,h]] |
| **785** | Rogioloxepane A | *L. microcladia* [667]; Synthesis [666[g,h], 668[d,g], 669[g]] |
| **786** | Rogioloxepane B | *L. microcladia* [667[d]] |
| **787** | Rogioloxepane C | *L. microcladia* [667[d]] |
| **788** | Neonipponallene | *L. nipponica* [670[d]] |
| **789** | Rogiolenyne D | *L. microcladia* [671[d]] |
| **790** | Rogiolenyne A | *L. microcladia* [671[d], 672] |
| **791** | Rogiolenyne B | *L. microcladia* [671[d], 672], *L. obtusa* [413] |
| **792** | Rogiolenyne C | *S. zimocca* in the vicinity of *L. microcladia* [163] |
| **793** | (3*E*)-Isoprelaurefucin | *L. nipponica* [673[a], 674[d]], *L. subopposita* [248]; Synthesis [675[g]] |
| **794** | (3*Z*)-Isoprelaurefucin | *L. subopposita* [248[a]], *L. nipponica* [199, 674[d]] |
| **795** | (3*E*)-Neoisoprelaurefucin | *L. obtusa* [413[a]], *L. okamurai* [100, 109] |
| **796** | Neoisoprelaurefucin | *L. nipponica* [676]; Synthesis [677[d,g]] |
| **797** | Laurenyne | *L. chondrioides* [73], *L. obtusa* [678[a,c], 679], *A. fasciata* [326], *A. punctata* [249]; Synthesis [680[b,d,g], 681[g], 682[e]] |
| **798** | (3*Z*)-Laurenyne | *L. nangii* [94, 310], *L. yonaguinensis* [513[a]], *A. dactylomela* [407] |
| **799** | (3*Z*)-Venustinene | *L. venusta* [683[a]], *A. fasciata* [326] |
| **800** | (*E*)-Rhodophytin | *Laurencia* sp. [684] |
| **801** | (*Z*)-Rhodophytin | *Laurencia* sp. [30, 684[b], 685[a]] |
| **802** | (*E*)-Chondriol | *Laurencia* sp. [684[d]] |
| **803** | (*Z*)-Chondriol | *Laurencia* sp. [30, 684[d], 685, 686[a], 687[b,c]] |
| **804** | (3*E*)-Venustin (Venustin B) | *L. venusta* [688] |
| **805** | (3*Z*)-Venustin | *L. venusta* [683] |
| **806** | Epoxy-rhodophytin | *Laurencia* sp. [684[c,d]] |
| **807** | (3*E*)-Epoxy-venustin (Venustin A) | *L. venusta* [683[d], 688] |
| **808** | (3*Z*)-Epoxy-venustin | *L. venusta* [683[c,d]] |
| **809** | (*E*)-Dihydrorhodophytin | *L. pinnatifida* [689, 690[a,d]] |
| **810** | (*Z*)-Dihydrorhodophytin | *L. filiformis* [420], *L. nangii* [94, 691], *L. pinnatifida* [689, 690[d]], *A. brasiliana* [692[a,c,d]] |
| **811** | Prelaureatin | *L. nipponica* [693[a]]; Synthesis [694–698[d,g], 699[g], 700[h], 701[e]] |
| **812** | (3*E*,12*R*,13*R*)-Pinnatifidenyne | *L. obtusa* [407[b,c,d], 702[a]] |
| ***ent*-812** | (3*E*,6*R*,7*R*)-Pinnatifidenyne | *A. dactylomela* [407[c,d]] |
| **813** | (*Z*)-*epi*-Dihydrorhodophytin | *L. filiformis* [420] |
| **814** | (3*E*)-Pinnatifidenyne | *L. chondrioides* [73], *L. obtusa* [593], *L. pinnatifida* [689[b,c,d], 703[a,c]], *A. dactylomela* [407]; Synthesis [704[g], 705[f,h]] |
| **815** | (3*Z*)-Pinnatifidenyne | *L. pinnatifida* [272, 689[b,c,d], 703[a,c]]; Synthesis [704[g], 705[f,h]] |

(continued)

**Table 4** (continued)

| No. | Name | Source of Isolation/Synthesis |
|---|---|---|
| ***ent*-815** | – | *L. nangii* [94, 118, 310[a,d]], *A. dactylomela* [118] |
| **816** | (3*Z*)-13-*epi*-Pinnatifidenyne | *L. claviformis* [107[a]], *A. fasciata* [326] |
| **817** | (*Z*)-Isodihydrorhodophytin | *L. obtusa* [435], *A. brasiliana* [692[a]] |
| **818** | (*E*)-Epoxy-isodihydrorhodophytin | *L. obtusa* [706[c,d]] |
| **819** | Laurencienyne | *L. obtusa* [321, 404, 411, 413, 707[a,c,d], 708] |
| **820** | Laurencienyne B | *L. obtusa* [411] |
| **821** | (3*E*)-13-Epilaurencienyne | *L. obtusa* [321[a], 702] |
| **822** | (3*Z*)-13-Epilaurencienyne | *L. obtusa* [702] |
| **823** | – | *L. obtusa* [709] |
| **824** | – | *L. obtusa* [702] |
| **825** | – | *L. obtusa* [702] |
| **826** | – | *Laurencia* sp. cf. *L. gracilis* [103] |
| **827** | – | *Laurencia* sp. cf. *L. gracilis* [103] |
| **828** | – | *Laurencia* sp. cf. *L. gracilis* [103] |
| **829** | Doliculol A | *D. auricularia* [710] |
| **830** | Doliculol B | *D. auricularia* [710] |
| **831** | Marilzallene | *L. marilzae* [711[d]] |
| **832** | (+)-4-Acetoxy-marilzallene | *L. marilzae* [711[c,d]] |
| **833** | Marilzallene B | *L. chondrioides* [73] |
| **834** | (–)-4-Acetoxy-marilzallene | *L. marilzae* [711] |
| **835** | Deacetyllaurencin | *L. nipponica* [609[a,d]], *Laurencia* sp. cf. *L. gracilis* [103]; Synthesis [694] |
| **836** | Laurencin | *L. glandulifera* [712[a], 713, 714[c,d]], *L. nipponica* [418, 606]; Synthesis [664[g], 696, 701[e], 715[g], 716[g], 717[g], 718[g], 719[g], 720[g], 721[g], 722[g,h], 723[g,h]] |
| **837** | Intricenyne | *L. intricata* [724[a]], *L. tryrsifera* [725[d]] |
| **838** | Laurepinnacin | *L. pinnata* [660]; Synthesis [661[e]] |
| **839** | Bermudenynol | *L. intricata* [726[c]]; Synthesis [727[d,g]] |
| **840** | Bermudenynol acetate | *L. intricata* [726[c]]; Synthesis [727[d,g]] |
| **841** | – | *L. tryrsifera* [725] |
| **842** | – | *L. glandulifera* [728] |
| **843** | – | *L. tryrsifera* [729] |
| **844** | – | *L. tryrsifera* [729[a]], *L. glandulifera* [728] |
| **845** | – | *L. glandulifera* [728] |
| **846** | – | *L. glandulifera* [728] |
| **847** | – | *L. glandulifera* [728] |
| **848** | – | *L. glandulifera* [728] |
| **849** | Poiteol | *L. poitei* [730[c,d]] |
| **850** | – | *L. tryrsifera* [725] |
| **851** | – | *L. tryrsifera* [725] |
| **852** | Laurallene | *L. nipponica* [77[b], 731[a]]; Synthesis [698[d,g], 700[g,h], 732, 733[g], 734[g]] |
| **853** | Pannosallene | *L. nangii* [94, 594], *L. pannosa* [735[a]]; Synthesis [734[b,d,g]] |

(continued)

**Table 4** (continued)

| No. | Name | Source of Isolation/Synthesis |
|---|---|---|
| **854** | Nipponallene | *L. nipponica* [670[d]] |
| ***ent*-854** | Epilaurallene | *L. nipponica* [77[a], 735[b], 736], *L. okamurai* [152] |
| **855** | – | *L. intricata* [631[a]], *Laurencia* sp. [105, 193] |
| **856** | – | *L. implicata* [91[b], 415[a]] |
| **857** | Laurendecumallene B | *L. decumbens* [611]; Synthesis [737[e]] |
| **858** | – | *L. implicata* [91] |
| **859** | – | *L. obtusa* [593[d]] |
| **860** | – | *L. obtusa* [318[a], 593[b,d]] |
| **861** | Aplyparvunin | *A. parvula* [291[c,d]] |
| **862** | Desepilaurallene | *L. okamurai* [152] |
| **863** | (*E*)-Laureatin | *L. nipponica* [736, 738[a]] |
| **864** | Laureatin | *L. nipponica* [77, 200, 739[a], 740[b], 741[b,d]], *L. okamurai* [109]; Synthesis [695, 742[g], 743[g], 744[e]] |
| **865** | (*E*)-Isolaureatin | *L. nipponica* [77, 736, 738[a]] |
| **866** | Isolaureatin | *L. nipponica* [77, 200, 740[b], 741[b,c,d], 745[a]]; Synthesis [743[g]] |
| **867** | (*Z*)-Chondrin | *Laurencia* sp. [684[d]] |
| **868** | (*Z*)-Chondrin | *Laurencia* sp. [684[d]] |
| **869** | Okamuragenin | *L. okamurai* [109] |
| **870** | (3*E*)-Laurefucin | *L. nipponica* [609, 746[a], 747[b,c,d]], *L. subopposita* [248], *Laurencia* sp. cf. *L. gracilis* [103]; Synthesis [705[f,h]] |
| **871** | (3*Z*)-Laurefucin | *L. nipponica* [138], *L. subopposita* [248[a]] |
| **872** | (3*E*)-Acetyllaurefucin | *L. nipponica* [609, 746[a], 747[b,d]], *L. subopposita* [248] |
| **873** | (3*Z*)-Acetyllaurefucin | *L. nipponica* [138], *L. subopposita* [248[a]] |
| **874** | Chlorofucin | *L. pannosa* [735], *L. snyderae* [730[a,c]], *Laurencia* sp. [193] |
| **875** | (3*Z*)-Chlorofucin | *L. elata* [221], *L. pannosa* [238[a]] |
| **876** | Bromofucin | *L. implicata* [415[a]], *L. pannosa* [735] |
| **877** | (3*Z*)-Bromofucin | *A. parvula* [748] |
| **878** | Laurefurenyne D | *Laurencia* sp. [619] |
| **879** | Laurefurenyne C | *Laurencia* sp. [619] |
| **880** | Laurefurenyne F | *Laurencia* sp. [619] |
| **881** | Laurefurenyne E | *Laurencia* sp. [619] |
| **882** | (3*E*)-Dehydrobromolaurefucin | *L. subopposita* [248[a]], *Laurencia* sp. cf. *L. gracilis* [103[b]]; Synthesis [705[e]] |
| **883** | (3*Z*)-Dehydrobromolaurefucin | *L. subopposita* [248] |
| **884** | (3*E*)-5-Hydroxy-dehydrobromolaurefucin | *Laurencia* sp. cf. *L. gracilis* [103] |
| **885** | Laurendecumenyne A | *L. decumbens* [611] |
| **886** | Laureoxanyne | *L. nipponica* [749] |
| **887** | Microcladallene A | *L. obtusa* [750[a]], *Laurencia* sp. [106] |

(continued)

**Table 4** (continued)

| No. | Name | Source of Isolation/Synthesis |
|---|---|---|
| **888** | Microcladallene C | *L. obtusa* [750[a]], *L. majuscula* [122] |
| **889** | Microcladallene B | *L. obtusa* [750]; Synthesis [751[b,d]] |
| **890** | – | *L. implicata* [415] |
| **891** | Brasilenyne | *A. brasiliana* [692[c,d]]; Synthesis [752[g], 753[g]] |
| **892** | Obtusenyne | *L. nangii* [310], *L. obtusa* [730, 754[a,c,d]]; Synthesis [755[e], 756[e], 757[g], 758[g], 759[g], 760[g]] |
| **893** | (+)-(3*E*)-12-Epiobtusenyne | *A. dactylomela* [372[c,d]] |
| **894** | (+)-(3*Z*)-12-Epiobtusenyne | *L. lageniformis* [310], *A. dactylomela* [372[a,d]] |
| **895** | (–)-(3*E*,12*R*,13*R*)-Obtusenyne | *L. chondrioides* [73], *L. pinnatifida* [689]; Synthesis [761[b,d]] |
| ***ent*-895** | (+)-(3*E*,6*R*,7*R*)-Obtusenyne | *A. dactylomela* [407[c,d]] |
| **896** | (–)-(3*Z*,12*R*,13*R*)-Obtusenyne | *L. pinnatifida* [689]; Synthesis [761[b,d]] |
| ***ent*-896** | (+)-(3*Z*,6*R*,7*R*)-Obtusenyne | *L. nangii* [94], *A. dactylomela* [407[a,d]] |
| **897** | – | *L. implicata* [415] |
| **898** | Isolaurallene | *L. nipponica* [762[a,c,d], 763]; Synthesis [734[g], 764[g], 765[g]] |
| **899** | Neolaurallene | *L. implicata* [91, 415], *L. intricata* [631, 763, 766[a,c,d]], *L. nangii* [94], *L. okamurai* [109], *L. saitoi* [480]; Synthesis [734[g]] |
| **900** | Itomanallene A | *L. intricata* [422]; Synthesis [734[b,g], 767[b,d,g]] |
| **901** | Laurendecumallene A | *L. decumbens* [611] |
| **902** | – | *L. implicata* [91] |
| **903** | Obtusallene II | *L. chondrioides* [73], *L. obtusa* [679[a,c,d], 769]; Synthesis [768[e]] |
| **904** | Obtusallene VIII | *L. obtusa* [770] |
| **905** | Obtusallene IX | *L. obtusa* [770] |
| **906** | Obtusallene IV (Dactylallene) | *L. marilzae* [711], *L. obtusa* [769[a]], *A. dactylomela* [771[a,c,d]]; Synthesis [768[e]] |
| **907** | 12-Epoxy-obtusallene IV | *L. marilzae* [711] |
| **908** | – | *L. marilzae* [711] |
| **909** | Obtusallene X | *L. marilzae* [711] |
| **910** | ObtusalleneVII | *L. obtusa* [770, 772[b]] |
| **911** | Obtusallene III | *L. chondrioides* [73], *L. obtusa* [679[a,c], 769] |
| **912** | Kasallene | *L. obtusa* [773] |
| **913** | Chondrioallene | *L. chondrioides* [73] |
| **914** | Obtusallene I | *L. obtusa* [679, 769, 774[a], 775[c,d], 776] |
| **915** | 10-Bromo-obtusallene I | *L. obtusa* [769, 776[a]] |
| **916** | ObtusalleneVI | *L. chondrioides* [73], *L. obtusa* [770[a], 772[b]] |
| **917** | Obtusallene V | *L. chondrioides* [73], *L. obtusa* [770[a], 772[b,c]] |
| **918** | Marilzabicycloallene A | *L. marilzae* [777] |
| **919** | Marilzabicycloallene C | *L. marilzae* [777] |
| **920** | Marilzabicycloallene D | *L. marilzae* [777] |
| **921** | Marilzabicycloallene B | *L. marilzae* [777] |
| **922** | Poitediene | *L. poitei* [778] |

(continued)

**Table 4** (continued)

| No. | Name | Source of Isolation/Synthesis |
|---|---|---|
| **923** | (*Z*)-Neolaurencenyne | *L. okamurai* [779[a]]; Synthesis [780[f]] |
| **924** | Neolaurencenyne | *L. composita* [31], *L. okamurai* [192, 204[a,f], 780] |
| **925** | (*Z*)-Laurencenyne | *L. composita* [31, 88], *L. nipponica* [138], *L. okamurai* [779[a]]; Synthesis [780[f], 781[f]] |
| **926** | Laurencenyne | *L. okamurai* [192, 204[a,f], 780], *L. composita* [31, 88], *A. punctata* [249]; Synthesis [781[f]] |
| **927** | (3*E*,6*Z*,9*Z*,12*E*)-Pentadeca-3,6,9,12-tetraen-1-yne | *L. majuscula* [416] |
| **928** | (3*E*,6*Z*,9*Z*,12*E*)-Pentadeca-3,6,9,12,14-pentaen-1-yne | *L. majuscula* [416] |
| **929** | (*E*)-12,13-Dihydrolaurediol | *L. nipponica* [782] |
| **930** | (*Z*)-12,13-Dihydrolaurediol | *L. nipponica* [782] |
| **931** | – | *L. pinnatifida* [689] |
| **932** | – | *L. pinnatifida* [689] |
| **933** | – | *L. pinnatifida* [689] |
| **934** | – | *L. pinnatifida* [689] |
| **935** | (3Z,9Z)-7-Chloro-6-hydroxy-12-oxo-pentadeca-3,9-dien-1-yne | *A. fasciata* [326] |
| **936** | (*Z*)-Adrienyne | *L. marilzae* [711] |
| **937** | (*E*)-Adrienyne | *L. marilzae* [711] |
| **938** | (*E*)-(12*Z*)-Laurediol | *L. nipponica* [782[a]], *L. okamurai* [109] |
| **939** | (*Z*)-(12*Z*)-Laurediol | *L. nipponica* [782] |
| **940** | – | *L. pinnatifida* [689] |
| **941** | (3*Z*,6*R*,7*R*,9*Z*,12*Z*)-6-Acetoxy-7-chloro-pentadeca-3,9,12-trien-1-yne | *L. pinnatifida* [689, 703[a,d]], *A. fasciata* [326]; Synthesis [783[b,g]] |
| **942** | (3*E*)-Laurediol | *L. nipponica* [738[a]], *L. obtusa* [593], *A. dactylomela* [178]; Synthesis [602[g], 784[g], 785[g], 786[g], 787[g]] |
| **943** | (3Z)-Laurediol | *L. nipponica* [738[a]], *L. obtusa* [593]; Synthesis [785[g]] |
| **944** | (3Z)-Acetyllaurediol | *L. nipponica* [138] |
| **945** | (3*E*)-Laurediol diacetate | *L. nipponica* [97, 670, 738[a]]; Synthesis [785[g]] |
| **946** | (3Z)-Laurediol diacetate | *L. nipponica* [138, 738[a]]; Synthesis [785[g]] |
| **947** | Prerogioloxepane | *L. microcladia* [667[d]] |
| **948** | (3Z,9Z,12*E*)-7-Acetoxy-6-chloro-pentadeca-3,9,12-trien-1-yne | *L. glandulifera* [607] |

[a]The isolation and structure elucidation of this compound was reported for the first time
[b]The structure of this compound has been revised
[c]The structure of this compound or of a derivative has been confirmed by X-ray crystallographic analysis
[d]The absolute stereochemistry of this compound has been established
[e]Partial synthesis
[f]Stereoselective total synthesis
[g]Enantioselective total synthesis
[h]Formal synthesis

### 3.4.1 Acetogenins Containing a Five-Membered Cyclic Ether (Tetrahydrofuran) Ring

To date, the group of acetogenins containing five-membered cyclic ethers includes a total of 65 compounds (**705–769**) that have been isolated from *Laurencia* species or sea hares feeding on them. These can be roughly divided into four subgroups, namely: (1) those containing one tetrahydrofuran ring (**705–729**); (2) those containing two isolated tetrahydrofuran rings (**730–738**); (3) those containing two fused tetrahydrofuran rings (**739–757**); and (4) maneonenes and isomaneonenes containing three tetrahydrofuran rings (**758–769**).

Acetogenins containing one tetrahydrofuran ring can feature either a 4,7-epoxy (**705–711**), a 6,9-epoxy (**712–720**), a 7,10-epoxy (**721–726**), a 9,12-epoxy (**727**), or a 10,13-epoxy (**728** and **729**) ring. The 4,7-epoxy acetogenins are characterized by either a bromoallene functionality (**705** and **707–711**) or a bromopropargylic unit (**706**), the 6,9-epoxy acetogenins possess either a (*E*)- or (*Z*)-enyne moiety (**712–716**) or a bromoallene functionality (**717–720**), whereas the 7,10-epoxy, the 9,12-epoxy and the 10,13-epoxy acetogenins exhibit a (*E*)- or (*Z*)-enyne moiety (**721–729**). Acetogenins containing two isolated tetrahydrofuran rings can either possess a bis(tetrahydrofuran-2-yl)methane (**730** and **731**) or a 2,2′-bis-tetrahydrofuran (**732–738**) core. The 2,2′-bis-tetrahydrofuran acetogenins (**732–738**) display a (*E*)- or (*Z*)-enyne terminus, while the bis(tetrahydrofuran-2-yl)methane acetogenins possess either a bromopropargylic (**730**) or a bromoallenic (**731**) unit. It is worth noting that the structures of elatenyne (**735**), isolated from *Laurencia elata* and *Laurencia decumbens* [196, 611], laurendecumbenyne B (**733**), isolated from *L. decumbens* [611], and compound **738**, obtained from *Laurencia majuscula* [416], were originally determined to contain a 2,7-dioxabicyclo[4.4.0]decane core. The structure of elatenyne (**735**) attracted the attention of many research groups, who, utilizing chemical and computational methods, finally succeeded in synthesizing the molecule corresponding to the spectroscopic data of the natural product, and revising the proposed structure to that of a 2,2′-bis-tetrahydrofuran derivative. However, the absolute stereochemistry is still unknown [221, 613, 615–618]. A similar revision was made to the structure of laurendecumbenyne B (**733**), which was established only on the basis of comparison of NMR data with those originally assigned for **735** [612]. Based on the same rationale, the proposed structure for **738** was also revised, albeit not yet confirmed by total synthesis [615]. Acetogenins containing two fused tetrahydrofuran rings can be differentiated into those with either a 2,6-dioxabicyclo[3.3.0] octane (**739–752**) or a 2,5-dioxabicyclo[2.2.1]heptane (**753–757**) substructure. The former can be further classified into 4,7:6,9-bisepoxy acetogenins featuring either a bromoallene functionality (**739–746** and **749**) or a bromopropargylic terminus (**747** and **748**) and to 7,10:9,12-bisepoxy acetogenins possessing a (*E*)- or (*Z*)-enyne moiety (**750–752**). All five 2,5-dioxabicyclo[2.2.1]heptane acetogenins exhibit a (*E*)- or (*Z*)-enyne terminus (**753–757**). Compounds **747–749** contain an additional tetrahydrofuran ring derived from etherification between C-9 and C-12, thus forming a bicyclic ketal motif. Maneonenes (**758–766**) and isomaneonenes (**767–769**) contain three tetrahydrofuran rings and all feature a (*E*)- or (*Z*)-enyne terminus. However, their structures are unique among all reported $C_{15}$ acetogenins in that the carbon chain has cyclized back on itself to create a monocarbocyclic (through the formation of bond between C-6 and C-11) and a bicarbocyclic (through the formation of bonds between C-5 and C-13, as well as between C-6 and C-11) framework, respectively. Maneonenes bear a chlorine atom at C-5 and a bromine atom at C-13, whereas isomaneonenes bear two bromine atoms at C-12 and C-13. The first members of these series, namely, (*Z*)-maneonenes A (**758**), B (**762**) and C (**760**), (*E*)-maneonene B (**765**), and isomaneonenes A (**767**) and B (**768**), were originally reported from a Hawaiian *Laurencia nidifica* population by the group of Erickson in 1976 [643, 644, 647]. The total synthesis of racemic **758**, **762** and **765** by Holmes et al. contributed to the structure elucidation of the series [646, 648]. In 2011, (12*E*)-(*Z*)-maneonene E (**761**) was reported twice by the same group from *L. papillosa* [101] and *L. obtusa* [645], both collected off the Saudi Arabia Red Sea coast at Jeddah, depicting though a different configuration at C-6 in the two reports. Moreover, in their second report, the authors also describe the isolation of (12*Z*)-*cis*-maneonene D (**764**) and (12*Z*)-(*Z*)-maneonene C (**766**), along with that of the known (*Z*)-maneonene A, for which, however, they incorrectly depicted the structure of the 6-*endo* isomer that is actually (*Z*)-maneonene C [645]. In agreement with the comment by Blunt and colleagues in 2013 [9] concerning the proposed structures, one has to admit that they are not very credible. Specifically, (12*Z*)-(*Z*)-maneonene D should have the same structure as either lembyne A (**763**) or (12*Z*)-(*Z*)-maneonene C, the semisynthetic product obtained after treatment of (*Z*)-maneonene C (**760**) with *p*-toluenesulfonic acid, which are thought to be C-5 epimers. Nonetheless, the NMR data of (12*Z*)-(*Z*)-maneonene D do not match either. In a similar manner, the structure depicted for (12*E*)-(*Z*)-maneonene E was, in fact, that of the known *cis*-maneonene A (**758**). Should (12*E*)-(*Z*)-maneonene E be the 6-*endo* isomer, then its NMR data do not match those reported for the C-5 epimers (12*E*)-lembyne (**759**) or *cis*-maneonene C (**760**).

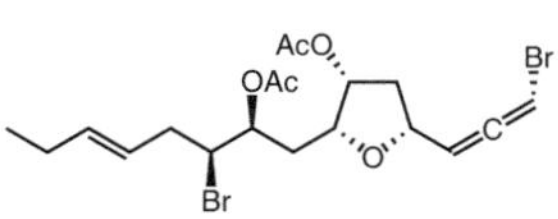

**705**

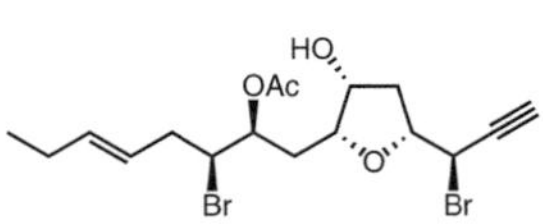

**706**

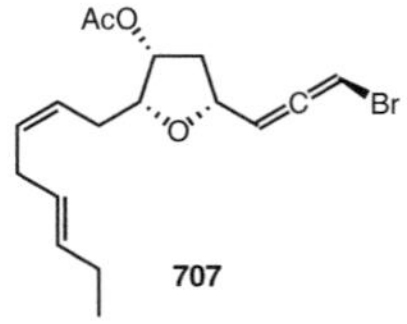

**707**

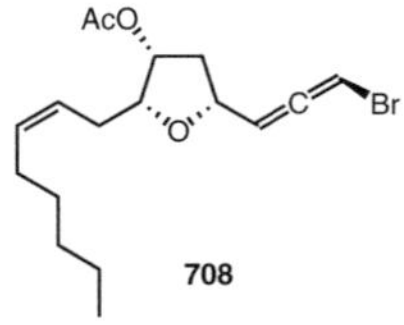

**708**

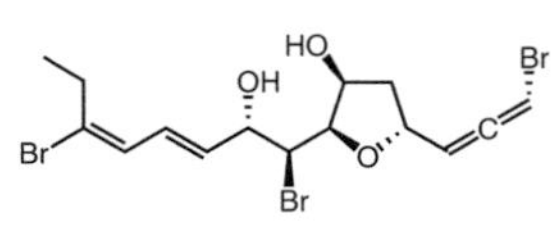

**709**

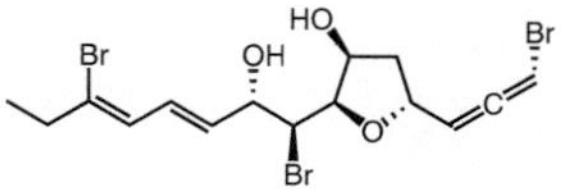

**710**

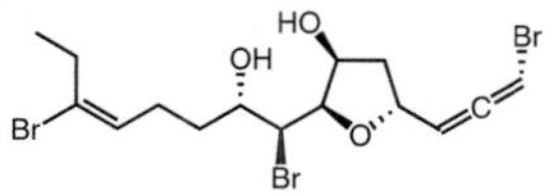

**711**

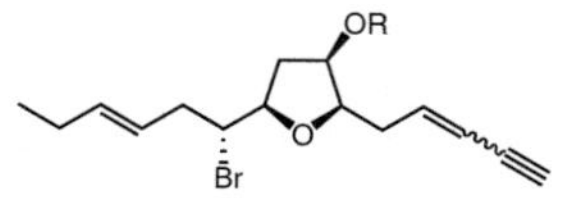

**712** (3*E*), R = H
**713** (3*Z*), R = H
**714** (3*E*), R = Ac
**715** (3*Z*), R = Ac

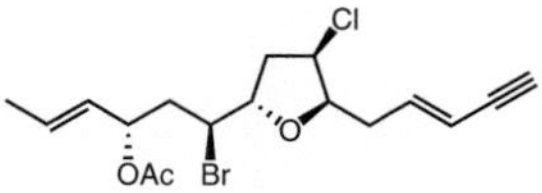

**716**

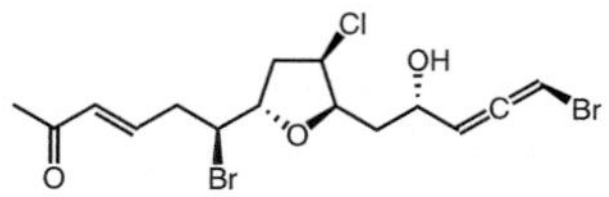

**717**

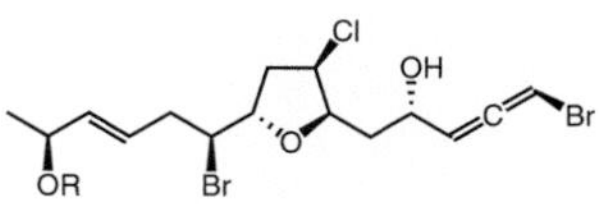

**718** R = H
**719** R = Me

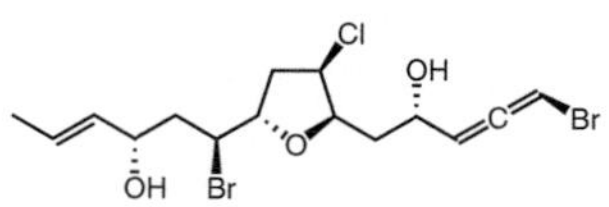

**720**

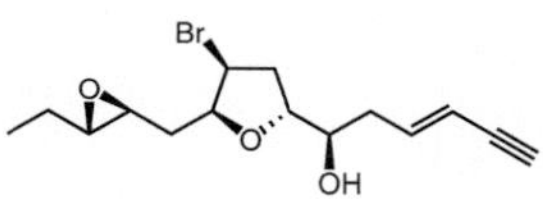

**721**

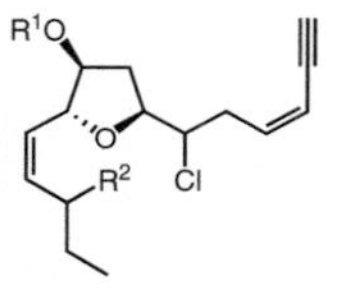

**722** $R^1$ = Ac, $R^2$ = Cl
**723** $R^1$ = H, $R^2$ = OH

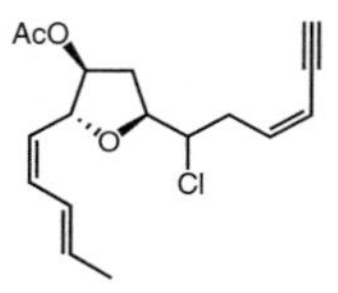

**724**

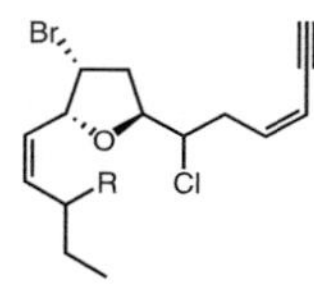

**725** R = Br
**726** R = OMe

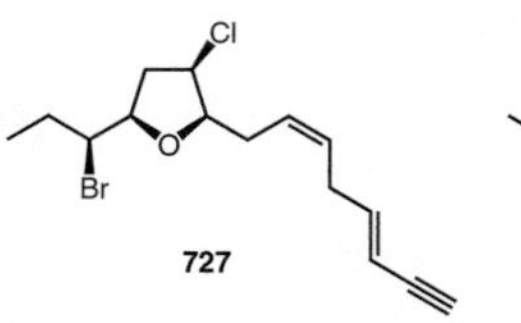

**727**

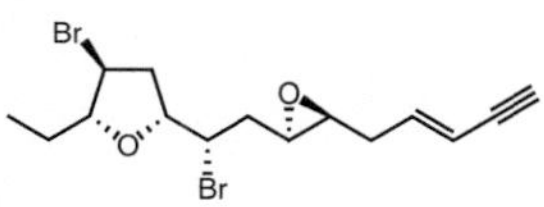

**728**

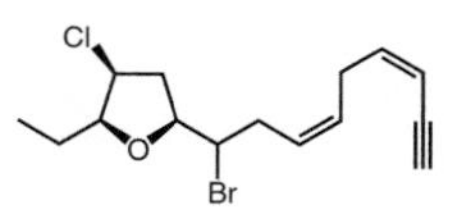

**729**

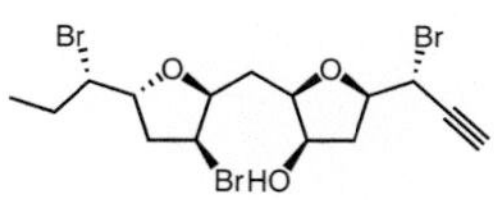

**730**

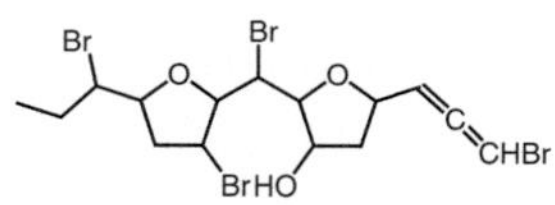

**731**

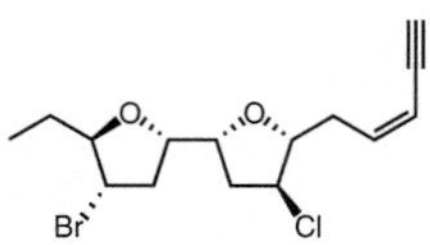

**732**

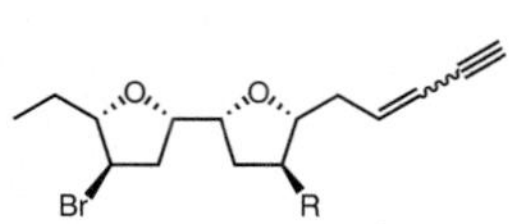

**733** (3*Z*), R = Cl
**734** (3*E*), R = Br

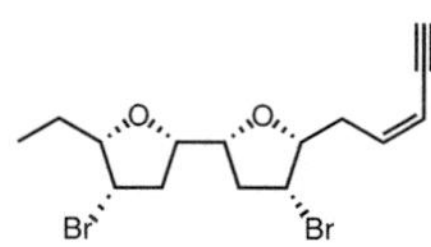

**735**

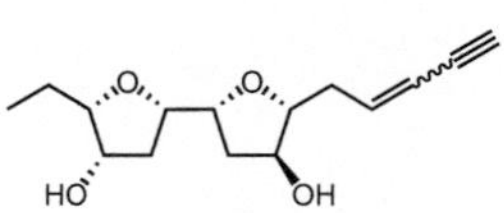

**736** (3*E*)
**737** (3*Z*)

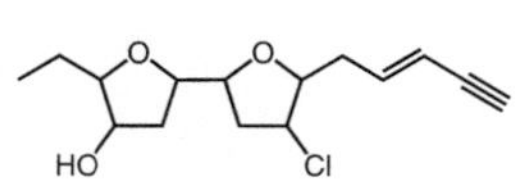

**738**

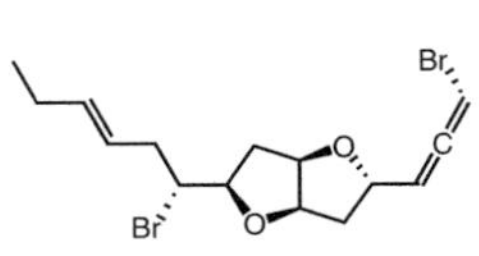

739

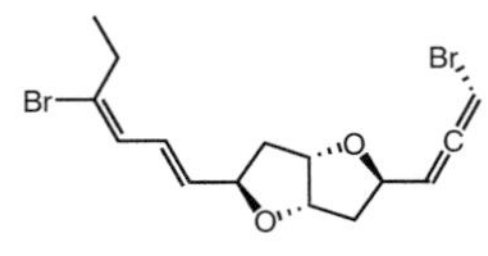

740

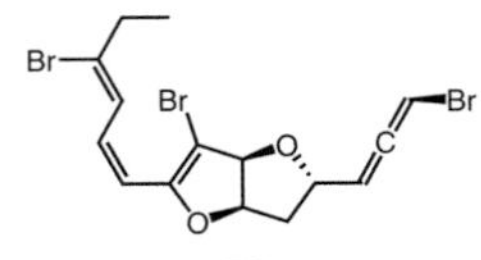

741

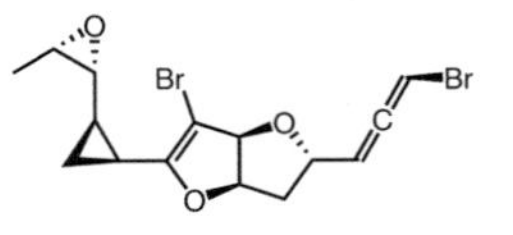

742

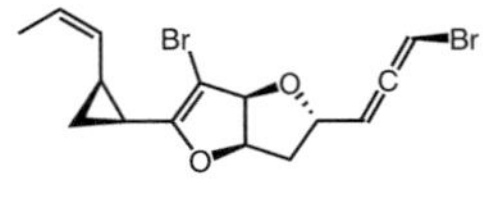

743

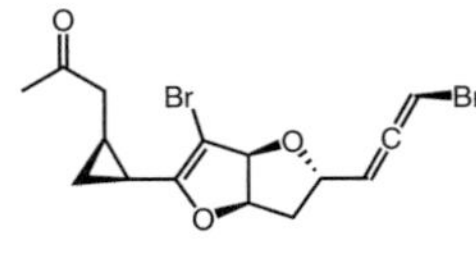

744

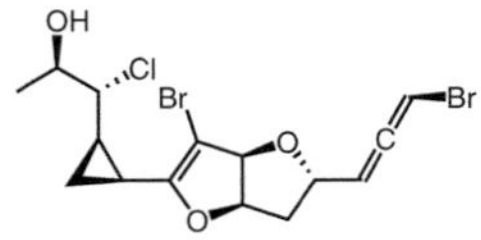

745

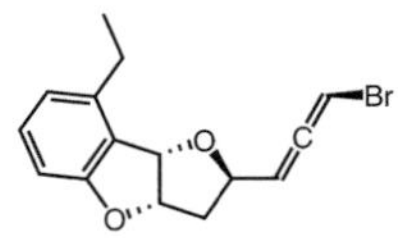

746

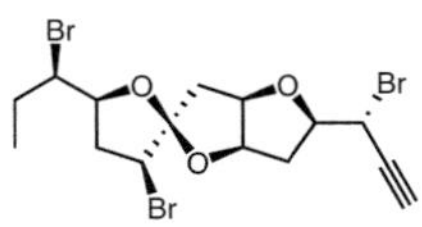

747

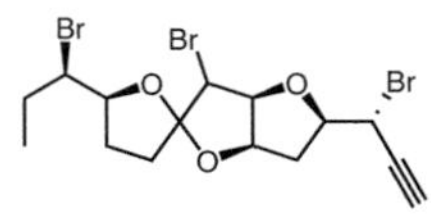

748

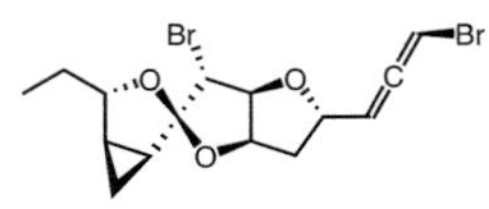

749

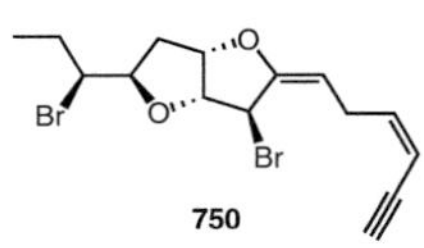

750

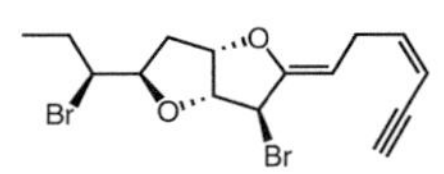

751

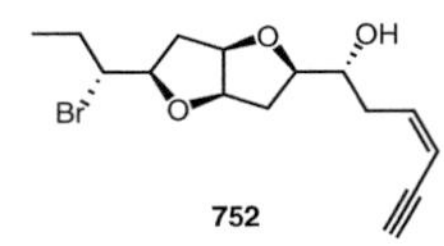

752

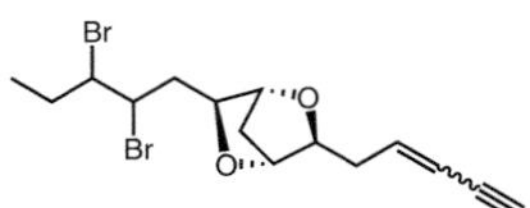

**753** (3*E*)
**754** (3*Z*)

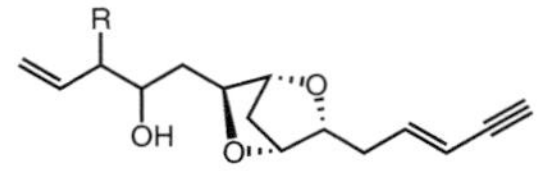

**755** R = Cl
**756** R = OH

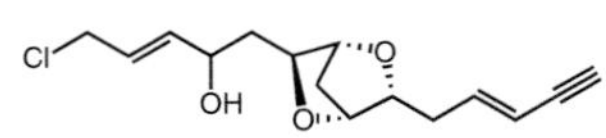

757

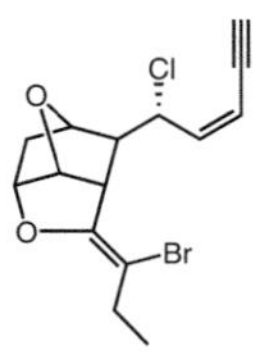

758

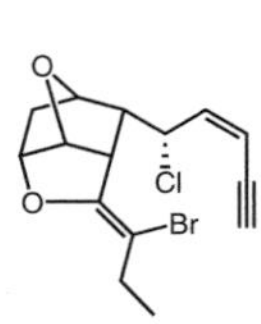

759

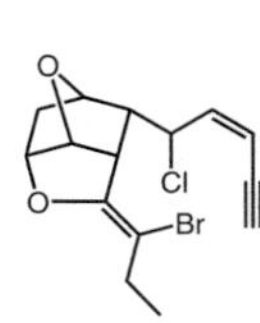

760

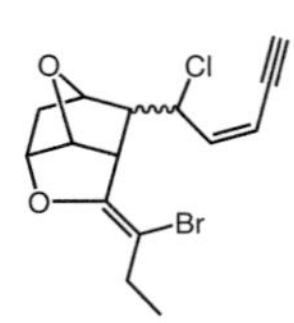

761

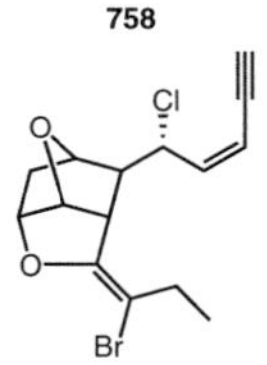

762

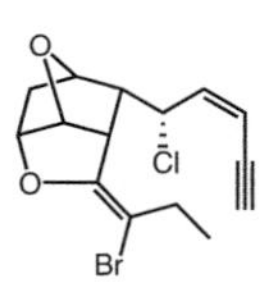

763

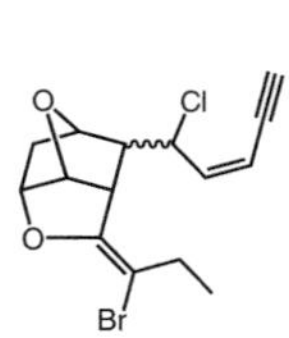

764

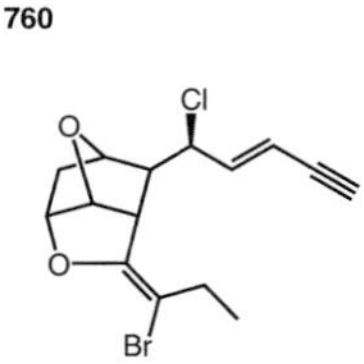

765

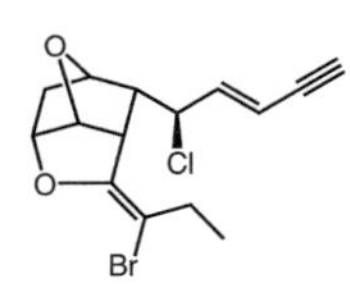

766

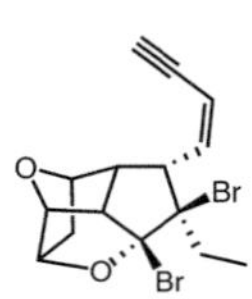

767

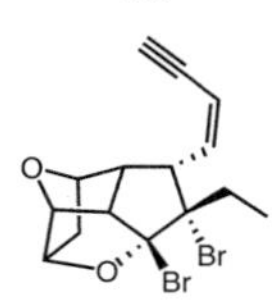

768

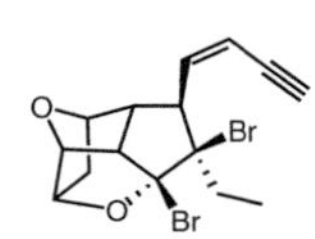

769

### 3.4.2 Acetogenins Containing a Six-Membered Cyclic Ether (Tetrahydropyran) Ring

The group of acetogenins containing six-membered cyclic ethers so far is rather small, including only 14 compounds (**770–783**) that have been reported from the genus *Laurencia* or mollusks feeding on these algae. These can be roughly divided into four subgroups, namely: (1) those containing one tetrahydropyran ring (**770–774**); (2) those containing isolated tetrahydrofuran and tetrahydropyran rings (**775–777**); (3) those containing fused tetrahydrofuran and tetrahydropyran rings (**778–781**); and (4) those containing two fused tetrahydropyran rings (**782** and **783**). Acetogenins containing one tetrahydropyran ring can feature either a 6,10-epoxy (**770** and **771**) or a 9,13-epoxy (**772–774**) ring. Nonetheless, all five display a (*E*)- or (*Z*)-enyne terminus. Acetogenins containing isolated tetrahydrofuran and tetrahydropyran rings present either a bromopropargylic functionality (**775** and **776**) or a terminal acetylene (**777**), but in all three cases isolated thus far C-6 was oxygenated and C-10 and C-12 were brominated. Acetogenins containing fused tetrahydrofuran and tetrahydropyran rings incorporate the 2,7-dioxabicyclo[4.3.0]nonane system in their structures through formation of either a 7,10:9,13-bisepoxy (**778–780**) or a 6,10:9,12-bisepoxy (**781**) core. Nonetheless, all four compounds are characterized by a (*E*)- or (*Z*)-enyne moiety. (3*E*)-Dactylomelyne (**782**) and its geometrical isomer **783**, isolated from the digestive glands of the sea hare *A. dactylomela* [372], are the only two metabolites featuring a 2,7-dioxabicyclo[4.4.0]decane substructure with a (*E*)- and (*Z*)-enyne terminus, respectively. Their structures have been secured by means of chemical synthesis [658, 659].

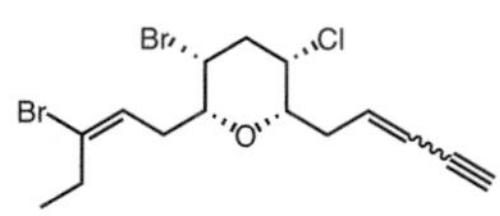

**770** (3*E*)
**771** (3*Z*)

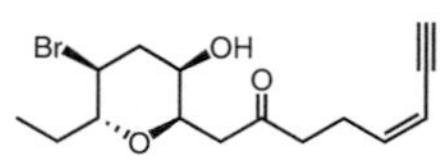

**772**

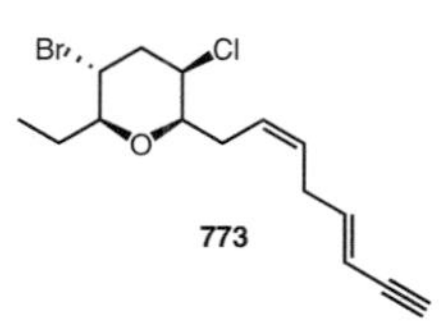

**773**

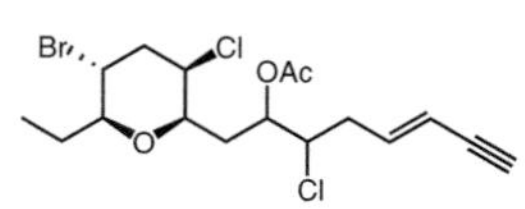

**774**

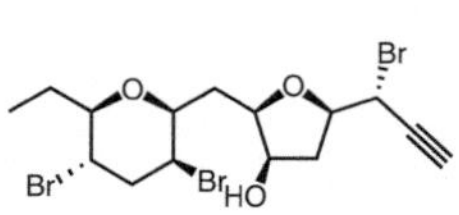

**775**

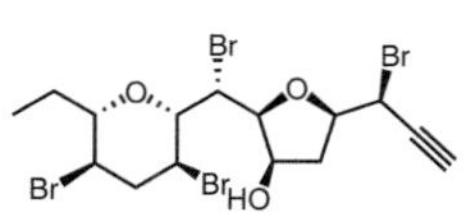

**776**

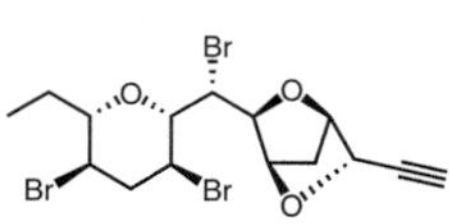

**777**

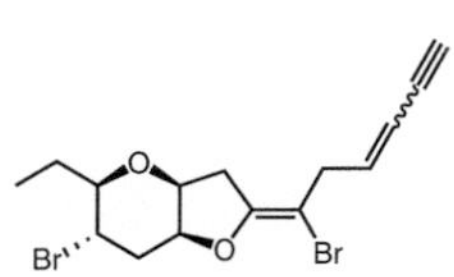

**778** (3*E*)
**779** (3*Z*)

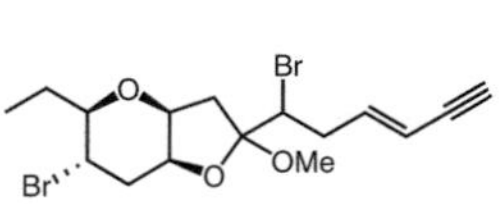

**780**

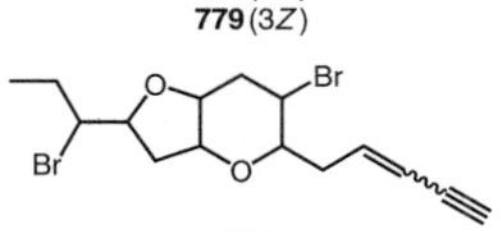

**781**

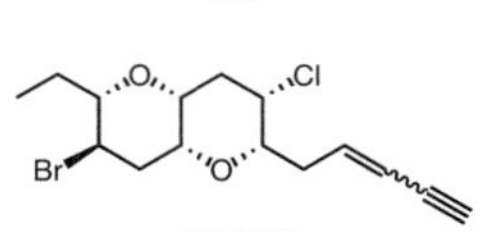

**782** (3*E*)
**783** (3*Z*)

### 3.4.3 Acetogenins Containing a Seven-Membered Cyclic Ether (Oxepane) Ring

To date, only 13 acetogenins containing seven-membered cyclic ethers (**784–796**) have been obtained from *Laurencia* species or their grazers. Their oxepane ring is formed in all cases by etherification between C-7 and C-12. Among them, three subgroups can be discerned, namely: (1) monocyclic acetogenins with a 1-bromopropanyl side chain attached to C-12 (**784–788**); (2) branched monocyclic acetogenins with bromomethyl and ethyl side chains attached to C-11 and C-12, respectively (**789–792**); and (3) acetogenins with a 1-bromopropanyl side chain attached to C-12 featuring a second ether ring formed between C-6 and C-9 (**793–796**). All these compounds display a (*E*)- or (*Z*)-enyne terminus, with the exception of **788** that bears a bromoallene moiety. While metabolites **784–788** and **793–796** bear a bromine atom at C-13, the branched acetogenins **789–792** have a Br substitution at C-15. Furthermore, compounds **784–787** and **789–792** have a chlorine atom at C-6, whereas the 6,9:7,12-bisepoxy acetogenins **793–796** display a second Br substitution at C-10. Rogiolenynes A (**790**), B (**791**), C (**792**), and D (**789**) can be singled out from all the other $C_{15}$ acetogenins isolated to date in that they are branched instead of having a straight $C_{15}$ chain. These were reported from *L. microcladia* and the sponge *Spongia zimocca*, co-occurring in the same small area of the Mediterranean Sea off the area Il Rogiolo, south of Livorno [163, 671, 672]. Guella and Pietra suggested that rogiolenyne A (**790**) could be transferred from the red alga to *S. zimocca* where it undergoes epoxide opening by $Cl^-$ to give rogiolenyne B (**791**), followed by acetylation to give rogiolenyne C (**792**) [672].

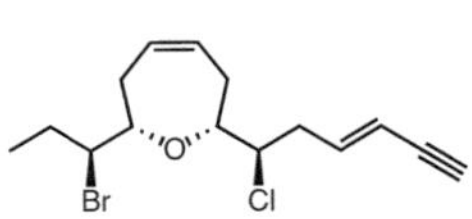


**784**

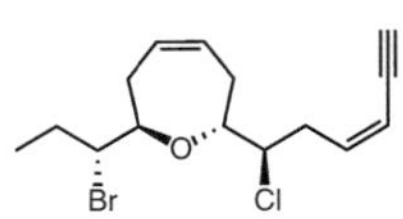


**785**

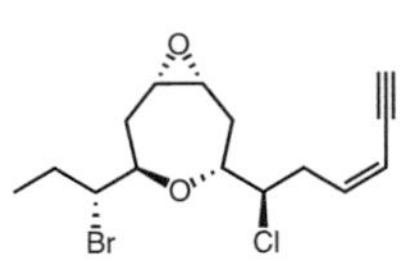


**786**

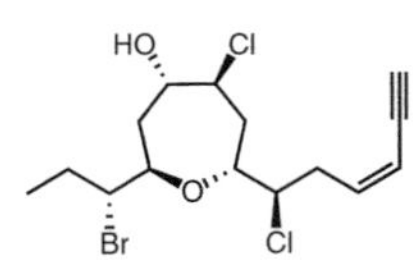


**787**

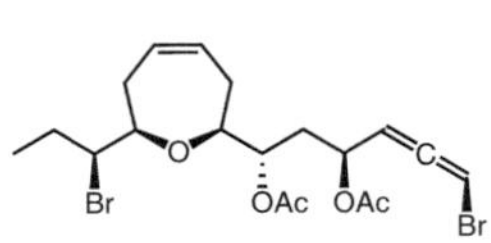


**788**

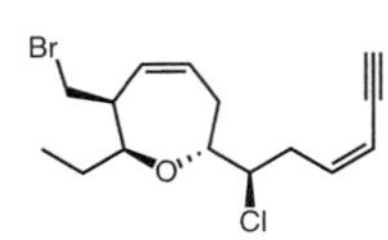


**789**

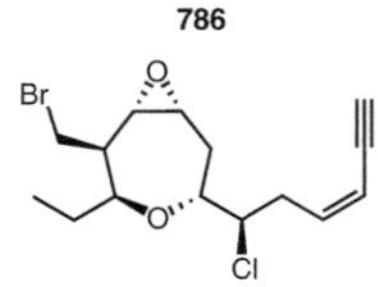


**790**

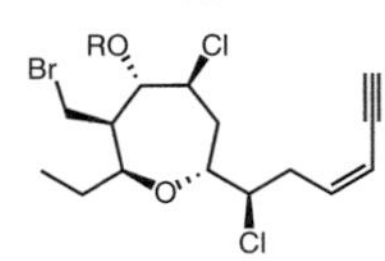


**791** R = H
**792** R = Ac

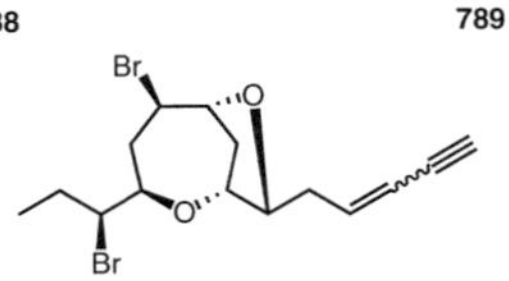


**793** (3*E*)
**794** (3*Z*)

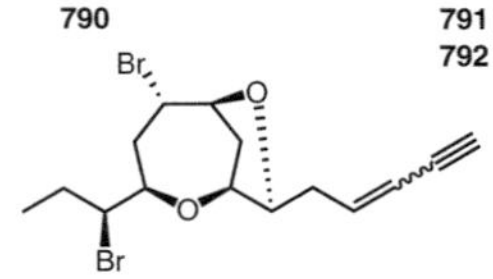


**795** (3*E*)
**796** (3*Z*)

### 3.4.4 Acetogenins Containing an Eight-Membered Cyclic Ether (Oxocane) Ring

The group of acetogenins containing eight-membered cyclic ethers is by far the largest group among all acetogenins isolated from red algae of the genus *Laurencia* or sea hares consuming them, encompassing a total of 93 derivatives (**797–889**). These can be roughly divided into six subgroups, namely: (1) those with a 6,12-epoxy, also known as laurenan-type acetogenins (**797–834**); (2) those with a 7,13-epoxy, also known as lauthisan-type acetogenins (**835–851**); (3) those with a 4,7:6,12-bisepoxy (**852–862**); (4) those with a 6,12-epoxy and a second epoxy system between C-7 and either C-9, C-10 or C-11 (**863–869**); (5) those with a 7,13-epoxy and a second epoxy system between C-6 and either C-9 or C-10 (**870–886**); and (6) those with a 4,10:9,13-bisepoxy (**887–889**) functionality. All laurenan-type acetogenins display a (*E*)- or (*Z*)-enyne terminus, with the exception of **831–834** that bear a bromoallene moiety. The majority of these compounds have a chlorine atom at C-7 and frequently they have a bromine atom at C-13. Similarly, all lauthisan-type acetogenins feature a (*E*)- or (*Z*)-enyne terminus. However, they often have a Cl substitution at C-6 and a Br substitution at C-12. All acetogenins possessing a 4,7:6,12-bisepoxy core bear a bromoallene functionality as terminus, with the exception of **859** and **860** having a bromopropargylic moiety and desepilaurallene (**862**) that is a $C_{12}$ derivative. In all cases, they exhibit a Br substitution at C-13. Among them, aplyparvunin (**861**), obtained from a Japanese *A. parvula* sample, possesses a cyclopropane ring fused with the eight-membered ether [291]. Its structure was determined by single-crystal X-ray diffraction analysis. Both acetogenins with a 6,12-epoxy and a second epoxy system between C-7 and either C-9, C-10, or C-11, as well as those with a 7,13-epoxy and a second epoxy system between C-6 and either C-9 or C-10 possess a (*E*)- or (*Z*)-enyne terminus, apart from the $C_{12}$ derivative okamuragenin (**869**).

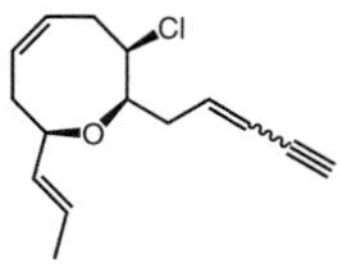

**797** (3*E*)
**798** (3*Z*)

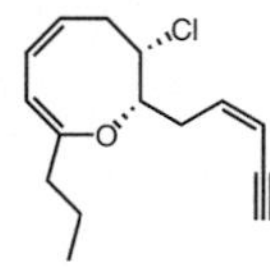

**799**

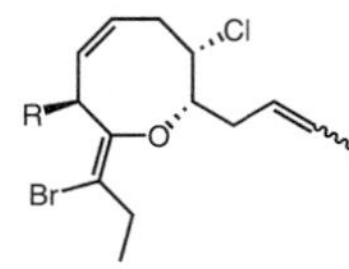

**800** (3*E*), R = H
**801** (3*Z*), R = H
**802** (3*E*), R = OH
**803** (3*Z*), R = OH

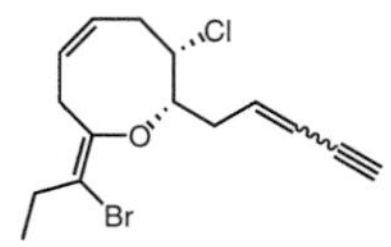

**804** (3*E*)
**805** (3*Z*)

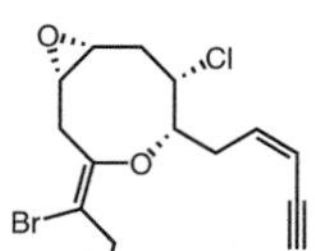

**806**

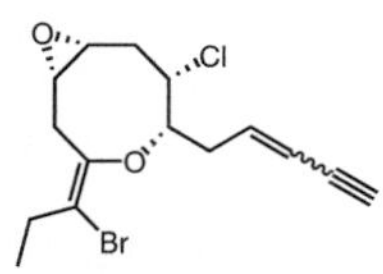

**807** (3*E*)
**808** (3*Z*)

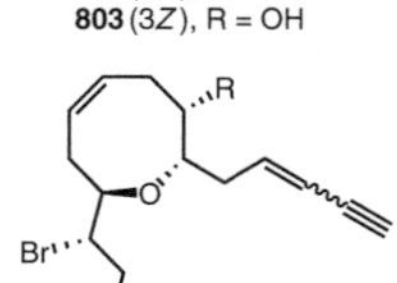

**809** (3*E*), R = Cl
**810** (3*Z*), R = Cl
**811** (3*Z*), R = OH

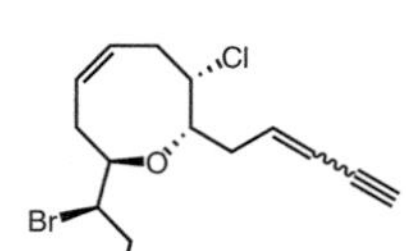

**812** (3*E*)
**813** (3*Z*)

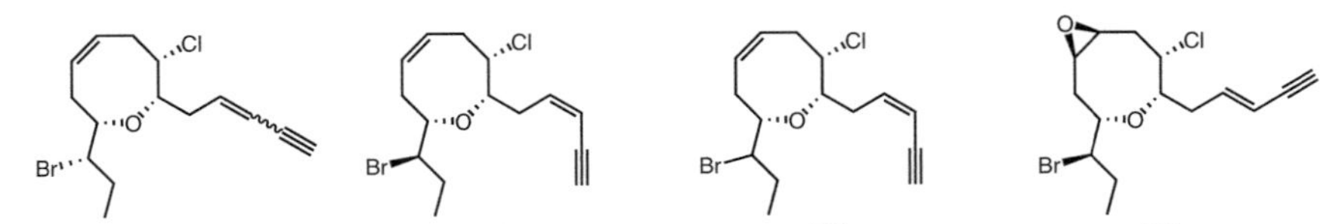

**814** (3*E*)
**815** (3*Z*)

**816**

**817**

**818**

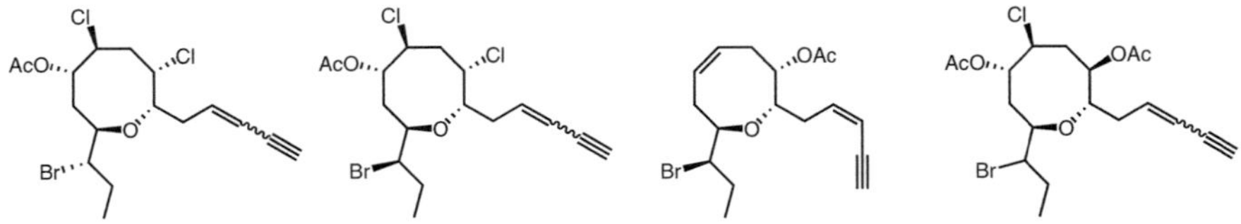

**819** (3*E*)
**820** (3*Z*)

**821** (3*E*)
**822** (3*Z*)

**823**

**824** (3*E*)
**825** (3*Z*)

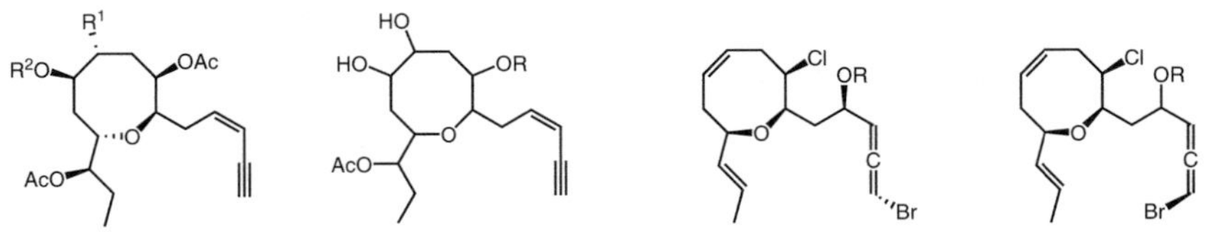

**826** R$^1$ = Cl, R$^2$ = Ac
**827** R$^1$ = OAc, R$^2$ = H
**828** R$^1$ = OH, R$^2$ = Ac

**829** R = H
**830** R = Ac

**831** R = H
**832** R = Ac

**833** R = H
**834** R = Ac

All members of the former subgroup, including **869**, have a Br substitution at C-13, whereas most members of the latter subgroup bear a Br substitution at C-12. The efficient isomerization of (*Z*)-enyne systems to (*E*)-enyne was exemplified by the chemical conversion of laureatin (**864**) to (*E*)-isolaureatin (**865**) [788]. Microcladallenes A (**887**), B (**889**), and C (**888**), initially isolated from *L. microcladia* collected off the French coast at Cape Ferrat [750], are the only examples of acetogenins incorporating a 4,10:9,13-bisepoxy core in their structures. All three bear a bromoallene moiety as the terminus and a bromine atom at C-12.

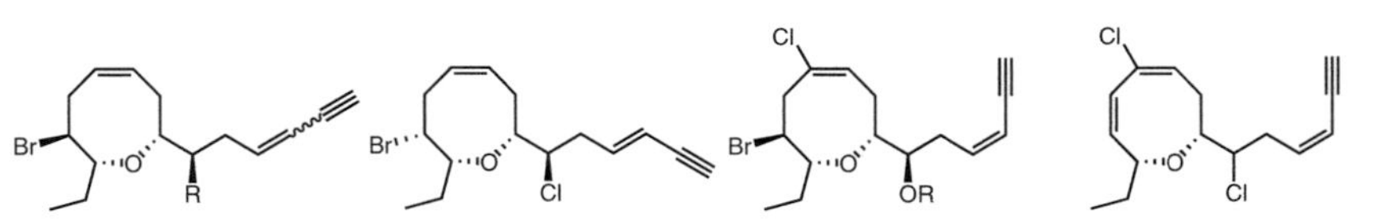

**835** (3*E*), R = OH
**836** (3*E*), R = OAc
**837** (3*Z*), R = Cl

**838**

**839** R = H
**840** R = Ac

**841**

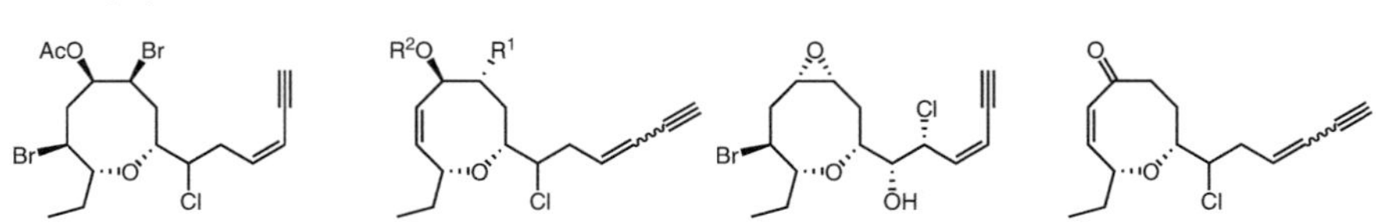

**842**

**843** (3*E*), R$^1$ = OH, R$^2$ = H
**844** (3*Z*), R$^1$ = OH, R$^2$ = H
**845** (3*Z*), R$^1$ = OH, R$^2$ = Ac
**846** (3*Z*), R$^1$ = OAc, R$^2$ = H
**847** (3*Z*), R$^1$ = OAc, R$^2$ = Ac
**848** (3*Z*), R$^1$ = Br, R$^2$ = Ac

**849**

**850** (3*E*)
**851** (3*Z*)

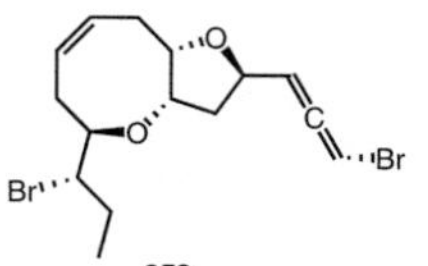

**852**

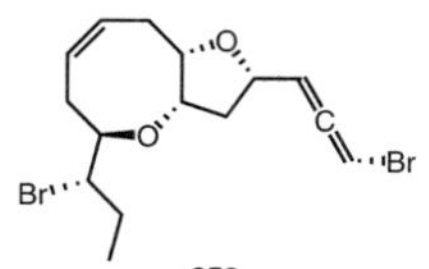

**853**

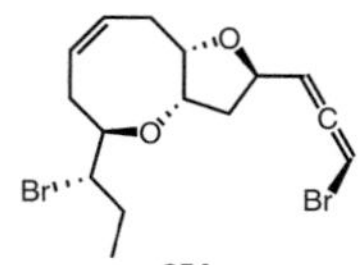

**854**

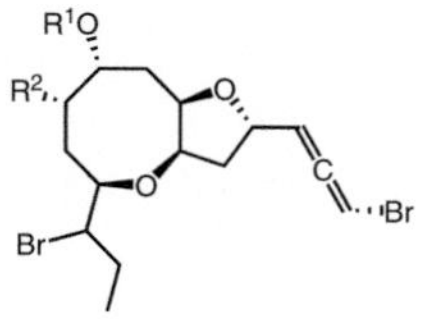

**855** $R^1$ = H, $R^2$ = Br
**856** $R^1$ = Ac, $R^2$ = Br
**857** $R^1$ = H, $R^2$ = OH

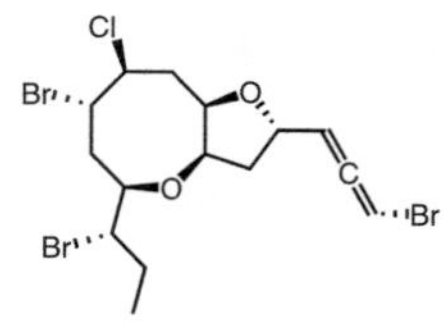

**858**

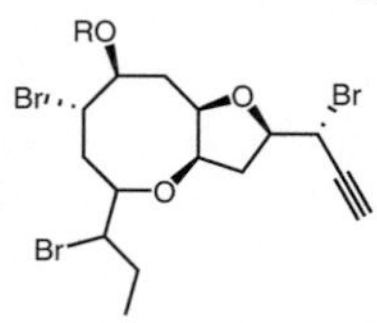

**859** R = H
**860** R = Ac

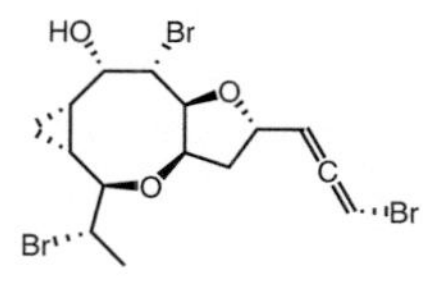

**861**

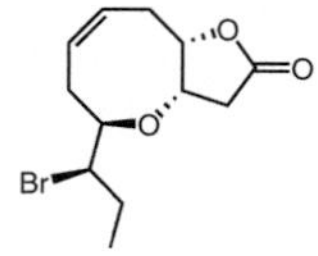

**862**

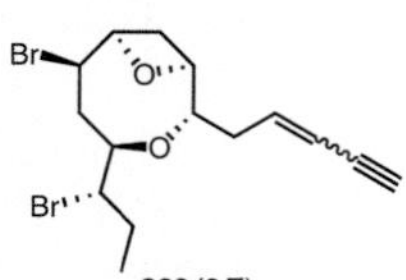

**863** (3*E*)
**864** (3*Z*)

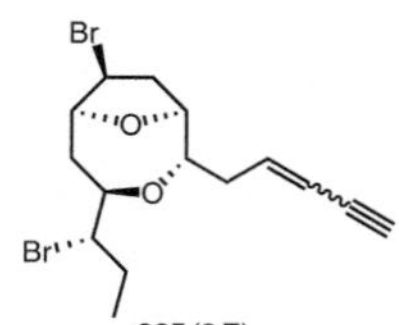

**865** (3*E*)
**866** (3*Z*)

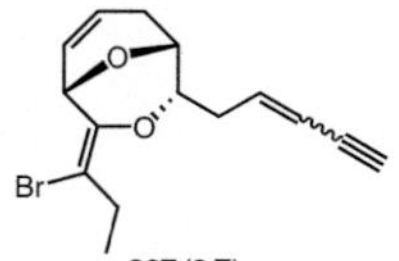

**867** (3*E*)
**868** (3*Z*)

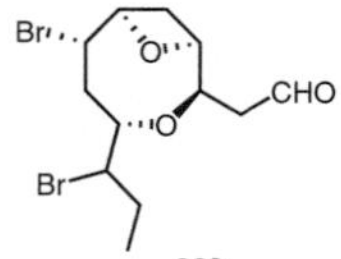

**869**

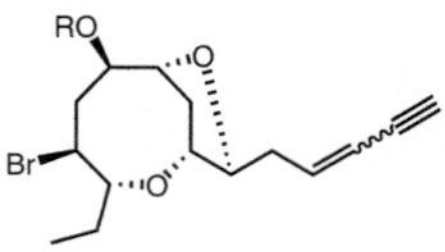

**870** (3*E*), R = H
**871** (3*Z*), R = H
**872** (3*E*), R = Ac
**873** (3*Z*), R = Ac

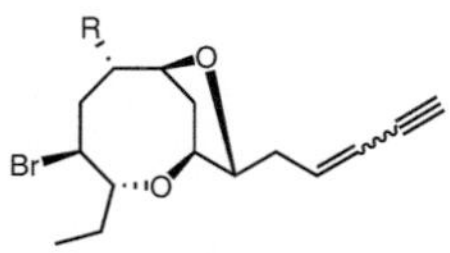

**874** (3*E*), R = Cl
**875** (3*Z*), R = Cl
**876** (3*E*), R = Br
**877** (3*Z*), R = Br

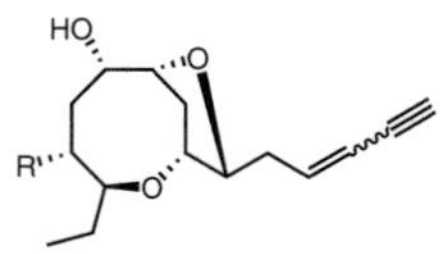

**878** (3*E*), R = OH
**879** (3*Z*), R = OH
**880** (3*E*), R = Br
**881** (3*Z*), R = Br

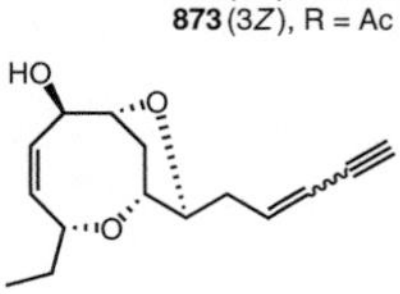

**882** (3*E*)
**883** (3*Z*)

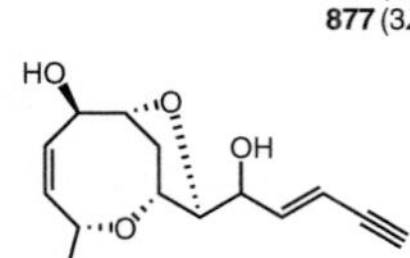

**884**

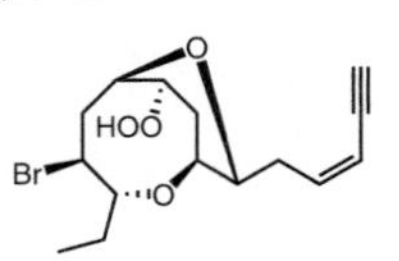

**885**

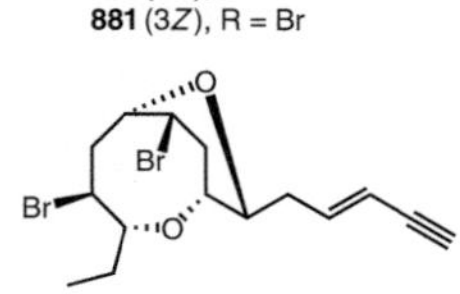

**886**

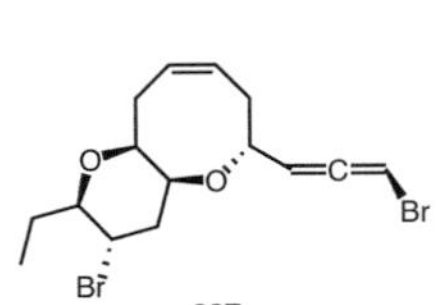
887

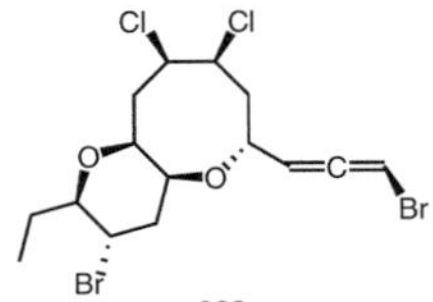
888

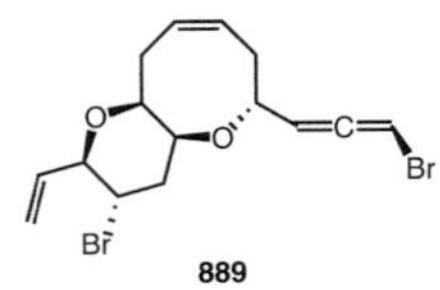
889

### 3.4.5 Acetogenins Containing a Nine- or Ten-Membered Cyclic Ether Ring

So far, twelve acetogenins containing a nine-membered cyclic ether unit (**890–896** and **898–902**) and only one containing a ten-membered cyclic ether (**897**) have been reported from *Laurencia* species and/or gastropods feeding on them. Among the nine-membered acetogenins, three subgroups can be discriminated, namely: (1) those containing a 5,12-epoxy ring (**890**); (2) those containing a 6,13-epoxy ring (**891–896**); and (3) those containing a 4,7:6,13-bisepoxy system (**898–902**). The first two subgroups feature a (*E*)- or (*Z*)-enyne terminus, while the third subgroup is characterized by a bromoallene functionality. With the exception of **890**, **891**, and **902**, all derivatives have a bromine atom at C-12, whereas compounds **891–896** possess a chlorine atom at C-7 as well. Noteworthy is obtusenyne (**892**), the first nine-membered acetogenin to be isolated, which was obtained from two Mediterranean samples of *Laurencia obtusa* almost simultaneously by two different research groups [730, 754]. Its first total synthesis was achieved in 1999 by the group of Murai [757], while three more synthetic protocols have also been reported [758–760]. Metabolite **897**, for which the relative configurations of its asymmetric centers have not been assigned yet, was isolated from *Laurencia implicata* collected in Florence Bay, Magnetic Island, and is the sole representative of the ten-membered acetogenins [415].

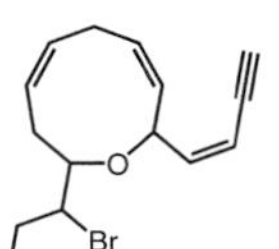
890

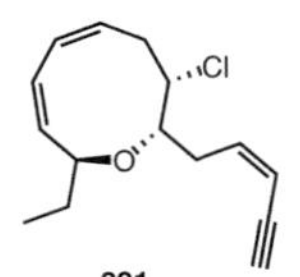
891

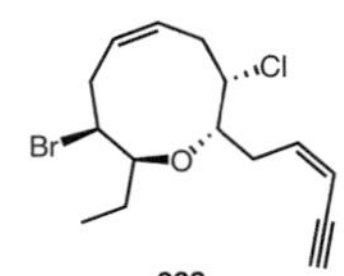
892

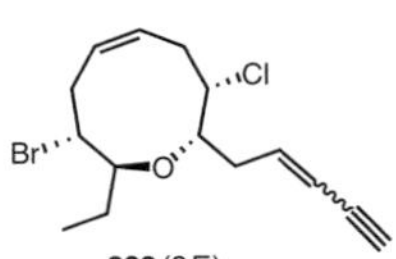
893 (3*E*)
894 (3*Z*)

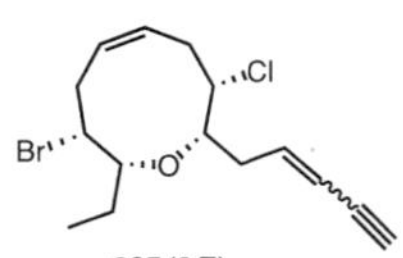
895 (3*E*)
896 (3*Z*)

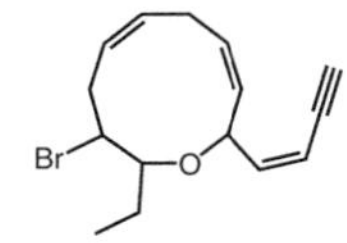
897

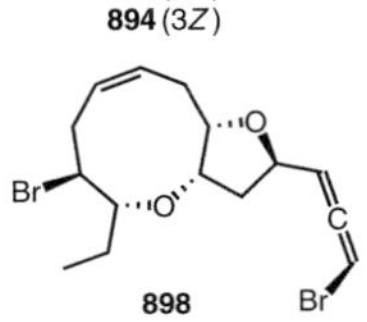
898

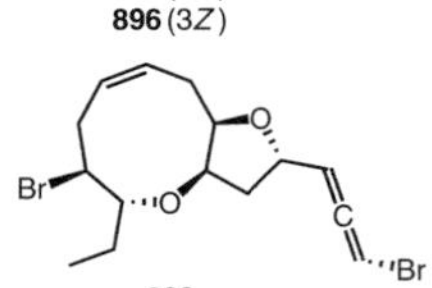
899

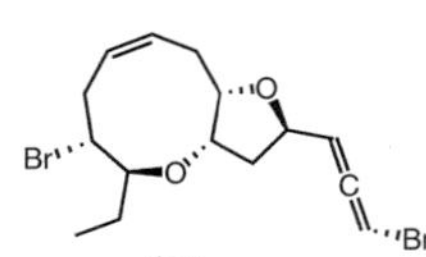
900

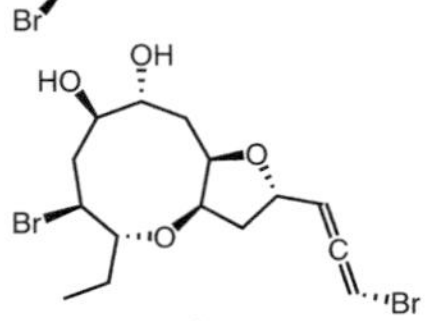
901

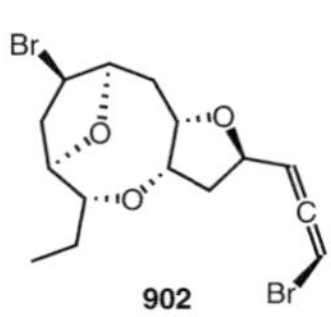
902

### 3.4.6 Acetogenins Containing a Twelve-Membered Cyclic Ether Ring

The group of acetogenins containing 12-membered cyclic ethers includes 20 compounds (**903–922**) that have been reported to date from the genus *Laurencia* or mollusks feeding on these algae. Metabolites **903–917**, known as obtusallenes, are characterized by a 4,14-epoxy core, a bromoallene functionality as one terminus at C-4 (with the exception of **908**) and a methyl group as the second terminus at C-14, and all of them incorporate one or two more ether rings into their structures. Among these, **903–909** feature a 6,9-epoxy ring, **910** presents a 9,12-epoxy ring, **911** contains a 10,13-epoxy ring, **912** and **913** have a 5,13-epoxy ring, and **914** and **915** display a 9,13-epoxy ring. Obtusallenes VI (**916**) and V (**917**) have two tetrahydrofuran rings as a result of etherification between C-6 and C-9, as well as between C-9 and C-12, thus leading to the formation of a *spiro*-ketal motif. The first member of this class, obtusallene I (**914**), was reported from a *Laurencia obtusa* population collected in Gökçeada Island, Aegean Sea, and its structure was determined by means of single crystal X-ray diffraction analysis [774, 775]. Obtusallene IV (**906**), also known as dactylallene, was isolated simultaneously from *A. dactylomela* from the Canary Islands and *L. obtusa* from Turkey [769, 771]. In the original report on the isolation of obtusallenes V (**917**), VI (**916**), and VII (**910**) from *L. obtusa*, the bromine and chlorine atoms were placed at C-7 and C-13, respectively, on the basis of analysis of the NMR spectroscopic data [770]. Several extensive investigations concerning the biosynthesis of these natural products were undertaken by Braddock and co-workers [768, 789, 790] that finally led to the reassignment of the position of the halogens at C-7 and C-13 on the basis of a $^{13}$C NMR chlorine induced isotopic shift, which was further confirmed by X-ray crystallographic analysis [772]. The latter study also predicted the existence of compounds with a 4,13-dioxabicyclo[5.5.1]tridecane ring system [772]. Subsequently, metabolites with this framework were isolated from *Laurencia marilzae* in 2011 and designated as marilzabicycloallenes A (**918**), B (**921**), C (**919**), and D (**920**) [777]. Poitediene (**922**), obtained from *Laurencia poitei* collected at Boca Grande Key, Florida [778], is a unique tricyclic acetogenin featuring a *cis*-dibromo-olefin functionality as terminus. Its structure, including absolute stereochemistry, was established on the basis of X-ray diffraction analysis.

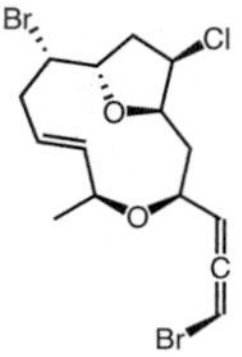
**903**

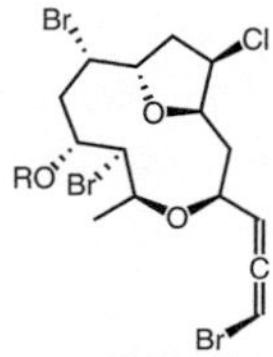
**904** R = H
**905** R = Ac

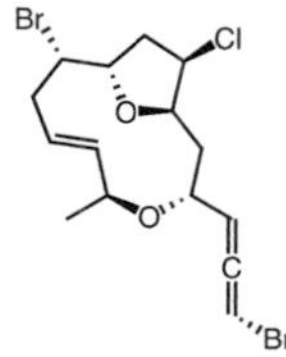
**906**

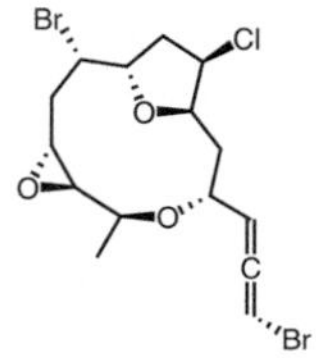
**907**

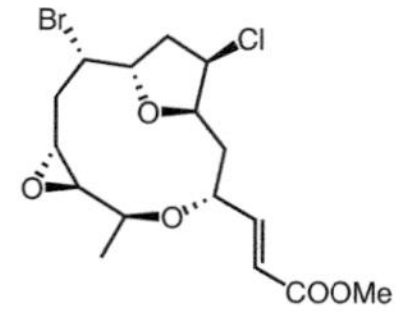
**908**

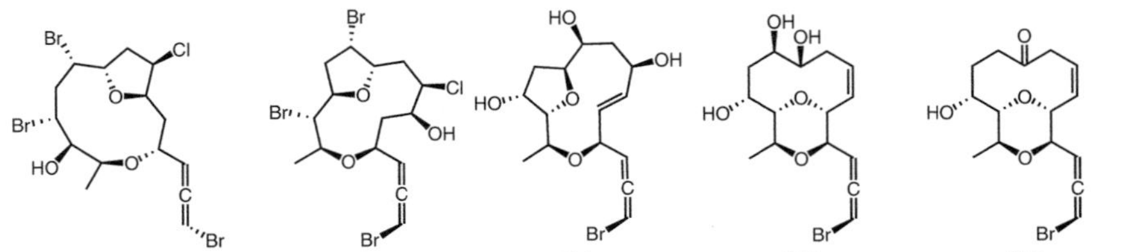

909 910 911 912 913

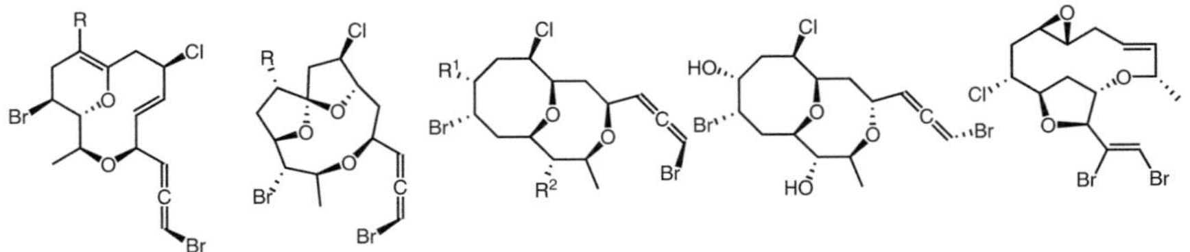

**914** R = H
**915** R = Br

**916** R = H
**917** R = Br

**918** $R^1 = R^2 = OH$
**919** $R^1 = OMe$, $R^2 = OH$
**920** $R^1 = R^2 = Cl$

**921**

**922**

### 3.4.7 Linear Acetogenins

Besides the cyclic ether-containing acetogenins, which constitute the majority of acetogenins isolated so far from *Laurencia* species and their grazers, 26 linear acetogenins (**923–948**) have been reported to date. All of them feature a (*E*)- or (*Z*)-enyne terminus, while most are characterized by multiple double bonds, but in all cases by a $\Delta^9$ functionality with (*Z*) configuration. Apart from compounds **923–928**, which do not have any heteroatoms in their structures, all the others bear oxygen and/or chlorine substitutions at C-6 and C-7. In particular, compounds **929**, **930**, **938**, **939**, and **942–946** are oxygenated at both C-6 and C-7, while **931–937**, **940**, and **941** are oxygenated at C-6 and chlorinated at C-7. Less frequently, chlorination at C-6 and oxygenation at C-7 is observed, as in the case of **947** and **948**. Metabolites **923–926**, found in Japanese populations of *L. okamurai* [192, 204, 779, 780], are considered to be biosynthetic precursors of various $C_{15}$ non-terpenoid metabolites found in *Laurencia* species. Their synthesis was reported by the Yamada group in 1981 and 1986 [204, 780], while another synthesis of laurencenyne (**926**) and its (*E*)-isomer **925** was later achieved by Holmeide et al. starting from eicosapentaenoic acid [781]. A Japanese *L. nipponica* population afforded a mixture of optical isomers ((6*R*,7*R*) and (6*S*,7*S*)) of (*E*)- (**942**) and (*Z*)-laurediol (**943**), along with their diacetates **945** and **946**, respectively [738].

**923** (3*E*)
**924** (3*Z*)

**925** (3*E*)
**926** (3*Z*)

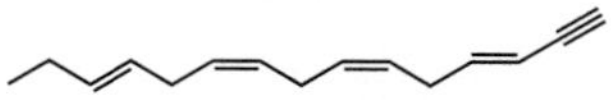

**927**

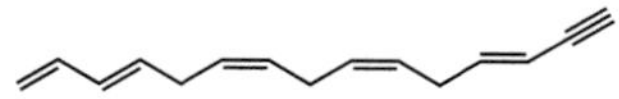

**928**

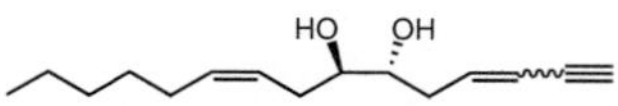

**929** (3*E*)
**930** (3*Z*)

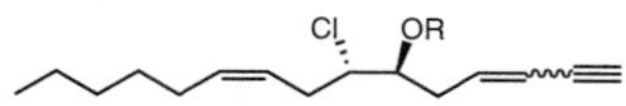

**931** (3*E*), R = H
**932** (3*Z*), R = H
**933** (3*E*), R = Ac
**934** (3*Z*), R = Ac

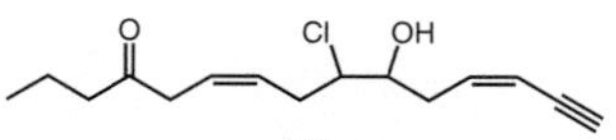

**935**

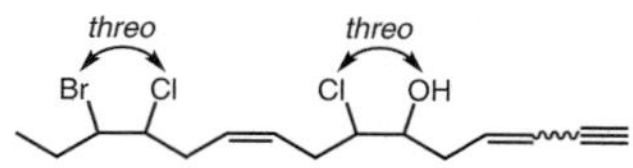

**936** (3*E*)
**937** (3*Z*)

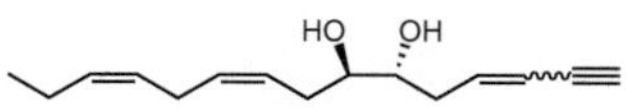

**938** (3*E*)
**939** (3*Z*)

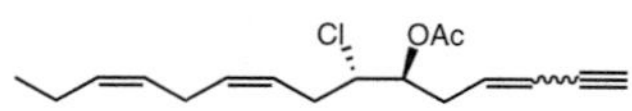

**940** (3*E*)
**941** (3*Z*)

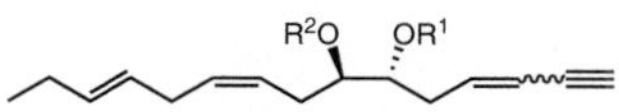

**942** (3*E*), $R^1 = R^2 = H$
**943** (3*Z*), $R^1 = R^2 = H$
**944** (3*Z*), $R^1 = H$, $R^2 = Ac$
**945** (3*E*), $R^1 = R^2 = Ac$
**946** (3*Z*), $R^1 = R^2 = Ac$

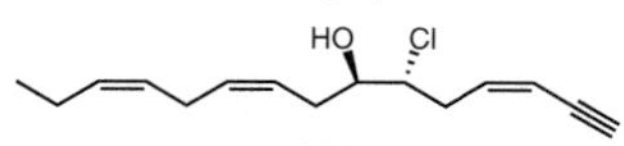

**947**

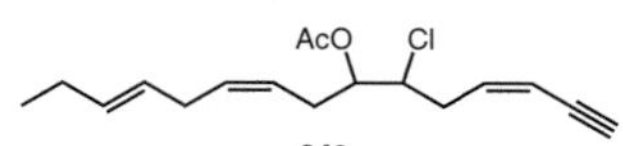

**948**

Biosynthesis studies were undertaken to understand the role of these compounds in the production of medium-sized ring ethers. The enzymatic syntheses of laurencin (**836**) (a lauthisan-type eight-membered acetogenin) from (3*E*,6*R*,7*R*)-laurediol and of prelaureatin (**811**) and laureatin (**864**) (laurenan-type eight-membered acetogenins) from (3*Z*,6*S*,7*S*)-laurediol, have been accomplished [694–697]. (*E*)-Adrienyne (**936**) and its (*Z*)-isomer **937** were found in *L. marilzae* from the Canary Islands. The relative configurations of the stereocenters C-6, C-7, C-12, and C-13 were determined using a configurational analysis based on coupling constants [711].

## 3.5 Indoles

To date, 35 indoles (**949–983**, Table 5) have been reported from *Laurencia* species and/or gastropods feeding on them.

All of these compounds contain at least two bromine atoms, with the exception of **972**, which is not brominated, and **973**, which is monobrominated. Indoles can be divided into three subclasses, namely: (1) monoindoles (**949–973**); (2) bisindoles

**Table 5** Indoles from *Laurencia* spp.

| No. | Name | Source of isolation/synthesis |
|---|---|---|
| **949** | *N*-Methyl-3,5-dibromo-indole | *L. similis* [791] |
| **950** | 3,5,6-Tribromo-indole | *L. similis* [118, 176[a], 431, 792], *C. papillosus* [342], *A. dactylomela* [118] |
| **951** | 2.3.6-Tribromo-indole | *L. similis* [176[a]] |
| **952** | 2,3,5,6-Tetrabromo-indole | *L. brongniartii* [793[a]], *L. complanata* [433], *L. saitoi* [794], *L. similis* [118, 176, 431, 432, 792, 795, 796], *C. papillosus* [342], *A. dactylomela* [118, 277]; Synthesis [797] |
| **953** | 2,4,6-Tribromo-indole | *L. brongniartii* [798] |
| **954** | 2,3,4,6-Tetrabromo-indole | *L. brongniartii* [798] |
| **955** | 3-Methylthio-2,4,6-tribromo-indole | *L. brongniartii* [798, 799] |
| **956** | 3-Methylsulfinyl-2,4,6-tribromo-indole | *L. brongniartii* [799] |
| **957** | 4,6-Dibromo-2-methylthio-indole | *L. brongniartii* [798[a], 799] |
| **958** | 4,6-Dibromo-2,3-di(methylthio)indole | *L. brongniartii* [798, 799] |
| **959** | Itomanindole B | *L. brongniartii* [798, 800[a]] |
| **960** | Itomanindole A | *L. brongniartii* [798–800[a]] |
| **961** | 4,6-Dibromo-2,3-di(methylsulfinyl)-indole | *L. brongniartii* [799] |
| **962** | 3-Methylthio-2,4,5,6-tetrabromo-indole | *L. brongniartii* [798, 799] |
| **963** | 2,3-Di(methylthio)-4,5,6-tribromo-indole | *L. brongniartii* [798, 799] |
| **964** | 4,5,6-Tribromo-2-methyl-sulfinylindole | *L. brongniartii* [801] |
| **965** | 2-Methylsulfinyl-3-methylthio-4,5,6-tribromo-indole | *L. brongniartii* [799] |
| **966** | *N*-Methyl-2,3,6-tribromo-indole | *L. brongniartii* [793[a]]; Synthesis [797, 802] |
| **967** | *N*-Methyl-2,3,4,6-tetrabromo-indole | *L. decumbens* [160] |
| **968** | *N*-Methyl-2,5-dibromo-indole | *L. similis* [431] |
| **969** | *N*-Methyl-2,3,5-tribromo-indole | *L. brongniartii* [793[a]], *L. complanata* [433], *L. decumbens* [160], *L. similis* [176, 431, 432, 792], *Laurencia* sp. [187], *A. dactylomela* [156, 270, 277]; Synthesis [797] |
| **970** | *N*-Methyl-3,5,6-tribromo-indole | *L. similis* [176[a]] |
| **971** | *N*-Methyl-2,3,5,6-tetrabromo-indole | *L. brongniartii* [793[a]], *L. complanata* [433], *L. decumbens* [160], *L. similis* [432, 792, 795, 796], *C. papillosus* [342], *A. dactylomela* [118, 277]; Synthesis [797, 802] |
| **972** | Indolyl-3-carbaldehyde | *L. tristicha* [290] |
| **973** | *N*-Methyl-5-bromo-1,3-dihydro-indole-2-one | *L. similis* [792] |
| **974** | 3,3′-bis-(2,5,6-Tribromo)indole | *L. similis* [791[a], 792] |
| **975** | 3,3′-bis-(4,6-Dibromo-2-methylthio)indole | *L. brongniartii* [798[a], 799] |
| **976** | 3,3′-bis-(2′-Methylsulfinyl-2-methylthio-4,6,4′,6′-tetrabromo)indole | *L. brongniartii* [799] |
| **977** | 3,3′-bis-(4,6-Dibromo-2-methylsulfinyl) indole | *L. brongniartii* [799] |
| **978** | 2,4,4′,6,6′-Pentabromo-2′-(methylsulfanyl)-1*H*,1′*H*-3,3′-bisindole | *L. brongniartii* [803] |

(continued)

**Table 5** (continued)

| No. | Name | Source of isolation/synthesis |
|---|---|---|
| **979** | 2,4,4′,6,6′-Pentabromo-2′,3-bis (methylsulfanyl)-1′*H*-1,3′-bisindole | *L. brongniartii* [803] |
| **980** | 2,4,4′,5′,6,6′-Hexabromo-2′,3-bis (methylsulfanyl)-1′*H*-1,3′-bisindole | *L. brongniartii* [803] |
| **981** | 2,4,4′,5,6,6′-Hexabromo-2′,3-bis (methylsulfanyl)-1′*H*-1,3′-bisindole | *L. brongniartii* [803] |
| **982** | 2,4,4′,5,5′,6,6′-Heptabromo-2′,3-bis (methylsulfanyl)-1′*H*-1,3′-bisindole | *L. brongniartii* [803] |
| **983** | Similisine A | *L. similis* [792[b]] |
| ***ent*-983** | Similisine B | *L. similis* [792[b]] |

[a]The isolation and structure elucidation of this compound was reported for the first time
[b]The absolute stereochemistry of this compound has been established

(**974–982**); and (3) *spiro*-trisindoles (**983**). Among these, monoindoles **955–965** and bisindoles **975–982** contain sulfur in their structures. Similisines A (**983**) and B (***ent*-983**), for which the structures and absolute configurations were determined by comprehensive analysis of their spectroscopic data, computer modeling and comparison of their experimental and theoretically calculated electronic circular dichroism (ECD) spectra, are a pair of enantiomeric polybrominated *spiro*-trisindoles isolated in 2013 from *L. similis* collected in the South China Sea [792].

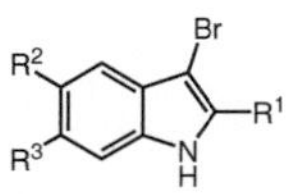

**949** $R^1 = R^3 = H, R^2 = Br$
**950** $R^1 = H, R^2 = R^3 = Br$
**951** $R^1 = R^3 = Br, R^2 = H$
**952** $R^1 = R^2 = R^3 = Br$

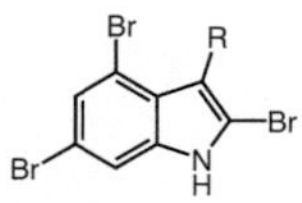

**953** R = H
**954** R = Br
**955** R = SMe
**956** R = SOMe

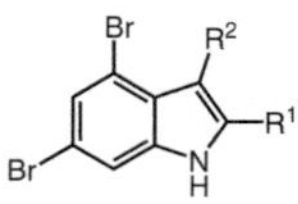

**957** $R^1$ = SMe, $R^2$ = H
**958** $R^1 = R^2$ = SMe
**959** $R^1$ = SMe, $R^2$ = SOMe
**960** $R^1$ = SOMe, $R^2$ = SMe
**961** $R^1 = R^2$ = SOMe

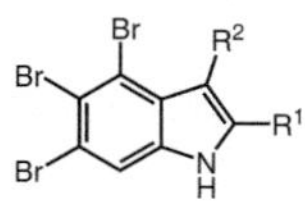

**962** $R^1$ = Br, $R^2$ = SMe
**963** $R^1 = R^2$ = SMe
**964** $R^1$ = SOMe, $R^2$ = H
**965** $R^1$ = SOMe, $R^2$ = SMe

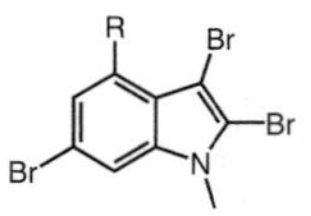

**966** R = H
**967** R = Br

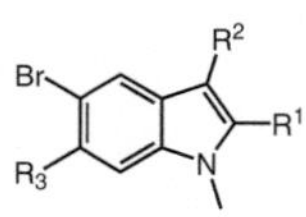

**968** $R^1 = Br, R^2 = R^3 = H$
**969** $R^1 = R^2 = Br, R^3 = H$
**970** $R^1 = H, R^2 = R^3 = Br$
**971** $R^1 = R^2 = R^3 = Br$

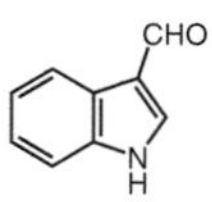

**972**

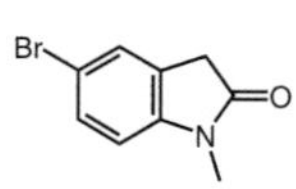

**973**

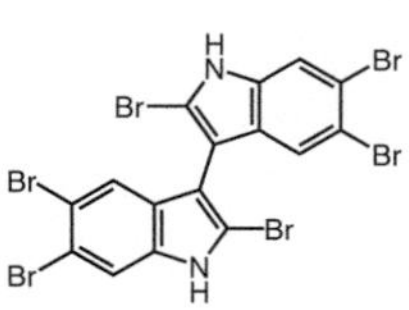

**974**

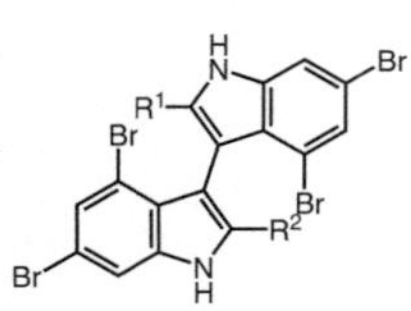

**975** $R^1 = R^2$ = SMe
**976** $R^1$ = SOMe, $R^2$ = SMe
**977** $R^1 = R^2$ = SOMe
**978** $R^1$ = Br, $R^2$ = SMe

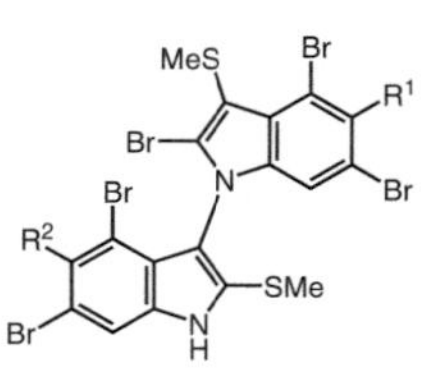

**979** $R^1 = R^2 = H$
**980** $R^1 = H, R^2 = Br$
**981** $R^1 = Br, R^2 = H$
**982** $R^1 = R^2 = Br$

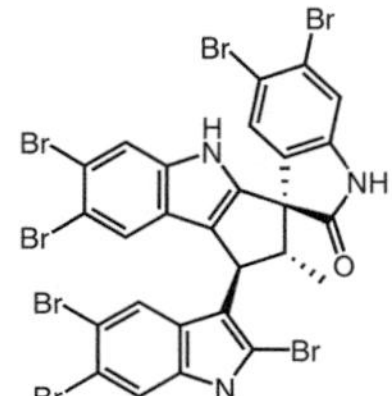

**983**

## 3.6 Aromatic Compounds

Only nine aromatic compounds (**984–992**, Table 6) have been isolated so far from red algae of the genus *Laurencia*.

These include four simple monoaryl derivatives (**984–987**), two diaryl derivatives (**988** and **989**), and three naphthalene derivatives (**990–992**). With the exception of **984**, **986**, and **987**, all other compounds are brominated. According to Blunt et al. the structures of **990–992** should be re-examined due to inconsistencies in the published NMR data [9].

**Table 6** Aromatic compounds from *Laurencia* spp.

| No. | Name | Source of isolation/synthesis |
|---|---|---|
| **984** | 4-Hydroxy-benzaldehyde | *L. papillosa* [804], *L. tristicha* [290], *A. dactylomela* [277] |
| **985** | – | *L. nipponica* [805] |
| **986** | 4-Methoxy-benzyl alcohol | *L. papillosa* [804] |
| **987** | 3-Hydroxy-4-methyl-acetophenone | *L. chilensis* [806] |
| **988** | 3′,5′,6′,6-Tetrabromo-2,4-dimethyldiphenyl ether | *L. similis* [807] |
| **989** | 2′,5′,6′,5,6-Pentabromo-3′,4′,3,4-tetramethoxybenzophenone | *L. similis* [807] |
| **990** | 2,5,6-Tribromo-3-bromoamino-7-bromomethylnaphthalene | *L. similis* [807] |
| **991** | 2,5,8-Tribromo-3-bromoamino-7-bromomethylnaphthalene | *L. similis* [807] |
| **992** | 1,2,5-Tribromo-3-bromoamino-7-bromomethylnaphthalene | *L. similis* [807] |

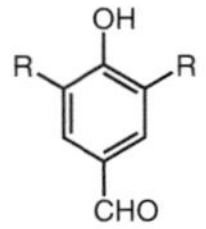

**984** R = H
**985** R = Br

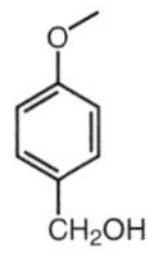

**986**

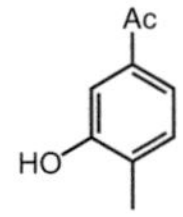

**987**

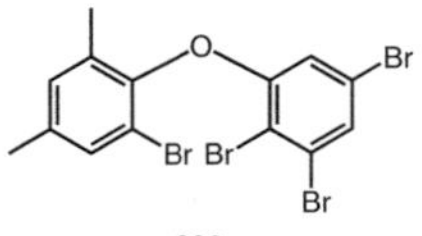

**988**

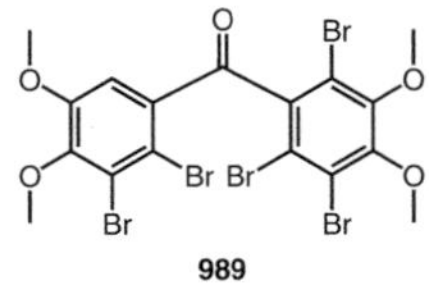

**989**

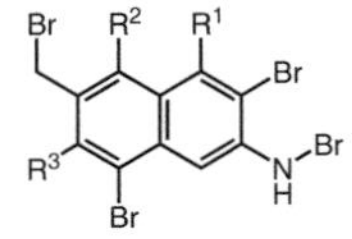

**990** $R^1 = R^2 = H, R^3 = Br$
**991** $R^1 = R^3 = H, R^2 = Br$
**992** $R^1 = Br, R^2 = R^3 = H$

## 3.7 Steroids

A total of 17 steroids (**993–1009**, Table 7) have been identified to date from species of the genus *Laurencia*.

Without exception, all steroids isolated thus far from *Laurencia* species are non-halogenated. Among them, 2a-oxa-2-oxo-5$\alpha$-hydroxy-3,4-dinorcholestane (**1009**), obtained from *L. obtusa* collected in the Lakshadweep Islands in India, represents the first example of a ring A bisnorsteroid isolated from a natural source [816].

**Table 7** Steroids from *Laurencia* spp.

| No. | Name | Source of isolation/synthesis |
|---|---|---|
| **993** | Cholesterol | *L. aldingensis* [373], *L. distichophylla* [264], *L. elata* [196], *L. karlae* [808], *L. majuscula* [809], *L. obtusa* [145], *L. okamurai* [60, 100], *L. tristicha* [290], *C. papillosus* [342] |
| **994** | Cholest-5-en-3$\beta$,7$\alpha$-diol | *L. okamurai* [100], *L. tristicha* [290] |
| **995** | β-Sitosterol | *L. tristicha* [290] |
| **996** | 24-Propylidene-cholest-5-en-3-ol | *L. papillosa* [810] |
| **997** | 3$\beta$,7$\beta$-Dihydroxy-5$\alpha$-cholestan-11-one | *L. papillosa* [811] |
| **998** | 3$\alpha$,6$\alpha$-Dihydroxy-5$\beta$-cholestan-12-one | *L. papillosa* [812] |
| **999** | Cholest-4-en-3-one | *L. obtusa* [98] |
| **1000** | 6$\beta$-Hydroxy-cholest-4-en-3-one | *L. papillosa* [812] |
| **1001** | (24*S*)-Stigmasta-4,25-diene-3$\beta$,6$\beta$-diol | *L. majuscula* [813] |
| **1002** | Stigmast-4-en-3$\alpha$,6$\beta$-diol | *L. saitoi* [794] |
| **1003** | Pinnasterol | *L. pinnata* [814[b]] |
| **1004** | Acetylpinnasterol | *L. pinnata* [814[a,b]] |
| **1005** | – | *L. pinnata* [815] |
| **1006** | – | *L. pinnata* [815] |
| **1007** | – | *L. pinnata* [815] |
| **1008** | – | *L. pinnata* [815] |
| **1009** | 2a-Oxa-2-oxo-5$\alpha$-hydroxy-3,4-dinorcholestane | *L. obtusa* [816] |

[a]The structure of this compound or of a derivative has been confirmed by X-ray crystallographic analysis

[b]The absolute stereochemistry of this compound has been established

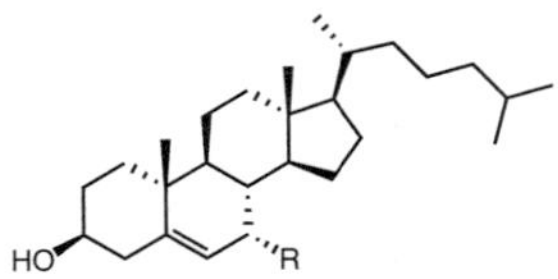

**993** R = H
**994** R = OH

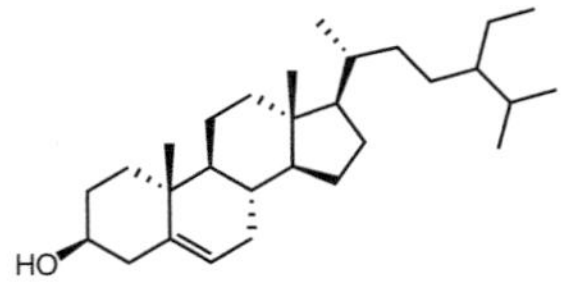

**995**

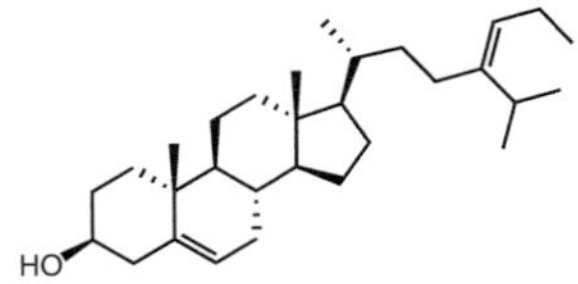

**996**

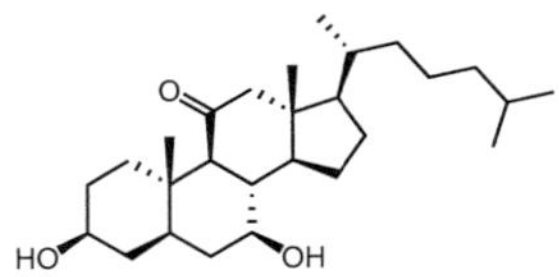

**997**

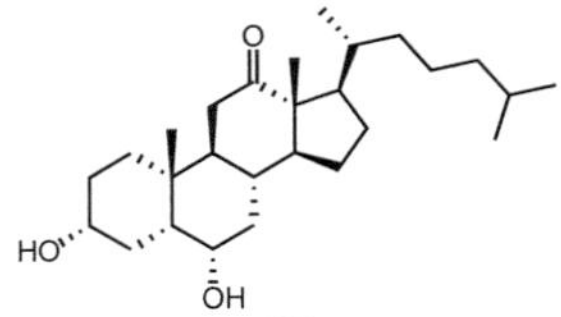

**998**

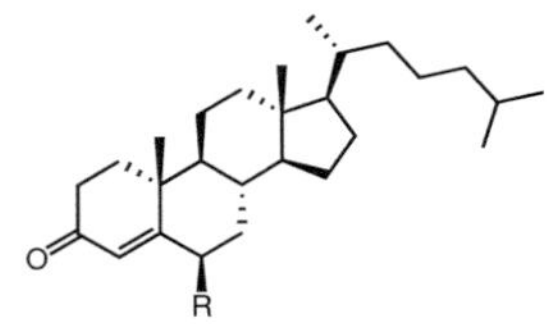

**999** R = H
**1000** R = OH

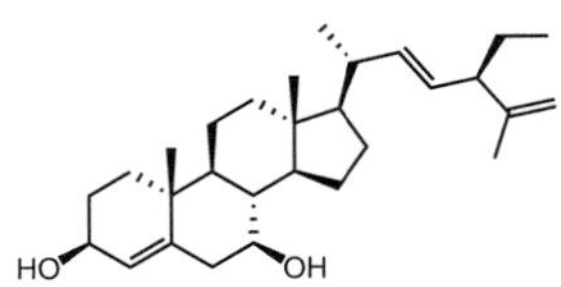

**1001**

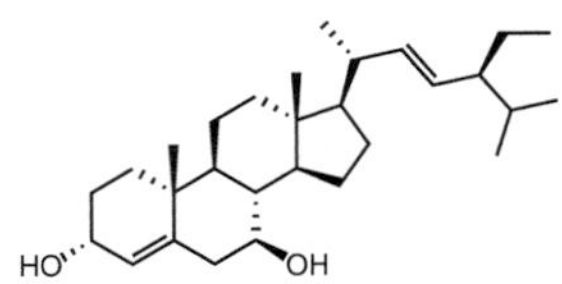

**1002**

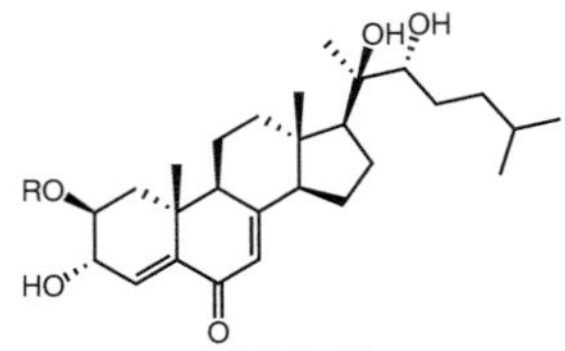

**1003** R = H
**1004** R = Ac

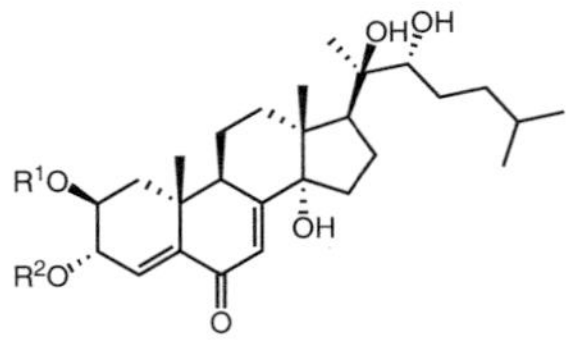

**1005** $R^1 = R^2 = H$
**1006** $R^1 = Ac, R^2 = H$
**1007** $R^1 = H, R^2 = Ac$

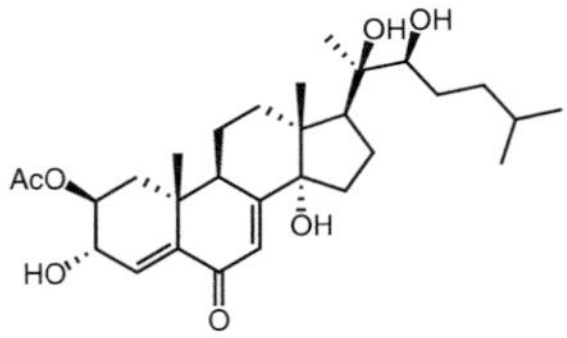

**1008**

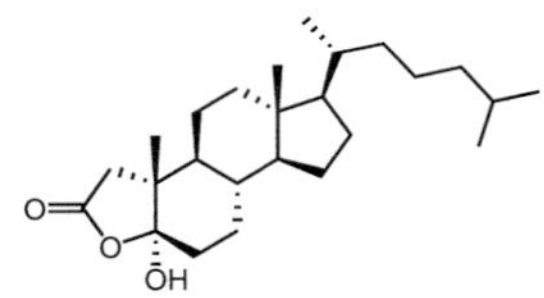

**1009**

## 3.8 Miscellaneous Metabolites

In addition to the *Laurencia* metabolites discussed above that can be assigned to distinct classes, such as sesquiterpenes, diterpenes, triterpenes, and acetogenins, 38 compounds isolated from *Laurencia* species feature atypical structures and cannot be classified into specific groups (**1010–1047**, Table 8).

Among them, the carotenoid zeaxanthin (**1010**) was isolated from *L. tristicha* [290]. Chilenones A (**1011**) and B (**1012**), a dimer and a trimer of 2-methyl-3(2*H*)-furanone, and the tetracyclic polyketal **1013** were isolated as naturally occurring racemates from *Laurencia chilensis* and their structures were elucidated on the basis of X-ray crystallographic analysis [817–819]. *Spiro*-bis-pinnaketal (**1014**) was isolated as a natural product from *Laurencia pinnatifida* [821], but was later proven to be an artifact derived from laurencione (**1015**), which was subsequently obtained as a labile mixture of tautomers from an extract of *Laurencia spectabilis* [822]. The related compounds **1016** and **1017** were isolated from *L. chilensis* [827]

**Table 8** Miscellaneous metabolites from *Laurencia* spp.

| No. | Name | Source of isolation/synthesis |
|---|---|---|
| **1010** | Zeaxanthin | *L. tristicha* [290] |
| **1011** | Chilenone A | *L. chilensis* [817[c]] |
| **1012** | Chilenone B | *L. chilensis* [818[c]] |
| **1013** | – | *L. chilensis* [819[c]]; Synthesis [820[e]] |
| **1014** | *spiro*-bis-Pinnaketal | *L. pinnatifida* [821[c], 822] |
| **1015** | Laurencione | *L. spectabilis* [822]; Synthesis [820[f], 823[f], 824[f,h], 825[f], 826[f,h]] |
| **1016** | 2-Methyl-3(2*H*)-dihydrofuranone | *L. chilensis* [827] |
| **1017** | 4-Hydroxy-2,3-dimethylnon-2-eno-4-lactone | *L. composita* [214] |
| **1018** | 5-Hydroxy-methylfurfural | *L. undulata* [828] |
| **1019** | *cis*-1-Methylcyclohexane-1,4-diol | *L. composita* [214] |
| **1020** | *trans*-1-Methylcyclohexane-1,4-diol | *L. composita* [214] |
| **1021** | Benkarlaol | *L. karlae* [829] |
| **1022** | 3*β*-Hydroxy-5*α*,6*α*-epoxy-β-ionone | *L. tristicha* [290] |
| **1023** | 3*β*-Hydroxy-5*β*,6*β*-epoxy-β-ionone | *L. tristicha* [290] |
| **1024** | 2-Bromo-γ-ionone | *L. saitoi* [335] |
| **1025** | – | *L. mariannensis* [74] |
| **1026** | Hexadecyl-1-*O*-*α*-L-arabinopyranoside | *L. majuscula* [830] |
| **1027** | *n*-Hexadecyl-*β*-D-pyranoside | *L. karlae* [808] |
| **1028** | Floridoside | *L. undulata* [831], *Laurencia* spp. [32] |
| **1029** | D-Isofloridoside | *L. pinnatifida* [832[a]], *L. undulata* [831] |
| **1030** | 1-*O*-Hexadecanoyl-3-*O*-*β*-D-galactopyranosylglycerol | *L. tristicha* [290] |
| **1031** | 1-*O*-Hexadecanoyl-3-*O*-*β*-D-galactopyranosylglycerol | *L. tristicha* [290] |
| **1032** | – | *L. pedicularioides* [833] |
| **1033** | – | *L. majuscula* [809] |
| **1034** | (2*Z*)-Chloro-pentadec-2-enal | *L. flexilis* [149]; Synthesis [834] |
| **1035** | (*E*)-2-((*E*)-Tridec-2-en-2-yl) heptadec-2-enal | *L. papillosa* [812] |
| **1036** | (*E*)-2-Tridecyl-2-heptadecanal | *L. filiformis* [89], *L. microcladia* [255], *L. obtusa* [145], *L. papillosa* [61[a]], *L. scoparia* [134], *L. undulata* [61[a]]; Synthesis [835] |
| **1037** | – | *L. nidifica* [836] |
| **1038** | Hybridalactone | *L. hybrida* [837[a], 838[c,d]]; Synthesis [839[g], 840[g], 841[g]] |
| **1039** | Papillamide | *L. papillosa* [842] |
| **1040** | – | *L. hybrida* [843] |
| **1041** | Pentadecanal | *L. filiformis* [89] |
| **1042** | (12*S*)-HEPE | *L. hybrida* [843, 844[b]], *L. spectabilis* [822] |
| **1043** | – | *L. okamurai* [845] |
| **1044** | – | *L. okamurai* [845] |

(continued)

**Table 8** (continued)

| No. | Name | Source of isolation/synthesis |
|---|---|---|
| **1045** | – | *L. okamurai* [845] |
| **1046** | – | *L. okamurai* [845] |
| **1047** | – | *L. okamurai* [845] |

[a]The isolation and structure elucidation of this compound was reported for the first time
[b]The structure of this compound has been revised
[c]The structure of this compound or of a derivative has been confirmed by X-ray crystallographic analysis
[d]The absolute stereochemistry of this compound has been established
[e]Partial synthesis
[f]Stereoselective total synthesis
[g]Enantioselective total synthesis
[h]Formal synthesis

and *Laurencia composita* [214], respectively, while 5-hydroxy-methylfurfural (**1018**) was recognized as a component of the edible alga *Laurencia undulata* [828]. The epimeric *cis*- (**1019**) and *trans*-1-methylcyclohexane-1,4-diol (**1020**) were also obtained from *L. composita* [214]. Benkarlaol (**1021**) was isolated from *Laurencia karlae* collected from the Nansha Islands in the South China Sea [829], the ionone derivatives **1022** and **1023** were obtained from *Laurencia tristicha* [290], 2-bromo-γ-ionone (**1024**) was found in *L. saitoi* [335], and the tetrahydropyran **1025** was extracted from *Laurencia mariannensis* collected from Hainan and Weizhou Islands of the People's Republic of China [74]. The glycolipids **1026** and **1027**, floridoside (**1028**) and D-isofloridoside (**1029**) and the related metabolites **1030–1032** were isolated from various species of *Laurencia*, all collected from mainland China, South Korea or India [290, 808, 830–833]. The iodolactone **1033** was isolated from *Laurencia majuscula* [809], and (*Z*)-2-chloro-pentadec-2-enal (**1034**) containing an unusual chloroenal functionality was obtained from *Laurencia flexilis* [149], and its synthesis was achieved in one step from commercial tridecanal [834], whereas the α,β-unsaturated aldehydes **1035** and **1036** have been isolated from a number of *Laurencia* species [61, 134, 145, 255, 812]. The latter metabolite is considered the aldol product of pentadecanal (**1041**), with which it was also isolated from *Laurencia filiformis* [89]. The sphingosine derivative **1037** was obtained from a Hawaiian *Laurencia nidifica* specimen [836], while the fatty acid amide **1039**, designated as papillamide, was isolated from a Japanese *Laurencia papillosa* [842]. Hybridalactone (**1038**), an unusual $C_{20}$ fatty acid-derived metabolite containing a cyclopropane and a macrolactone ring, was isolated from *Laurencia hybrida* and its structure, including the absolute stereochemistry, was established on the basis of analysis of its spectroscopic data, X-ray diffraction analysis, molecular mechanics calculations, chemical derivatizations, and total synthesis [837–841]. A number of fatty acid derivatives (**1040** and **1042–1047**) have been isolated from species of *Laurencia* [822, 843, 844]. Finally, common fatty acids, such as palmitic acid, eicosanoic acid, arachidonic acid and eicosapentaenoic acid, have been registered as constituents of the genus *Laurencia* [145, 373, 809, 846].

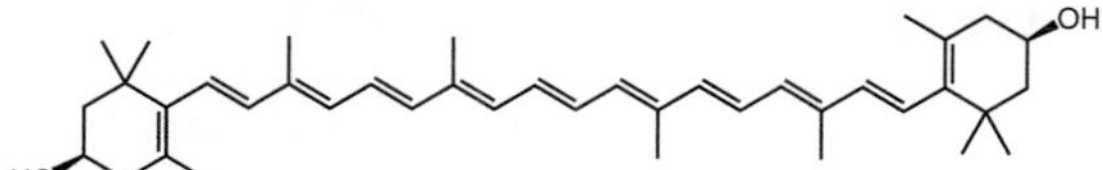

**1010**

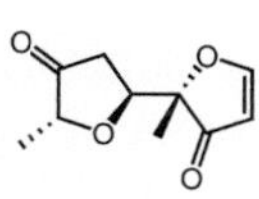

**1011**

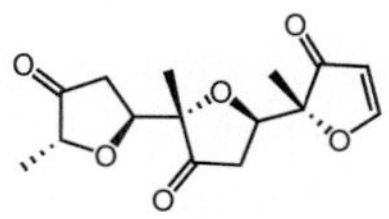

**1012**

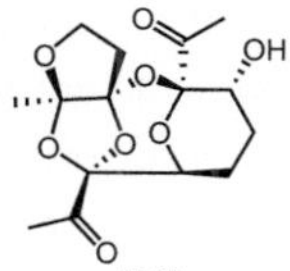

**1013**

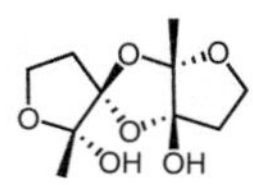

**1014**

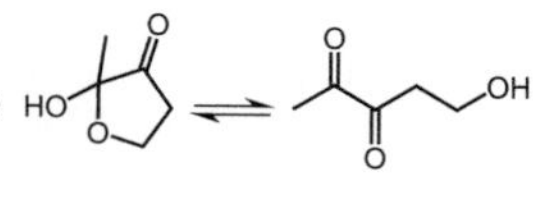

**1015**

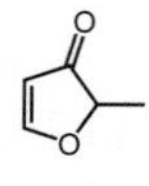

**1016**

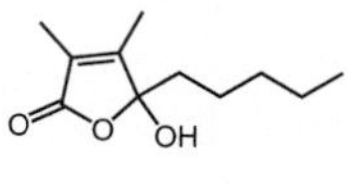

**1017**

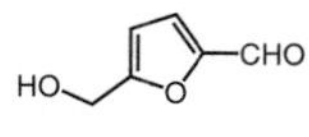

**1018**

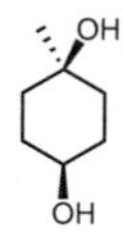

**1019**

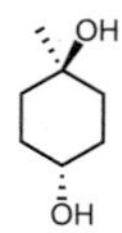

**1020**

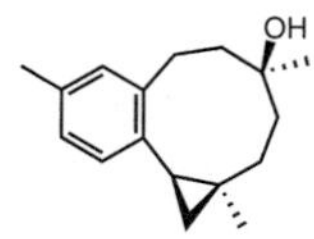

**1021**

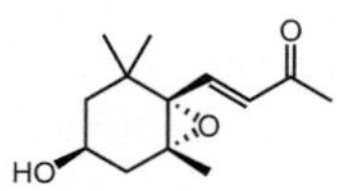

**1022**

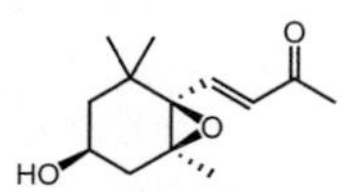

**1023**

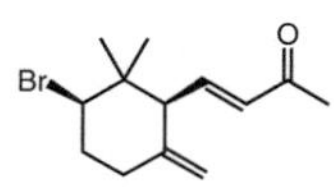

**1024**

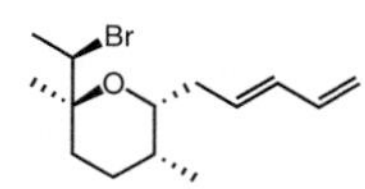

**1025**

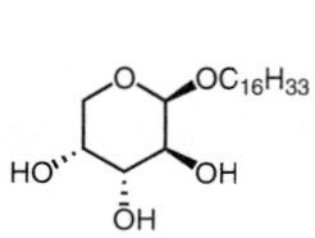

**1026**

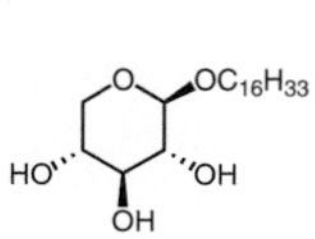

**1027**

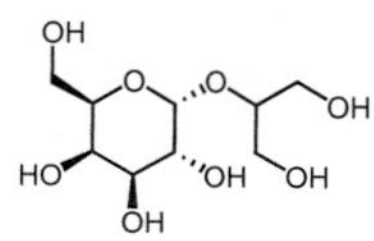

**1028**

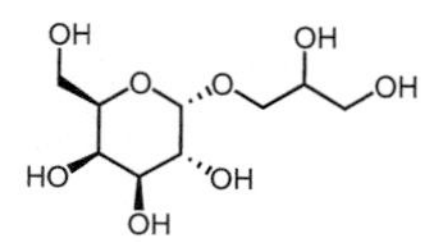

**1029**

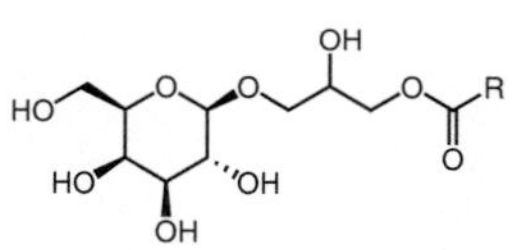

**1030** R = $C_{15}H_{31}$
**1031** R = $C_{17}H_{35}$

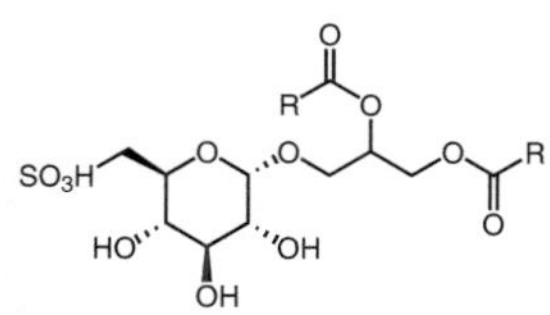

**1032** R = $C_{15}H_{31}$

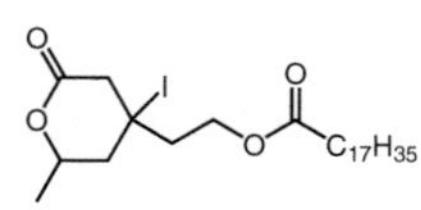

**1033**

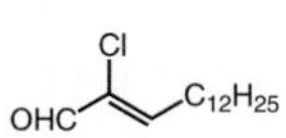

**1034**

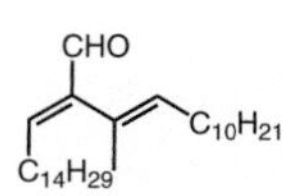

**1035**

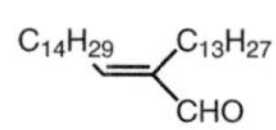

**1036**

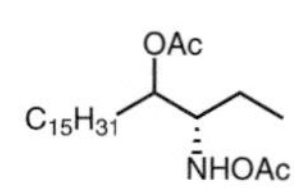

**1037**

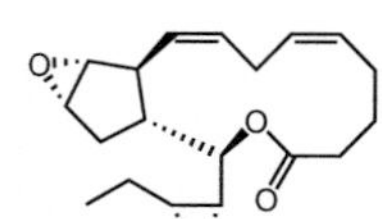

**1038**

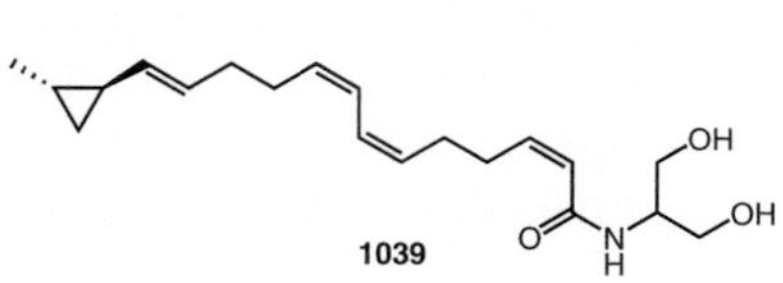

**1039**

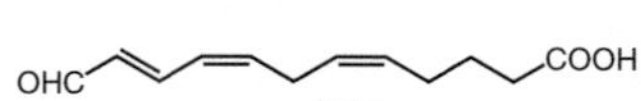

**1040**

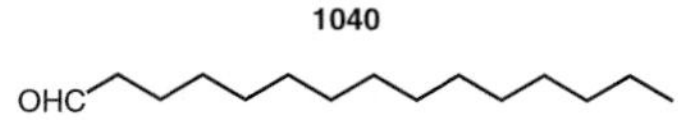

**1041**

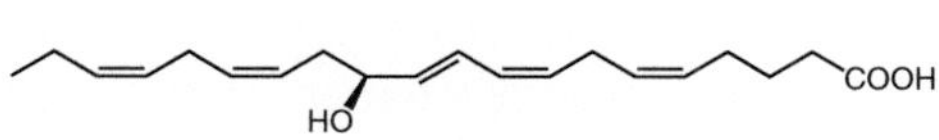

**1042**

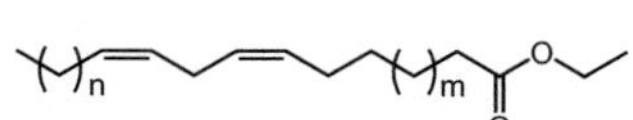

**1044** n = 4, m = 5
**1045** n = 5, m = 4

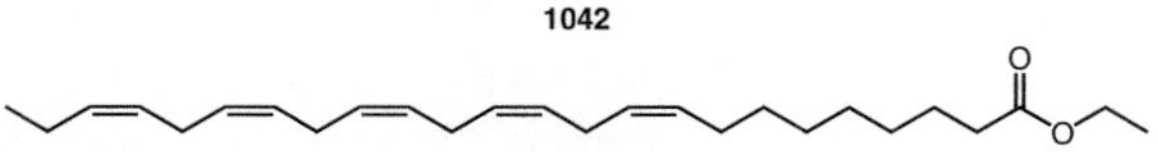

**1043**

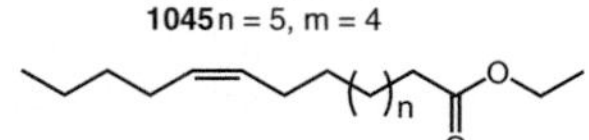

**1046** n = 8
**1047** n = 6

# 4 Biological Activity

The identification of natural products that can become lead molecules towards the development of new drugs requires investigation of the highest possible number of chemical entities in as many pharmacological activity screening assays as possible. In prescreening bioassays, large numbers of initial samples are tested so it can be determined whether or not some of them possess the desired type of activity. The detection of biological activity can be performed by either general screening bioassays or specialized screening bioassays. Depending on the strategy adopted, either a general screening, which can pick up many different effects, or a specific assay focused on finding effects against a specific disease, is selected. Specialized screening bioassays can be subdivided according to the target organisms that are used. These can be lower organisms e.g. microorganisms, insects, mollusks, protozoa, helminthes, and isolated subcellular systems, such as enzymes, receptors, organelles, isolated intact cells of human or animal origin, isolated organs or vertebrates, or whole animals [847]. Although the chemistry of *Laurencia* species has been investigated thoroughly, the biological activity evaluation of the isolated secondary metabolites has not been studied in a systematic way. The activity of *Laurencia* metabolites is organized below according to the biological target, regardless of the bioassays used for the evaluation.

## *4.1 Cytotoxic Activity*

Cancer remains a most serious health threat to mankind and is considered as one of the leading causes of death worldwide. Natural products frequently possess highly selective and specific biological activities and are recognized as a promising reservoir of bioactive compounds for modern medicine, particularly for the treatment of cancer. In vitro cell proliferation assays are the most widely used assays for evaluating preliminary anticancer activity of both natural products and synthetic derivatives. These assays give an indication of whole cell cytotoxicity, but in order to determine the exact molecular target specialized assays are needed. Although these screens have assisted in the bioguided isolation of many active compounds, they are considered non-specific for the discovery of the next generation of drugs. Fortunately, a large number of specific biochemical and genetics-based screens are now available routinely [848–850].

The chamigrane ma'iliohydrin (**36**) displayed cytotoxicity in the NCI 60-cell line human tumor screen at a mean panel $GI_{50}$ concentration of $10^{-5}$ *M*, with the greatest activity observed for the NCI/ADR-RES breast cancer cell line ($GI_{50} = 10^{-8}$ *M*) [148]. A dechloro-hydroxy analogue of hurgadol (**32**) was found to be weakly cytotoxic against HL-60 and THP-1 leukemia cells with $IC_{50}$ values of 102 and 152 μ*M*, respectively [144]. Dactylone (**39**) was proven effective at non-toxic doses as a cancer preventive agent, initiating early apoptosis in HL-60 and THP-1 cells [144, 851].

The chamigranes cartilagineol (**31**), **44**, **46**, elatol (**59**), isoobtusadiene (**64**) and mailione (**131**) exhibited remarkable cytotoxicity against P-388, A-549, HT-29 and MEL-28 cancer cell lines, but especially against HT-29 cells [142]. The cytotoxic properties of **44**, its geometric isomer (**46**), and isoobtusadiene (**64**) were evaluated against the NCI 60-cell line panel and were proven cytotoxic to specific cell lines of the colon cancer subpanel at concentrations 10–100-fold lower than the mean cytotoxic concentration observed for the other tumor subpanels [154].

The chamigrane sesquiterpenes elatol (**59**), allolaurinterol acetate (**152**), isoobtusol acetate (**50**), **44**, **46**, and the indole derivative **969** were tested for cytotoxicity against P-388 cells, but their potency was rather weak [156]. It is worth noting that the activity of compounds **44**, **46**, and **59** on the P-388 cancer cell line was inconsistent with previously published values.

Obtusadiene (**63**) and isoobtusadiene (**64**) were devoid of any discernible cytotoxic activity against the KB cancer cell line [131]. The chamigranes deschlorelatol (**56**), its acetate **57**, elatol (**59**), its acetate **60**, as well as the bisabolanes caespitol (**337**), and 8-acetylcaespitol (**338**) were evaluated against HeLa, HEP-2 and non-tumor Vero cells and **59** proved to be the most active of the tested compounds, while **337** was slightly active and **56** and **57** were inactive [86].

Pacifenol (**105**), cycloelatanenes A (**119**) and B (**118**), and the acetogenin (3*Z*)-chlorofucin (**875**) displayed negligible cytotoxic activity when tested [221]. The inhibition of cytokinesis in the embryos of the sea urchin *Tetrapygus niger*, used as a preliminary test for an indication of antineoplastic activity, revealed that 4,10-dibromo-3-chloro-α-chamigrene (**5**), prepacifenol (**88**), deoxyprepacifenol (**87**), and claviol (**124**) had no effect on the cytokinesis, while **6** and pacifenol (**105**) displayed only weak inhibition [87]. Chemical transformation of pacifenol (**105**) afforded a number of derivatives that were also evaluated for their cytotoxicity [852].

Among the sesquiterpenes johnstonol (**113**), **163–165**, aplysin (**195**), 10-hydroxy-aplysin (**196**), 10-hydroxy-bromo-*epi*-aplysin (**198**), 10-hydroxy-*epi*-aplysin (**199**), laurentristich-4-ol (**246**), debromoaplysinol (**202**), aplysinol (**204**), debromo-*epi*-aplysinol (**210**), aplysin-9-ene (**201**), *epi*-aplysinol (**211**), **222**, cyclolauren-2-ol (**228**), laurebiphenyl (**239**), gossonorol (**319**), **320**, **327**, **328**, and the linear diterpene (*E*)-phytol (**641**), only laurebiphenyl (**239**) showed moderate cytotoxicity against the tested cancer cell lines (A-549, BGC-823, BEL-7402, HCT-8, and HeLa), while **202** and **210** were selectively cytotoxic against the HeLa cell line [232, 290, 295].

Laurepoxyene (**185**), 10-hydroxy-aplysin (**196**), 3β-hydroperoxy-aplysin (**197**), 10-hydroxy-*epi*-aplysin (**199**), 3α-hydroperoxy-3-*epi*-aplysin (**200**), aplysinol (**204**), 10-bromoisoaplysin (**207**), 8,10-dibromoisoaplysin (**208**), laurokamurene A (**242**), (5*S*)-5-acetoxy-β-bisabolene (**307**), and *seco*-laurokamurone (**469**) were evaluated for their cytotoxicity against the HL-60 and A-549 human cancer cell lines, but only **196** and **197** showed moderate levels of inhibitory activity against A-549 cells [275].

Aplysin (**195**), isolaurinterol (**161**), aplysinal (**209**), and laurinterol (**231**) were found to be cytotoxic against the A-549, SK-OV-3, SK-MEL-2, XF-498, and HT-15 cancer cell lines [268]. Noteworthy were the inconsistent results for aplysin (**195**) on the A-549 cell line in comparison to earlier published data. Aplysin (**195**) was reported to act as a sensitizer for TRAIL (TNF-related apoptosis-inducing ligand) using several TRAIL-resistant cancer cell lines and to potentiate TRAIL-induced apoptosis in A-549 and MCF-7 cells [853].

Debromoallolaurinterol (**149**), *iso*-laurenisol (**174**), bromolaurenisol (**175**), **184**, laurinterol (**231**), perforenone A (**377**), 3-*epi*-perforenone A (**379**), perforenol B (**384**), **393**, **421**, and (*E*)-2-tridecyl-2-heptadecenal (**1036**) were evaluated against five human cancer cell lines (K-562, MCF-7, PC-3, HeLa, A-431) and Chinese hamster ovary (CHO) cells [255]. Moreover, the cytotoxicity of bromolaurenisol (**175**), laurinterol (**231**), **240**, and **421** was evaluated against the NSCLC-N6 and A-549 cancer cell lines [255]. A *L. okamurai* extract containing laurinterol (**231**) was reported to induce apoptosis in melanoma cells [854]. Debromolaurinterol (**226**), laurinterol (**231**) and its acetate **232** were moderately cytotoxic against HeLa cells, while debromolaurinterol acetate (**227**) was inactive [309]. Allolaurinterol (**151**) and its acetate **152**, isolaurenisol (**176**) and its acetate **177** were tested for cytotoxicity against the P-388 and BSC-1 cancer cell lines, and showed that the moderate levels of activity exhibited by **151** and **176** were even lower for their acetates **152** and **177** [260].

The laurane sesquiterpene **217**, bromolaurinterol (**235**), 8-iodolaurinterol (**236**), the dimer **238**, *ent*-$\alpha$-bromocuparene (***ent*-415**), and $\alpha$-isobromocuparene (**417**) were devoid of any significant activity against five human cancer cell lines (HT-29, MCF-7, PC-3, HeLa, and A-431) [262]. 10-Hydroxy-isolaurene (**181**), and the rearranged metabolites **242** and **422** were cytotoxic towards the NSCLC-N6 and A-549 lung cancer cell lines, with **181** and **242** exhibiting the highest levels of activity [274].

The cytotoxic activity against KB, HEP-G2 and MCF-7 cancer cell lines of the compounds laur-2-ene-3,12-diol (**167**), **245**, **420**, and kahukuen-10-ol B (**633**) was evaluated. Kahukuen-10-ol B (**633**) was active against all cell lines, while the cuparane **420** exhibited moderate activity against KB cells, laur-2-ene-3,12-diol (**167**) exhibited moderate activity towards KB and MCF-7 cells, and **245** was inactive [271]. Isolauraldehyde (**183**) and 8,11-dihydro-12-hydroxy-isolaurene (**191**) showed very promising activity in an in vitro model of Ehrlich ascites carcinoma [98]. *epi*-Obtusane (**51**) and oculiferane (**428**) exhibited high levels of cytotoxic activity against the PC-3, A-549, MCF-7, HEP-G2, and HCT-116 cell lines [167].

The snyderanes α-snyderol (**247**) and its acetate (**248**), and **256**, **258**, **264**, and **265** did not exhibit cytotoxic activity against a panel of cell lines (KB, LU1, Col 2, LNCaP) [319]. Aplysistatin (**276**) and its 6$\beta$-hydroxy derivative **277** were cytotoxic the against KB cell line [343, 855] and the former was also active against P-388 cells [343]. Luzonensol (**252**), **254**, isopalisol (**261**), palisadin B (**266**), 5$\beta$-acetoxy-palisadin B (**268**), 4-hydroxy-palisadin C (**272**), aplysistatin (**276**), palisadin A (**278**), and the aristolanes **451**, **452**, **454**, **457–459**, **461**, and **462** were proven inactive against BEL-7402 cells [325]. Aplysistatin (**276**), **454**, and **460** were inactive against K-562, SGC-7901, and SMMC-7721 cells, the aristolane derivative **451** was found cytotoxic, whereas (*E*)-phytol (**641**) showed only weak activity against SGC-7901 cells [342]. The aristolanes ***ent*-451** and **453** displayed cytotoxic effects against selected cancer cell lines, while compounds **454** and **455** were inactive [432]. The bisabolane-related sesquiterpene majapolene A (**362**) displayed modest activity in the NCI 60-cell line cytotoxicity screen [390].

Obtusol (**28**), elatol (**59**), and compound **92** were cytotoxic against the HM02, HEP-G2 and MCF-7 cancer cell lines, while isoobtusol acetate (**50**), puertitol B acetate (**312**), caespitane (**333**), caespitol (**337**), 8-acetylcaespitol (**338**), caespitenone (**349**), furocaespitane (**352**), laucapyranoid A (**372**), 3-*epi*-perforenone A (**379**), **381**, 4-hydroxy-1,8-*epi*-isotenerone (**397**), isopinnatol B (**553**), dactylomelol (**600**), and dactylopyranoid (**631**) were inactive in this regard [50, 399]. Elatol (**59**) was also active against L-6 muscle myoblast cells with a *MIC* value of 3.3 μg/cm$^3$ [50]. Obtusol (**28**), isoobtusol (**49**), and its acetate **50** and the bisabolane derivatives caespitol (**337**), isocaespitol (**346**), and its acetate **347** displayed low cytotoxicity against the HeLa-229 cancer cell line, with isoobtusol acetate (**50**) being the most active ($LD_{50} = 4.5$ μg/cm$^3$) [384]. Semisynthetic derivatives of **49** were prepared and tested for their cytotoxicity against the A-549 and RD cancer cell lines but without evidence of improved cytotoxicy in comparison to the natural compound [856].

Caespitane (**333**), caespitol (**337**), 8-acetylcaespitol (**338**), **342**, and caespitenone (**349**) were found to be moderately active against the HT-29, MCF-7 and A-431 cell lines, exhibiting activity with $IC_{50}$ values lower than 20 μ*M*, whereas metabolites **335**, **341** and **356** displayed weak activity and **343**, 8-desoxyisocaespitol (**345**), isocaespitol (**346**), **348**, the lactol **354**, and **355** were inactive [370]. Deodactol (**334**) showed moderate cytotoxic activity against L-1210 cells [371].

In vivo and in vitro studies were undertaken to clarify the mechanism of action of elatol (**59**). It seems that it promotes a delay in the cell cycle, probably in the G1/S transition, activating the apoptotic process, but further studies are needed to confirm this [857]. Another study has reported potent cytotoxic properties for isoobtusol (**49**), (+)-elatol (**59**), and palisadin A (**278**) [166].

The eudesmane-related $C_{12}$ terpene teuhetenone A (**449**) and the *seco*-eudesmane chabrolidione B (**448**) were found inactive against the HEP-G2, HEP-3B, MCF-7, MDA-MB-231, and A-549 cancer cell lines [858]. In another study, teuhetenone A (**449**) exhibited higher cytotoxic activity against the MCF-7 cancer cell line in comparison to the eudesmane **440** and chabrolidione B (**448**) [423].

The diterpenes parguerol (**514**) and its 16-acetate (**517**), deoxyparguerol (**525**), isoparguerol (**540**), and its 16-acetate (**541**) exhibited cytotoxic activity against the P-388 cancer cell line [156]. The cytotoxicity of deacetylparguerol (**513**), parguerol 16,19-diacetate (**519**), parguerol triacetate (**521**), 2-deacetoxy-deoxyparguerol (**523**), **524**, deoxyparguerol (**525**), **532**, **539**, isoparguerol 7,16-diacetate (**542**), and the A-*seco*-parguerane **544** was evaluated against the HeLa and P-388 cancer cell lines and only **519**, **521**, **525**, **542**, and **544** exhibited activity, while **544** was found cytotoxic against the B-16 cancer cell line ($IC_{50} = 0.78$ μg/cm$^3$) [475, 479, 483]. Parguerol 7,16-diacetate (**520**) and parguerol triacetate (**521**) were found to inhibit P-glycoprotein in multidrug resistant human cancer cells [859].

The cytotoxicity against Ehrlich carcinoma tumor cells of parguerol 16-acetate (**517**), parguerol 7,16-diacetate (**520**), deoxyparguerol (**525**) and its 7-acetate (**526**), isoparguerol (**540**), its 16-acetate (**541**) and 7,16-diacetate (**542**) were evaluated. The cytotoxicity of the isoparguerol derivatives was determined as being more potent than that of the parguerol derivatives, while the deoxyparguerol derivatives showed the lowest activity [860]. Virtual screening of the SeaWeed Metabolite Database (SWMD) resulted in the identification of four potential lead compounds with the parguerane skeleton that showed good binding with β-tubulin [861].

Luzodiol (**590**) and the dactylomelanes **592–599** were found inactive when tested against five human tumor cell lines (HeLa, MCF-7, A-431, HEP-G2, and A-549) [515]. The obtusane diterpene **609** exhibited cytotoxicity against the KB ($ED_{50} = 4.5$ μg/cm$^3$) and PS ($ED_{50} = 10$ μg/cm$^3$) cell lines in NCI in vitro bioassays [519]. The rearranged obtusane laurenditerpenol (**613**) exhibited potent inhibition of HIF-1-mediated hypoxic signaling in breast tumor cells [521]. Two analogues featuring the 7-oxabicyclo[2.2.1]heptane ring system present in the natural product were prepared and their weak inhibition of HIF-1 activation indicated the importance of the cyclohexenol part of laurenditerpenol (**613**) [862]. The diterpenes prevezols B-E (**617–620**) were tested for their cytotoxicity against five human tumor cell lines (MCF-7, PC-3, HeLa, A-431, and K-562) and prevezols B (**617**) and C (**618**) were found to be the most potent, while prevezol E (**620**) was inactive [528]. Dactyloditerpenol acetate (**635**) was tested for cytotoxicity against the A-2058 and DU-145 cancer cell lines, but displayed weak activity as the observed reductions of cell viability were only 67% and 77%, respectively [178]. Viridiols A (**624**) and B (**625**) showed significant cytotoxicity against the P-388, A-549, HT-29, and MEL-28 cancer cell lines, while only the former exhibited slight activity on RNA synthesis and neither compound had any inhibitory effect on protein and DNA synthesis [534].

The triterpenes thyrsiferol (**646**), its 23-acetate (**647**), 15(28)-anhydrothyrsiferyl diacetate (**651**), 15-anhydrothyrsiferyl diacetate (**657**), and magireols A (**665**), B (**666**), and C (**667**) displayed cytotoxicity against the P-388 cancer cell line, with compound **647** being the most active [544, 557]. Thyrsiferol (**646**) also showed cytotoxicity against the A-549, HT-29 and MEL-28 cancer cell lines [548]. Thyrsiferyl 23-acetate (**647**) was found to potently and specifically inhibit serine/threonine protein phosphatase 2A (PP2A) without affecting other phosphatases (PP1, PP2B, PP2C, PTP) and to induce apoptosis in various T- and B-leukemia cells [863, 864]. Thyrsiferol (**646**) and 15(28)-anhydrothyrsiferyl 23-acetate (**650**) also induced apoptosis, but they did not inhibit PP2A [864]. Furthermore, compound **646** inhibited mitochondrial respiration and HIF-1 activation [542]. Teurilene (**699**) did not show any cytotoxicity against P-388 cells [544], but was found to be cytotoxic against KB cells ($IC_{50} = 7.0$ μg/cm$^3$) [865].

The mechanisms of growth inhibition by the cytotoxic compound dehydrothyrsiferol (**649**) have been studied extensively [548, 563, 568, 569]. In particular, **649** was shown to circumvent multidrug resistance mediated by P-glycoprotein [866] and to cause growth inhibition in KB cancer cells without inducing apoptosis [867], but did not modulate multidrug resistance-associated protein 1 (MRP1) resistance [868]. Its effects on estrogen-dependent and -independent breast cancer cells were evaluated [869] and its observed inhibition activity on VLA integrins was considered as a potent tool in controlling metastasis in breast cancer cells [870]. Furthermore, **649** showed inhibition of PP2A [871]. The antimitotic activities of a synthetic analogue, 7,11-*epi*-thyrsiferol, were evaluated and a structure-activity relationship (SAR) model was developed for such compounds [872].

16-Hydroxy-dehydrothyrsiferol (**652**), 10-*epi*-dehydrothyrsiferol (**655**), 10-*epi*-15,16-dehydrothyrsiferol (**658**), dehydrovenustatriol (**660**), 15,16-dehydrovenustatriol (**662**), isodehydrothyrsiferol (**673**), thyrsenols A (**677**) and B (**678**), and predehydrovenustatriol acetate (**692**) were proven to possess significant cytotoxicity against P-388, A-549, HT-29, and MEL-28 cancer cells [548, 560, 568]. 16-*epi*-Hydroxy-dehydrothyrsiferol (**653**), martiriol (**681**), pseudodehydrothyrsiferol (**685**), and dioxepan-dehydrothyrsiferol (**698**) exhibited moderate cytotoxic activities against the same panel of cancer cell lines [558].

The PP2A inhibition activity of the triterpenes **652**, **653**, **655**, **658**, **660**, **662**, **673**, **677**, **678**, **683-686**, **688**, and **698** was evaluated and the most potent metabolites were found to be 16-hydroxy-dehydrothyrsiferol (**652**) and thyrsenol B (**678**) [871].

22-Hydroxy-15(28)-dehydrovenustatriol (**661**), iubol (**674**), 15-dehydroxythyrsenol A (**679**), 1,2-dehydropseudodehydrothyrsiferol (**682**), prethyrsenol A (**689**) 13-hydroxy-prethyrsenol A (**690**), and secodehydrothyrsiferol (**693**) had significant cytotoxicity against the Jurkat, MM-144, HeLa and CADO-ES-1 cell lines [563, 569]. Since metabolites **679**, **689**, and **690** have limited water solubility, four semisynthetic analogues modulating the solubility of the lead compound, dehydrothyrsiferol (**649**), were prepared and found to maintain the same levels of activity [569].

Callicladol (**672**) showed cytotoxicity against P-388 murine leukemia cells [566]. Aplysiols A (**648**) and B (**668**) displayed significant cytotoxic activity against P-388 tumor cells with $IC_{50}$ values of 0.6 and 6.6 mg/cm$^3$, respectively [543]. Metabolites **680**, **694**, and **695** did not show any discernible cytotoxicity against the HS-578-T and T-47D cell lines [570]. The triterpenes saiyacenols A (**669**) and B (**670**) exhibited cytotoxicity against Jurkat, MM-144, HeLa, and CADO-ES-1 cells [555]. Aplysqualenol A (**675**) exhibited potent cytotoxic properties against the MCF-7, NCI-H460 and SF-268 cancer cell lines [567, 873]. The triterpenes aurilol (**696**) and auriculol (**703**) exhibited cytotoxicity against HeLa $S_3$ cells [573, 589].

The acetogenins **722**, **723**, **726**, and **948** were evaluated for their cytotoxicity against the HT-29, MCF-7, PC-3, HeLa, and A-431 human tumor cell lines, but were devoid of any significant activity [607]. The need for the development of a computational (in silico) prescreening method for activity prediction led Brogi and co-workers to the creation of a ligand-based pharmacophore for MCF-7 cells, which was able to predict with accuracy the cytotoxicity levels for an array of compounds against this cell line [874].

When tested, the laurenan-type acetogenin (3*Z*)-dihydro-rhodophytin (**810**) was found to be devoid of cytotoxicity against a a panel of normal and tumor cell lines [691]. Elatenyne (**735**), laurendecumbenynes A (**885**) and B (**733**), and laurendecumallenes A (**901**) and B (**857**) were not active against the A-549 cancer cell line [611, 612]. Laurefurenynes A (**737**), B (**736**), C (**879**), D (**878**), E (**881**), and F (**880**) were tested for cytotoxicity against the murine L-1210, Colon 38, CFU-GM, H-116, H-125, and human CFU-GM cell lines and only laurefurenynes C (**879**) and F (**880**) were moderately cytotoxic, but without showing any selectivity [619].

Marilzallene (**831**), (+)-4-acetoxy-marilzallene (**832**), obtusallene IV (**906**), 12-epoxy-obtusallene IV (**907**), **908**, marilzabicycloallenes A (**918**), and B (**921**) and (*E*)-adrienyne (**936**) were inactive against the A-2780, HBL-100, HeLa, SW-1573, T-47D, and WiDr solid tumor cell lines [711, 777].

Only 2,3,5,6-tetrabromoindole (**952**), among the simple bromoindoles **952**, **966**, **969**, and **971**, showed cytotoxic activity against the L-1210 tumor cell line [793]. Metabolites **952** and **971** exhibited significant cytotoxicity against the K-562, SGC-7901 and SMMC-7721 tumor cell lines, while compound **950** displayed strong cytotoxic activity only for the K-562 cell line, whereas cholesterol (**993**) was found to be inactive [342]. The monoindoles **952**, **969**, and **971**, the bisindole **974**, and the *spiro*-trisindoles similisines A (**983**) and B (***ent*-983**) did not exhibit cytotoxicity against eight human cancer cell lines (MCF-7, BGC-823, HeLa, A-549, HCT-8, HEP-G2, HL-60, Jurkat) at a concentration of 10 μg/cm$^3$, while metabolite **973** showed very weak cytotoxic activity only against the HL-60 ($IC_{50} = 35.1$ μ*M*) and Jurkat ($IC_{50} = 53.3$ μ*M*) cell lines [792]. Metabolite **969** was not active against P-388 cell line [156]. When compounds **952**, **969**, and **971** were evaluated for their cytotoxic effects against selected cancer cell lines, only **952** was found active [432]. Tetrabromoindole **952** and the sterol stigmast-4-en-3*α*,6*β*-diol (**1002**) were devoid of cytotoxic properties [794]. When the indoles **955–958**, **960–963**, **965**, and **975–977** were evaluated against the HT-29 and P-388 cancer cell lines, compound **976** exhibited cytotoxicity against both cancer cell lines, while **977** was active only against the P-388 cell line [799]. The cytotoxic activity of 4,5,6-tribromo-2-methylsulfinylindole (**964**), along with other known related compounds was evaluated [801].

4-Hydroxy-benzaldehyde (**984**) and 4-methoxy-benzyl alcohol (**986**) were found inactive when tested against the KB cancer cell line [804]. Indolyl-3-carbaldehyde (**972**), 4-hydroxy-benzaldehyde (**984**), cholesterol (**993**), β-sitosterol (**995**), zeaxanthin (**1010**), the ionone derivatives **1022** and **1023**, and the glycosides **1030** and **1031** did not display cytotoxicity, when evaluated against several human tumor cell lines (A-549, BGC-823, BEL-7402, HCT-8, and HeLa), whereas cholest-5-en-3*β*,7*α*-diol (**994**) was cytotoxic against all these cell lines [290].

(24*S*)-Stigmasta-4,25-diene-3*β*,6*β*-diol (**1001**) when tested was found to be cytotoxic against P-388 cell line ($IC_{50} = 1.2$ μg/cm$^3$) [813]. The arabinoside **1026** exhibited significant antitumor activity in the LOVO and BEL-7402 cell lines and additional studies were performed to clarify the mechanism of this action [830]. (*E*)-2-((*E*)-Tridec-2-en-2-yl)-heptadec-2-enal (**1035**) demonstrated moderate cytotoxic activity against several cell lines in which it was tested (CHO, K-562, MCF-7, PC-3, HeLa, A-431) [255], whereas papillamide (**1039**) did not exhibit cytotoxic properties against the P-388 and B-16 cancer cell lines [842].

## *4.2 Antibacterial and Antifungal Activity*

The already emerged and increasing resistance to antibiotics is becoming a global threat to public health. Only 80 years after their introduction, humans are now facing the possibility of a future without effective antibiotics. Methicillin-resistant *Staphylococcus aureus* alone infects more than 94,000 people and kills yearly 19,000 in the U.S., more deaths than those caused by HIV/AIDS, Parkinson's disease, emphysema, and homicides combined [875].

A large number of *Laurencia* metabolites have been evaluated against different bacterial and fungal strains exhibiting various levels of activity. Many of these metabolites, isolated from independent sources, were submitted to antimicrobial activity evaluation assays utilizing different methods and/or different strains of microorganisms.

Among the *Laurencia* sesquiterpenes, elatol (**59**) can be singled out as the metabolite tested for its antimicrobial activity in the highest number of investigations, but the reported results are frequently contradictory, as even against the same strains variable levels of activity have been reported over the years [50, 81, 135, 157, 165, 166, 855, 876, 877].

Isoobtusol (**49**), isoobtusol acetate (**50**), and obtusol (**175**) are also found as common constituents in several *Laurencia* species and their antibacterial and antifungal activities have been reported in a number of articles [50, 135, 155, 384, 876]. Evaluation of the related sesquiterpenes allolaurinterol (**151**) [166, 248, 259, 260, 855, 877], allolaurinterol acetate (**152**) [152, 260], debromoallolaurinterol (**149**) [248], debromoallolaurinterol acetate (**150**) [152], isolaurinterol (**161**) [62, 201, 267, 279, 877], laurinterol (**231**) [62, 105, 137, 309, 877], debromolaurinterol (**226**) [309, 855, 878], and isodebromolaurinterol (**229**) [201, 267, 279] has been performed in several, albeit fragmented, studies, which do not allow direct comparison of the structural features with the reported antimicrobial activities.

The bisabolane sesquiterpenes caespitane (**333**) [50, 135], caespitol (**337**) [50, 384], 8-acetylcaespitol (**338**) [50], isocaespitol (**346**) [384], caespitenone (**349**) [50, 135], and furocaespitane (**352**) [50] have been evaluated for their antibacterial and/or antifungal activities in a number of investigations.

Furthermore, a significant number of *Laurencia* sesquiterpenes including (–)-(10*R*)-bromo-α-chamigrene (**1**) [81], laurecomin C (**2**) [84], **7** [84], laurokamin C (**12**) [79], 10-bromo-9-hydroxy-α-chamigrene (**14**) [81], dendroidiol (**34**) [146], laurecomin B (**37**) [84], 10-bromo-β-chamigren-8-ol (**41**) [152, 275], **44** [62, 155] and its geometric isomer **46** [62, 155], obtusane (***ent*-47**) [146], rogiolol (**48**) [162], rigidol (**53**) [81], (+)-(10*S*)-10-bromo-β-chamigrene (**55**) [81], ***ent*-56** [62], obtusadiene (**63**) [131], isoobtusadiene (**64**) [131], **76** [62], chamigrane **81** [187], **82** [152, 275], deoxyprepacifenol (**87**) [84, 201, 267, 279], prepacifenol (**88**) [878], **92** [50], laurecomin D (**94**) [84], yicterterpene A (**111**) [229], yicterterpene B (**112**) [229], laurokamin B (**120**) [79], pannosane (**137**) [238], laurokamin A (**139**) [79], laurecomin A (**140**), pannosanol (**142**) [84], laurene (**148**) [152], isolaurenisol (**176**) [260], **177** [260], **181** [201, 267, 279], 12-hydroxy-isolaurene (**182**) [275], isolauraldehyde (**183**) [275], laurepoxyene (**185**) [275], 8,11-dihydro-12-hydroxy-isolaurene (**191**) [275], 10-hydroxy-aplysin (**196**) [201, 267, 279], **197** [152, 275], **200** [152, 275], debromoaplysinol (**202**) [201, 267, 279], aplysinol acetate (**205**) [201, 267, 279], laureperoxide (**206**) [201, 267, 279], 10-bromo-isoaplysin (**207**) [201, 267, 275, 279], **208** [275], filiformin (**214**) [201, 267, 279], filiforminol (**218**) [166, 257], **227** [309], isodebromolaurinterol (**229**) [201, 267, 279], **232** [309], cyclolaurenol (**233**) [305], cyclolaurenol acetate **234** [305], laurokamurene A (**242**) [201, 267, 275, 279], laurokamurene B (**241**) [201, 267, 279], laurokamurene C (**243**) [275], aplysistatin (**276**) [339, 342, 855], **277** [855], palisadin A (**278**) [166, 337, 339], palisadin B (**266**) [337, 339], 5*β*-acetoxy-palisadin B (**268**) [337, 339], 12-hydroxy-palisadin B (**269**) [337, 339], isodactyloxene A (**302**) [74], 3-acetoxy-(*E*)-γ-bisabolene (**305**) [81], puertitol B acetate (**312**) [50], isocaespitol acetate (**347**) [384], majapolene A (**362**) [193], majapolene B (**364**) [193], majapolene A acetate (**363**) [193], majapolene B acetate (**365**) [193], tiomanene (**366**) [193], laucapyranoid A (**372**) [50], 3-*epi*-perforenone A (**379**) [399], **381** [399], 4-hydroxy-1,8-*epi*-isotenerone (**397**) [399], α-isobromocuparene (**417**) [201, 267, 279], cupalaurenol (**418**) [187, 305, 877], cupalaurenol acetate (**419**) [305], the cuparene-type ether **421** [201, 267, 279], **424** [152], **434** [84], **435** [84], cycloeudesmol (**446**) [878], **451** [342], **454** [342], **460** [342], **490** [81, 146, 458], **492** [458], and dactylomelatriol (**504**) [135] were evaluated for their antibacterial and/or antifungal properties, exhibiting variable activity levels.

The aristolanes ***ent*-451** and **453** displayed good antibacterial activity against antibiotic-resistant bacteria, while compounds **454** and **455** when tested in the same assay were inactive [432]. The eudesmane **440** and the *seco*-eudesmane chabrolidione B (**448**) exhibited significant antifungal activity, while teuhetenone A (**449**) was less active in this regard [423].

The diterpenes deacetylparguerol (**513**), parguerol (**514**), **524**, deacetylisoparguerol (**539**), and isoparguerol (**540**) were tested for their antibacterial activity against seven pathogenic bacterial strains (*Escherichia coli*, *Staphylococcus aureus*, *S. albus*, *Bacillus cereus*, *Vibrio parahemolyticus*, *Vibrio anguillarum*, and *Pseudomonas putida*). None of these compounds showed evident antibacterial activity [476]. The labdanes **560** and **561** were proven inactive against several pathogenic bacteria (*E. coli*, *Staphylococcus epidermis*, *S. aureus*, *Salmonella* sp., and *Pseudomonas* sp.) [106].

Moreover, the diterpenes isopinnatol B (**553**) [50], **556** [105], 10-acetoxy-angasiol (**585**) [187], dactylomelol (**600**) [50], dactylopyranoid (**631**) [50],

10-hydroxy-kahukuene B (**633**) [74], aplysiadiol (**638**) [187], and (*E*)-phytol (**641**) [342] have been evaluated for their antibacterial and antifungal properties.

Neoirietetraol (**589**), when tested against six bacterial strains (*Alcaligenes aquamarinus*, *Alteromonas* sp., *Azomonas agilis*, *Erwinia amylovora*, *E. coli*, and *Halococcus* sp.), exhibited weak activities only against *A. aquamarinus* and *E. coli* [513]. Dactyloditerpenol acetate (**635**) exhibited marginal antibacterial activity against *Mycobacterium tuberculosis* [178]. Laurencianol (**626**) displayed antibacterial activity against *Bacillus subtilis* and *E. coli* [535].

The triterpene thyrsiferol (**646**) and its 23-acetate (**647**), when tested, did not show any antimicrobial activity [545]. Aplysqualenol A (**675**) did not exhibit any inhibitory activity against *M. tuberculosis* [567].

The $C_{15}$ acetogenins **842** and **844–848** were examined for their antistaphylococcal activity against a panel of multidrug- and methicillin-resistant *S. aureus* strains, displaying variable levels of activity [728]. (3*Z*)-Dihydrorhodophytin (**810**) exhibited better antibacterial activity than that of cefuroxime against six foodborne pathogens [691].

The antibacterial and/or antifungal activities of the acetogenins laurenmariallene (**741**) [74], chinzallene (**749**) [105], (12*E*)-lembyne A (**759**) [62], lembyne A (**763**) [165], (3*Z*)-laurenyne (**798**) [513], (*Z*)-chondriol (**803**) [878], ***ent*-854** [152], desepilaurallene (**862**) [152], and (3*Z*)-chlorofucin (**875**) [238] have also been evaluated in a number of studies.

Among the simple bromoindoles **952**, **966**, **969**, and **971**, only 2,3,5,6-tetrabromoindole (**952**) showed activity against *B. subtilis* and *Saccharomyces cerevisiae* [793]. Metabolites **950**, **952**, **971**, and cholesterol (**993**) did not show activity against *S. aureus* and MRSA strains [342], although **952** was previously reported as displaying a wide spectrum of activity against Gram-positive antibiotic-resistant bacteria [877]. When the indoles **952**, **969**, and **971** were tested against antibiotic-resistant clinical bacteria, only **952** was proven active [432]. The antimicrobial activity of a *Laurencia complanata* extract containing the bromoindoles **952**, **969**, and **971** and the aristolane debilone (**456**) could be, at least to some extent, attributed to the presence of compound **952** [433]. The antibacterial activity of 4,5,6-tribromo-2-methylsulfinylindole (**964**), along with a series of other previously reported related compounds has been examined [801].

3-Hydroxy-4-methyl-acetophenone (**987**) has shown moderate antimicrobial activity against *Staphylococcus aureus*, *Escherichia coli*, *Flavobacterium* sp., *Alcaligenes* sp., and *Candida albicans* [806].

6$\beta$-Hydroxycholest-4-en-3-one (**1000**) showed poor antifungal activity against *C. albicans* ($MIC = 2$ mg/cm$^3$), but was not active against *Aspergillus fumigatus* and *Aspergillus flavus* [812]. (*E*)-2-((*E*)-Tridec-2-en-2-yl)-heptadec-2-enal (**1035**) showed antifungal activity against *A. fumigatus* and *A. flavus*, with *MIC* values of 0.2 and 1 mg/cm$^3$, respectively [812].

The unstable metabolites **1040** and 12-*S*-HEPE (**1042**) were reported to possess antimicrobial activity [843]. The antifungal activity of the fatty acid ethyl esters **1043–1047** was evaluated and proven to be only moderate [845].

## 4.3 Antiviral Activity

There is an increasing need for molecules with antiviral activity since the treatment of viral infections with the drugs now available often leads to viral resistance. Therefore, the evaluation of antiviral activities and the potential therapeutic value of marine metabolites are receiving intense interest among virologists.

The chamigranes nidificene (**29**) and nidifidienol (**73**) have exhibited significant antiviral activity against the HSV-1 virus (*Herpes simplex* virus type 1) with $IC_{50}$ values of 1.5 and 2.4 μg/cm$^3$, respectively. Pacifenediol (**102**), dehydroxyprepacifenol epoxide (**85**), and pacifenol (**105**) were found moderately active, while **5**, **6**, **18**, **79**, prepacifenol epoxide (**86**), deoxyprepacifenol (**87**), prepacifenol (**88**), and jonhstonol (**113**) did not show any discernible activity [95].

The sesquiterpenes perforenone (**376**), perforenone D (**380**), perforenol (**382**), debromoperforatone (**387**), perforatone (**388**), and the related **390–392** were found to be inactive against a wide variety of different RNA and DNA viruses [393]. 3-*epi*-Perforenone A (**379**), **381** and the rearranged cycloperforane 4-hydroxy-1,8-*epi*-isotenerone (**397**) were tested for their ability to inhibit the reverse transcriptase enzyme of the HIV-1 virus (human immunodeficiency virus type 1), but were found to be devoid of activity [399].

The chamigranes obtusol (**28**), cartilagineol (**31**), isoobtusol (**49**) and its acetate **50**, elatol (**59**), **92**, the bisabolanes puertitol B acetate (**312**), caespitane (**333**), caespitol (**337**), 8-acetylcaespitol (**338**), caespitenone (**349**), furocaespitane (**352**), laucapyranoid A (**372**) and the diterpenes isopinnatol B (**553**), dactylomelol (**600**), and dactylopyranoid (**631**) were proven inactive when evaluated for the inhibition of HIV-1 reverse transcriptase [50].

The triterpenes thyrsiferol (**646**), its 23-acetate (**647**), and venustatriol (**659**) exhibited antiviral properties against the VSV (vesicular stomatitis virus) and HSV-1 viruses [547]. Aplysqualenol A (**675**) exhibited significant activity against the HSV-1, HSV-2 (*Herpes simplex* virus type 2), HCMV (human cytomegalovirus), EBV (Epstein-Barr virus), and VZV (varicella zoster virus) viruses [567].

## 4.4 Activity Against Parasites and Their Vectors

Many natural products with diverse molecular structures have revealed potency against parasitic tropical diseases and/or their vectors and represent interesting lead structures for the development of new and urgently needed therapeutic interventions for the control of parasitic infections [879].

The snyderanes α-snyderol (**247**) and its acetate (**248**), **256**, **258**, **264**, and **265** were evaluated for antimalarial activity in vitro, but only **258** was weakly active against the D6 and W2 clones of *Plasmodium falciparum* [319].

The chamigrane deoxyprepacifenol (**87**) and the acetogenins laureatin (**864**) and isolaureatin (**866**) showed insecticidal activity against *Culex pipiens pallens* mosquito larvae, while pacifenol (**105**) was inactive [200].

Obtusol (**28**), cartilagineol (**31**), isoobtusol (**49**) and its acetate (**50**), elatol (**59**), **92**, puertitol B acetate (**312**), caespitane (**333**), caespitol (**337**), 8-acetylcaespitol (**338**), caespitenone (**349**), furocaespitane (**352**), laucapyranoid A (**362**), isopinnatol B (**553**), dactylomelol (**600**), dactylopyranoid (**631**), as well as 3-*epi*-perforenone A (**379**), **381**, and 4-hydroxy-1,8-*epi*-isotenerone (**397**) were tested for nematicidal activity against *Caenorrhabditis elegans*, antiplasmodial activity against *P. falciparum*, and antitrypanosomal activities against *Trypanosoma brucei rhodesiense* and *Trypanosoma cruzi* [50, 399]. Obtusol (**28**), elatol (**59**), **92** and caespitol (**337**) showed nematicidal activity and only elatol (**59**) exhibited antitrypanosomal activity to *T. cruzi* ($IC_{50} = 0.92$ μg/cm$^3$, $MIC = 3.3$ μg/cm$^3$). None of the tested compounds exhibited any antiplasmodial activity [50, 399].

Elatol (**59**) showed activity against the promastigote and intracellular amastigote forms of *Leishmania amazonensis* ($IC_{50} = 4.0$ and 0.45 μ*M*, respectively, after 72 h), while it was not toxic to macrophages [172]. Its enantiomer ***ent*-59** was found to be effective as an antitrypanosomal agent without cytotoxicity to mammalian cells [880–882]. Elatol (**59**) has also displayed larvicidal activity against the dengue mosquito (*Aedes aegypti*) [883]. Furthermore, **59**, while moderately toxic, showed a high degree of repellent activity against the spider mite *Tetranychus urticae* in laboratory experiments [173].

Obtusol (**28**) and ***ent*-59** displayed in vitro and in vivo leishmanicidal activity against both the promastigote and the amastigote forms of *L. amazonensis*, while the triquinane **490** had moderate activity. None of them promoted enhanced nitric oxide production by macrophages and they had very low toxicity against two different populations of mammalian cells [126].

Obtusane (***ent*-47**), isorigidol (**35**), **42**, **44** and its geometric isomer **46**, majusculone (**130**), mailione (**131**), and (*E*)-2-tridecyl-2-heptadecenal (**1036**) demonstrated moderate in vitro anthelmintic activity against the parasitic stage of *Nippostrongylus brasiliensis*, while obtusol (**28**), (+)-(3*Z*)-bromomethylidiene-10-*β*-bromo-β-chamigrane (**43**) and its geometric isomer **45** and scopariol (**136**) were found inactive [134]. Among the β-bisabolanes **313**, **315**, and **316** and the triquinanes **490** and **491**, only **313** showed weak anthelmintic activity against the parasitic stage of *N. brasiliensis* ($EC_{50} = 0.11$ m*M*) [366].

The anthelmintic activities against earthworms (*Allolobophora caliginosa*) of parguerol 16-acetate (**517**), parguerol 7,16-diacetate (**520**), deoxyparguerol (**525**), its 7-acetate (**526**), isoparguerol (**540**), its 16-acetate (**541**), and its 7,16-diacetate (**542**) were evaluated [860]. The parguerane and isoparguerane derivatives were found to be more effective than the deoxyparguerane derivatives and displayed higher levels of anthelminthic activity in comparison to mebendazole, which was used as a reference compound.

Squalene (**701**) has shown activity against *P. falciparum* and *T. brucei rhodesiense*, but not against *T. cruzi* and *Leishmania donovani*, while it displayed low cytotoxicity to mammalian cells [884]. The triterpenes aplysqualenols A (**675**) and B (**676**) showed moderate antiplasmodial activity against *P. falciparum* [567].

The seven-membered acetogenin isolaurepinnacin (**784**) and the lauthisan-type eight-membered acetogenin laurepinnacin (**838**) exhibited insecticidal activity [660]. Moreover, the laurenan-type eight-membered acetogenins, (3*E*,12*R*,13*R*)-pinnatifidenyne (**812**), (3*E*)-13-epilaurencienyne (**821**) and its (3*Z*)-isomer (**822**), and **824** and its (3*Z*)-isomer (**825**) were assayed for their toxicity to *Pheidole pallidula* ants. The (*Z*)-isomers **822** and **825** exerted strong toxicity with a noticeable knockdown effect on the first day of application, while the (*E*) isomers **812**, **821**, and **824** showed toxicity that escalated gradually [702]. (12*E*)-(*Z*)-Maneonene E (**761**) showed larvicidal activity against the larvae of both the confused flour beetle (*Tribolium confusum*) and the mosquito (*C. pipiens*) [101].

The simple aromatic compounds 4-hydroxy-benzaldehyde (**984**) and 4-methoxybenzyl alcohol (**986**) were isolated from the $CH_2Cl_2$ extract of *L. papillosa*, which demonstrated significant and selective in vitro antimalarial activity. 4-Hydroxy-benzaldehyde (**984**) showed, albeit weak, selective in vitro antimalarial activity against *P. falciparum*, while **986** was inactive. Inspection by NMR spectroscopy of the other active fractions resulting from the initial chromatographic separation of the extract showed the presence of typical *Laurencia* compounds, suggesting that additional metabolites might also possess antimalarial activity [804].

(*E*)-2-((*E*)-Tridec-2-en-2-yl)-heptadec-2-enal (**1035**) demonstrated moderate in vitro anthelmintic activity against the parasitic stage of *N. brasiliensis* ($ED_{50} < 100\ \mu M$) [134].

## 4.5 Anti-inflammatory Activity

Many of the inflammatory diseases are becoming common in aging societies throughout the world. Several clinically used anti-inflammatory drugs suffer from the disadvantages of both side effects and often a high cost. Natural products, as the largest chemical library of diverse structures, offer great hope for the identification of bioactive lead compounds for the development of new drugs for the treatment of inflammatory diseases [885, 886].

The chamigranes dendroidiol (**34**), obtusane (***ent*-47**) and debromoelatol (**72**), and the triquinane **490** were evaluated for their potential anti-inflammatory activity, with ***ent*-47**, **72**, and **490** showing inhibitory effects on NO release from lipopolysaccharide (LPS)-activated RAW 264.7 macrophages, while **72** moderately affected TNF-α production [146].

Palisol (**260**), palisadins A (**278**) and B (**266**), 5*β*-hydroxy-palisadin B (**267**), 5*β*-acetoxy-palisadin B (**268**), and pacifigorgiol (**485**) exhibited anti-inflammatory activity using LPS-stimulated RAW 264.7 cells [330, 451]. Aplysistatin (**276**) and 5*β*-hydroxy-palisadin B (**267**) inhibited NO and PGE2 production via inhibition of iNOS and COX-2 [330, 451]. The latter reduced the release of the pro-inflammatory cytokines TNF-α, IL-1β, and IL-6 [451], and was effective in vivo in the LPS-induced zebra fish embryo model [887]. The cell viability of some of these compounds was tested with a lactate dehydrogenase (LDH) cytotoxicity assay [146, 330, 451].

The snyderanes palisadins A (**278**) and B (**266**), 5*β*-acetoxy-palisadin B (**268**) and the aristolanes **451** and **453** when tested did not inhibit bee venom-derived phospholipase A2 [888].

Pacifenol (**105**) displayed anti-inflammatory properties and was shown to be an effective inhibitor of the degranulation process in human neutrophils, inhibiting in vitro eicosanoid release [889].

The effects of (*Z*)-maneonenes D (**764**) and E (**761**) in the apoptosis on blood neutrophils were studied and it was suggested that they may be involved in the regulation of programmed cell death in the initiation and propagation of inflammatory responses [645].

The tricyclic brominated diterpenoid neorogioltriol (**623**) showed anti-inflammatory activity in vivo using the carrageenan-induced rat paw edema model and in vitro effects on LPS-treated RAW 264.7 macrophages by inhibition of LPS-induced NF-κB activation and TNF-α production [890]. Moreover, a 1 mg/kg dose of neorogioltriol (**623**) proved to be as effective as a 300 mg/kg dose of aspirin in reducing inflammation. The exact mechanism of action of neorogioltriol (**623**) needs to be clarified and the target(s) in the cell identified.

Dactyloditerpenol acetate (**635**) exhibited significant in vitro anti-neuro-inflammatory activity through inhibition of thromboxane $B_2$ and superoxide anion generation from *E. coli* LPS-activated rat brain microglia, with concomitant low short-term toxicity [178].

4,5,6-Tribromo-2-methylsulfinylindole (**964**) and other related compounds isolated from *L. brogniartii* were evaluated for their anti-inflammatory activity [801].

(12*S*)-HEPE (**1042**), which is produced from the oxidation of EPA (eicosapentaenoic acid), is recognized as a potent inhibitor of platelet aggregation and mediator of inflammation, stimulates degranulation by neutrophils and inhibits the activity of ATPases [844]. Floridoside (**1028**) was found to exhibit anti-inflammatory properties in neurons [891].

## *4.6 Miscellaneous Biological Activities*

Obtusol (**28**), cartilagineol (**31**), isoobtusol (**49**) and its acetate **50**, elatol (**59**), **92**, puertitol B acetate (**312**), caespitane (**333**), caespitol (**337**), 8-acetylcaespitol (**338**), caespitenone (**349**), furocaespitane (**352**), laucapyranoid A (**372**), 3-*epi*-perforenone A (**379**), **381**, 4-hydroxy-1,8-*epi*-isotenerone (**397**), isopinnatol B (**553**), dactylomelol (**600**), as well as dactylopyranoid (**631**) were evaluated for the inhibition of tyrosine kinase p56lck, but were proven inactive [50, 399].

The metabolites **4**, deschloroelatol (**56**), elatol (**59**), and their acetates **57** and **60**, caespitol (**337**), and 8-acetylcaespitol (**338**) were evaluated for their acetylcholinesterase inhibition potential using computational techniques. Among them, acetylelatol (**60**) was found to be the most promising [892].

The lauranes debromolaurinterol (**226**) and laurinterol (**231**) and the acetogenin aplysiallene (**740**) exhibited inhibitory activity against $Na^+/K^+$-ATPase [308]. Aplysin (**195**) has shown a protective effect on hepatic injury induced by ethanol in rats [893]. The triquinanes **490** and **492** exhibited weak antioxidant activity when evaluated in the DPPH radical-scavenging assay [458]. The diterpenes parguerol (**514**) and isoparguerol (**540**) induced neurite outgrowth in rat pheochromocytoma (PC-12) cells at concentrations of 25 and 50 μ*M*, respectively, while deoxyparguerol 7,16-acetate (**651**) was proven inactive even at a 100 μ*M* concentration [478].

The acetogenins dactylyne (**771**), laurencin (**836**), laureatin (**864**) and isolaureatin (**866**) but not (3*E*)-laurefucin (**870**) exhibited a prolongation of pentobarbital-induced hypnosis in rats without having any other apparent effects [894, 895]. Due to the non-toxic nature of dactylyne (**771**), a theoretical study of this metabolite and its stereoisomers has been conducted, so that details of its action could be understood [897]. Dactylyne (**771**) was also proven to be an inhibitor of cytochrome P450 [896].

The tricyclic brominated diterpenoid neorogioltriol (**623**) was evaluated for its analgesic activity using a writhing test in mice and a formalin test in rats. A dose-dependent antinociceptive response was observed in the writhing test at 0.5 and 1 mg/kg with an $IC_{50}$ value of 12.5 mg/kg. The potent analgesic activity of **623** was estimated to be 400 times higher than that of aspirin, which was used as a standard [533]. Furthermore, significant analgesic properties were demonstrated for the related prenylated bisabolane, *O*-11,15-cyclo-14-bromo-14,15-dihydrorogiol-3,11-diol (**637**), which was able to signal directly to primary afferents, through a mechanism dependent on the activation of opioid receptors [538].

An acetone extract of *L. nipponica* yielded the bromophenol **985**, which showed weak inhibition ($IC_{50} = 76.6$ μ*M*) of glucose 6-phosphate dehydrogenase (G6PD). G6PD inhibitors are considered as a potential therapy for obesity [805].

The highly brominated aromatic compounds **988** and **989** were identified as inhibitors of the protein tyrosine phosphatase 1B (PTP1B) with $IC_{50}$ values of 3.0 and 2.7 μg/$cm^3$, respectively. Moreover, the brominated naphthalene derivatives **990–992** were identified in the same extract, but showed weak inhibition of PTP1B [807]. Pinnasterol (**1003**), acetylpinnasterol (**1004**) and the related steroids **1005–1008** have shown activity as molting hormones [814, 815]. 5-Hydroxymethylfurfural (**1018**), floridoside (**1028**) and D-isofloridoside (**1029**) have exhibited antioxidant activity [828, 831].

## 5 Ecological Functions

Irrespective of their taxon of origin, the chemicals that play a prominent role in interspecific interactions are rarely the same substances used by an organism to meet the daily challenges of living, such as respiration, digestion, excretion, or, in the case of plants, photosynthesis. Despite their name, secondary metabolites are by no means of secondary importance to the life of the organisms and in most cases are restricted to certain organisms where they fulfill specific roles, thus being essential to their survival. The functions of marine and terrestrial natural products have obvious analogies as they primarily act as toxins, antipredator defense substances, and pheromones. Furthermore, chemical ecology studies have also highlighted their importance in the prevention of fouling, competition for space and nutrients, and mediation of larval settling and metamorphosis.

## *5.1 Antifeedant Activity and Toxicity to Marine Organisms*

The ecological significance of marine metabolites in the chemical defense systems of the producing organisms are frequently proposed on the basis of repellence, feeding inhibition, and/or toxicity exerted on various marine animals, especially those connected to their food chain.

The chamigrane elatol (**59**) and the laurane isolaurinterol (**161**) showed potent antifeedant activities in feeding assays conducted on coral reefs and in the laboratory, while aplysin (**195**), which is a 2,7-epoxy analogue of isolaurinterol (**161**), was inactive [898]. In a more recent study, besides elatol (**59**), the sesquiterpenes isoobtusol (**49**) and palisadin A (**278**) showed equally potent antifeedant properties [166]. Additionally, elatol (**59**) reduced the feeding activity of the crab *Pachygraphus transversus* and the sea urchin *Lytechinus variegatus* [899], but surprisingly was found palatable to the black sea urchin *Echinometra lucunter* [175].

The chamigrane **84** and the rearranged chamigrane **143** were toxic to the reef damselfish *Pomacentrus coeruleus* at concentrations of 15 $\mu g/cm^3$ [188]. Cyclolaurenol (**233**), cupalaurenol (**418**) and their respective acetates (**234** and **419**) showed ichthyotoxic activity against guppies at concentrations of 5–10 ppm [305].

Parguerol (**514**), parguerol 19-acetate (**515**), parguerol 7-acetate (**516**), **517**, **519–521**, **523**, **525**, **527**, **535–538**, **540**, **542**, and **543** exhibited weak to moderate antifeedant activity against young abalone (*Haliotis discus hannai*) and young sea urchins (*Stronglyocentrotus nudus* and *Stronglyocentrotus intermedius*), with the exception of parguerol triacetate (**521**), 2-deacetoxy-deoxyparguerol (**523**), and deoxyparguerol 16-acetate (**527**), which showed potent activity against *H. discus hannai*. Furthermore, **521** and **527** were active against *S. nudus* and *S. intermedius* [477].

The triterpenes aplysiols A (**648**) and B (**668**), in a toxicity assay, were lethal to *Gambusia affinis* at a concentration of 10 ppm and active against the goldfish *Carassius auratus*, in a feeding-deterrence test, at a concentration of 50 $\mu g/cm^2$ [549].

The acetogenins (3*Z*)-laurenyne (**798**) and (−)-(3*E*,6*R*,7*R*)-pinnatifidenyne (***ent*-812**) showed antifeedant activity against *C. auratus* at 50 $\mu g/cm^2$ concentrations, while brasilenol (**398**), (*E*)-pinnatifidenyne (**814**), (+)-(3*E*, 6*R*,7*R*)-obtusenyne (***ent*-895**), and (+)-(3*Z*,6*R*,7*R*)-obtusenyne (***ent*-896**) were inactive [407].

Aplyparvunin (**861**) showed potent ichthyotoxicity against the mosquito fish with a lethal concentration of 3 ppm within 24 h [291].

The acetogenins panacene (**746**) [78], brasilenyne (**891**), (*Z*)-dihydrorhodophytin (**810**) and (*Z*)-isodihydrorhodophytin (**817**) [221], as well as the sesquiterpenes brasudol (**432**) and isobrasudol (**433**) [419] exhibited variable levels of inhibition of fish feeding activity. Pacifigorgiol (**485**), a highly rearranged sesquiterpene, exhibited toxicity towards the reef-dwelling fish *Eupomacentrus leucostictus* [900].

Dactylallene (**906**) exhibited significant toxicity against the mosquito fish (*G. affinis*) at 10 ppm and deterred the goldfish (*C. auratus*) at a concentration of 100 μg/cm$^2$ of food pellets [771].

## *5.2 Brine Shrimp Toxicity*

The toxicity of marine metabolites and subsequently the ability to offer chemical protection against predators has been estimated frequently by lethality exerted on the brine shrimp *Artemia salina*. Further to this type of preliminary indication, documentation of ecological functions then requires field assays with specialized predators. The brine shrimp lethality assay was also for many years considered as a useful tool for the preliminary assessment of cytotoxicity to cancer cells. Since its introduction in 1982 by McLaughlin and co-workers [901], this test has been used for the bioguided isolation of in vivo-active natural products.

Elatol (**59**) and the chamigrane **92** exerted 100% mortality to brine shrimp after 24 h of incubation (0.5 mg/cm$^3$), whereas isopinnatol B (**553**) made the animals hyperactive. Obtusol (**28**), caespitane (**333**), and caespitol (**337**) showed low levels of toxicity, while isoobtusol acetate (**50**), puertitol B acetate (**312**), 8-acetylcaespitol (**338**), caespitenone (**349**), furocaespitane (**352**), laucapyranoid A (**372**), dactylomelol (**600**), and dactylopyranoid (**631**) were found to be inactive [50].

The epoxy derivative **82** of okamurene E debromoallolaurinterol acetate (**150**), and allolaurinterol acetate (**152**) exhibited potent activity in the brine shrimp assay, **41**, **424**, and epilaurallene (***ent*-854**) showed weak activity, whereas laurene (**148**) and desepilaurallene (**862**) were found inactive [152].

Laurecomins A (**140**) and B (**37**), **434**, and **435** displayed potent brine shrimp toxicity, but laurecomins C (**2**), D (**94**), and **7** displayed moderate or weak activity [84].

Pacifenol (**105**) displayed 90% toxicity to brine shrimp (23 μg/cm$^3$), while 2,10-dibromo-3-chloro-α-chamigrene (**5**), deoxyprepacifenol (**87**), and prepacifenol acetate (**89**) showed moderate activity [84, 89, 109]. A number of semisynthetic derivatives of pacifenol (**105**) were also assayed in a similar manner [852].

Yicterterpene A (**111**), with a chlorination at C-9, exhibited higher levels of inhibitory activity against the marine zooplankton *A. salina* than yicterterpene B (**112**), featuring a bromine at the same site [229].

In a comparative study, β-snyderol (**249**) was found as the most active metabolite, among laurencienyne (**819**) and (3*E*)-13-epilaurencienyne (**821**), which exhibited the lowest toxicity to *A. salina* [321].

Debromoallolaurinterol (**149**) exhibited potent toxicity to *A. salina* ($LD_{50} = 1.8\ \mu M$), while the sesquiterpenes **6**, **122**, laurene (**148**), debromofiliformin (**213**), filiformin (**214**), 12-bromo-filiformin (**216**), okamurenes A (**331**), B (**332**), C (**322**), D (**321**), and E (**10**) and α-isobromocuparene (**417**) and the $C_{15}$ acetogenins **795**, laureatin (**864**), okamuragenin (**869**), neolaurallene (**899**), and (*E*)-(12*Z*)-laurediol (**938**) showed moderate or weak toxicity [109].

Deacetylparguerol (**513**), parguerol (**514**), **524**, deacetylisoparguerol (**539**), and isoparguerol (**540**) were also evaluated for their toxicity to *A. salina*, but only parguerol (**514**) and isoparguerol (**540**) showed marginal toxicity at a concentration of $0.5\ \mu M$ [476].

Prevezols A (**616**) and B (**617**) were proven toxic to *A. salina* ($LC_{50} = 6.35$ ppm) [527] and so was neoirietetraol (**589**) ($LC_{50} = 40.1\ \mu M$), while (3*Z*)-laurenyne (**798**) showed a weaker effect ($LC_{50} = 467.0\ \mu M$) [513].

## *5.3 Antifouling Activity*

Biofouling represents one of the most serious problems in the modern maritime industries and its control is one of the biggest challenges of marine biotechnology. When a surface is submerged in seawater, a biofilm forms composed of microorganisms rapidly covering the surface that is subsequently colonized by algae and invertebrates, culminating in its corrosion. Settlement of marine invertebrates on the hulls of ships results in reduction of speed, increased fuel consumption, and increased air pollution. Marine natural products have attracted significant interest as candidate molecules towards the development of new efficient and biodegradable antifouling marine paints [902, 903].

The sesquiterpenes 10-bromo-9-hydroxy-α-chamigrene (**14**), deschlorelatol (**56**), allolaurinterol (**151**), 3-acetoxy-(*E*)-γ-bisabolene (**305**), and the aristolane **490** exhibited moderate antialgal activity against the unicellular alga *Chlorella fusca*, while (–)-(10*R*)-10-bromo-α-chamigrene (**1**), rigidol (**53**), (+)-(10*S*)-10-bromo-β-chamigrene (**55**), and elatol (**59**) were inactive [81, 259]. Elatol (**59**) and deschloroelatol (**56**) strongly inhibited the settlement of the barnacle *Balanus amphitrite* cyprids and the bryozoan *Bugula neritina* larvae. However, their strong toxicity restricts their commercial development as they severely affect the survival rate of *B. amphitrite* nauplii [81, 904]. In another study from the same group, obtusol (**28**), caespitane (**333**), and caespitol (**337**) showed low levels of antialgal activity (*C. fusca*), while isoobtusol acetate (**49**), elatol (**59**), **135**, puertitol B acetate (**312**), 8-acetylcaespitol (**338**), caespitenone (**349**), furocaespitane (**352**), laucapyranoid A (**372**), isopinnatol B (**553**) and dactylomelol (**600**), and dactylopyranoid (**631**) were inactive [50]. 3-*epi*-Perforenone A (**379**), **381** and 4-hydroxy-1,8-*epi*-isotenerone (**397**) were devoid of any activity against the alga

*C. fusca* [399]. Yicterterpene A (**111**), chlorinated at C-9, exhibited higher levels of inhibitory activity against a marine phytoplankton (*Heterosigma akashiwo*) than its brominated analogue yicterterpene B (**112**) [229].

The triterpenes dehydrothyrsiferol (**649**), saiyacenols A (**669**), B (**670**) and C (**691**), and 28-hydroxy-saiyacenol B (**671**) showed antifouling activity against benthic diatoms (*Phaeodactylum tricornutum*, *Cylindrotheca* sp., *Navicula* cf. *salinicola*) and a macroalgal zoospore (*Gayralia oxysperma*). Comparison of their chemical structures in relation to the levels of activity indicated that the hydroxy group at C-28 is an important structural element for exhibition of activity [564].

The antifouling activity of omaezallene (**709**) and its congeners **710** and **711** was evaluated against the barnacle *B. amphitrite*. Omaezallene (**709**) and **710** showed potent settlement inhibition against *B. amphitrite* cyprids with $ED_{50}$ values of 0.22 and 0.30 μg/cm$^3$, respectively, while the activity of **711** was moderate ($ED_{50} = 1.5$ μg/cm$^3$). Intricatetraol (**700**) and aplysiallene (**740**) exhibited much weaker activity. The ecotoxicity of omaezallene (**709**) against the marine crustacean *Tigriopus japonicus* was quite low ($LC_{50} = 2.5$ μg/cm$^3$ in 48 h), especially when compared to that of commercial antifoulants [584].

## 6 Concluding Remarks

Algae, ranging in size from microscopic forms to gigantic representatives, form the basis of a complex food web involving all organisms living within the marine ecosystem. Through the years, humans have used them as food, as colorants, as fertilizers, as ingredients in folk medicinal recipes, as producers of polysaccharides, and, in the last few decades, as a source of unique metabolites with pharmacological potential.

In its current status, the genus *Laurencia*, one of the most diverse genera of marine algae, is comprised of 146 taxonomically accepted species and along with the genera *Chondrophycus*, *Osmundea*, *Palisada*, *Yuzurua*, and *Laurenciella* is included in the so called "*Laurencia* complex". The genus *Laurencia* seems to be actively evolving within the "*Laurencia* complex", considering the number of species and infrageneric categories that have been established and are constantly added on the basis of their morphological features with phylogenetic significance [22].

Even though species of *Laurencia* have been among the most well-known seaweeds in local communities for many centuries and various edible species, such as *L. undulata* and *L. nidifica* (known as "mane'one'o" in Hawaii) [828, 905], were consumed throughout the Pacific Ocean, the genus owes its reputation to the amazing treasure trove of the structurally diverse and unique metabolites that have been constantly isolated from representatives of this genus over the last few decades. The continuous interest of marine chemists, ecologists, and biologists in this genus is reflected upon by the increasing number of scientific articles reporting the isolation and evaluation of biological properties of *Laurencia* metabolites. The frequently unique structures and the significant levels of bioactivity that they often exhibit render them challenging targets for synthetic chemists that through their efforts for enantioselective syntheses have discovered new synthetic methodologies and have confirmed or revised the proposed structures for the natural products (see Fig. 8).

The observed extraordinary chemical variation of *Laurencia*, even within morphologically homogeneous species and populations, has been the subject of numerous studies that have provided frequently contradictory explanations. Nevertheless, this chemical variation could not be ascribed only to environmental conditions or genetic variables.

The co-evolution of the algae with marine grazing invertebrates has enhanced the biosynthesis of complex secondary metabolites that act as chemical defenses to protect the algae from predators. These metabolites, in such cases, are sequestered through the food chain in the mollusks feeding on *Laurencia*, supplementing their chemical armory. For this reason, in the present chapter, metabolites isolated from *Aplysia* sea hares (invertebrates feeding on *Laurencia* most frequently) with postulated origin from *Laurencia* have been included.

Altogether, 1047 secondary metabolites, often featuring new carbocycles, have been isolated from *Laurencia* and *Aplysia* species, and have been organized in the present contribution according to their chemical class, covering the literature up to August 2015 (see Fig. 9).

The metabolites included were classified into sesquiterpenes, diterpenes, triterpenes, acetogenins, indoles, aromatic compounds, steroids, and miscellaneous compounds. Among the isolated metabolites, sesquiterpenes comprise the most abundant class with 512 representatives, followed by cyclic ether-containing and linear $C_{15}$ acetogenins encompassing 244 chemical structures. Diterpenes and triterpenes include 133 and 59 compounds, representing the third and fourth chemical classes in abundance (see Fig. 10).

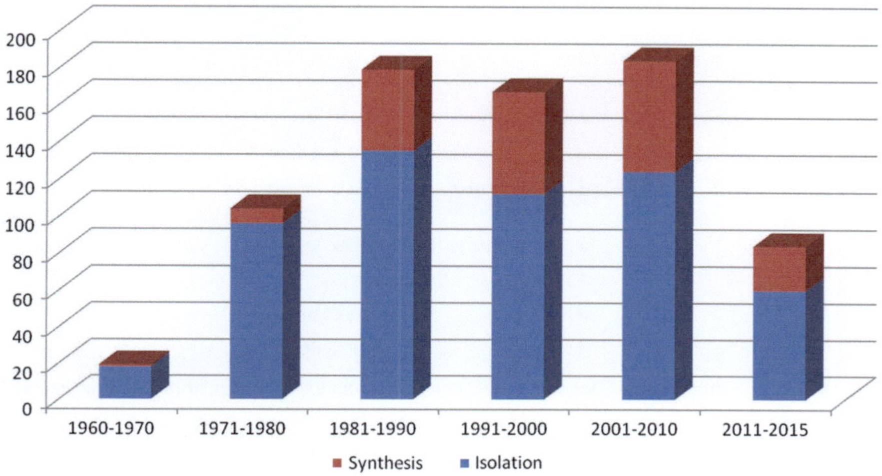


**Fig. 8** Number of publications on the isolation and synthesis of *Laurencia* metabolites per decade

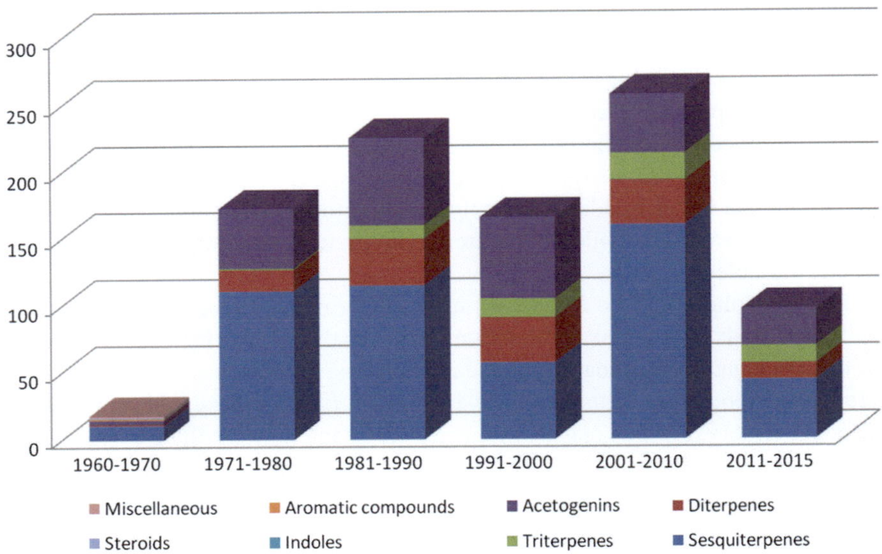


**Fig. 9** Number of metabolites isolated per decade from *Laurencia* species according to their chemical class

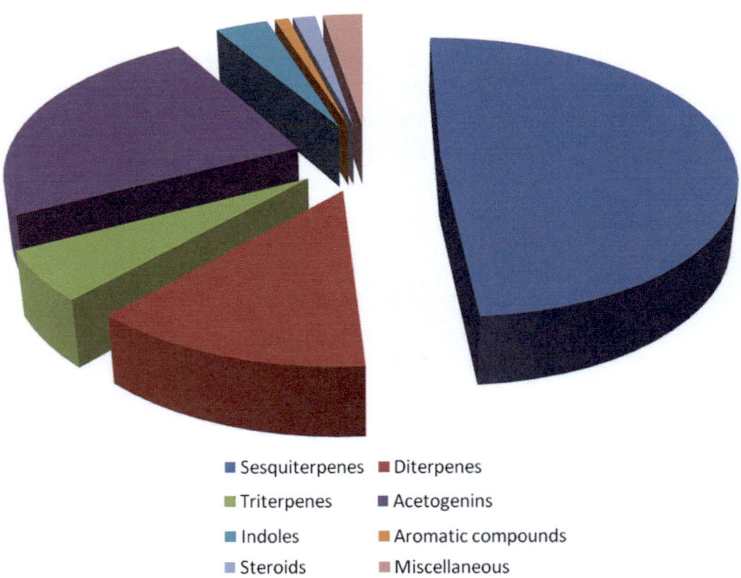


**Fig. 10** Contribution of the different chemical classes to the total number of metabolites isolated from *Laurencia* species

The majority of metabolites are halogenated, with brominated compounds being more abundant than chlorinated ones, regardless of the relative concentration of bromide and chloride ions in the sea water.

Despite the fact that there are now available sophisticated chromatographic and spectroscopic techniques allowing the isolation and structure elucidation of minor metabolites in sub-milligram amounts, the astonishing number of already isolated *Laurencia* metabolites renders efforts towards the discovery of new metabolites constantly more copious. This difficulty highlights the necessity for integrated platforms, including analytical techniques, software and databases for the accelerated dereplication of known metabolites in the extracts and the detection of potentially new natural products at the early stages of phytochemical analysis.

Even though a relatively high number of *Laurencia* metabolites have been evaluated for a wide spectrum of bioactivities and many of them displayed notable levels of activity, the chemical wealth of the genus has not been systematically assessed. The unparalleled chemical ensemble of *Laurencia* metabolites represents a rich database with unique structures for in silico structure-activity studies for the design of new drugs. This represents a treasure trove of compounds available to the pharmaceutical industry, which through currently existing automated high-throughput in vitro screens can examine thousands of compounds or extracts in very short periods of time.

## References

1. WHO (2013) WHO traditional medicine strategy: 2014–2023. World Health Organization, Geneva
2. Farnsworth NR, Akerele O, Bingel AS, Soejarto DD, Guo Z (1985) Medicinal plants in therapy. Bull World Health Org 63:965
3. Pascolutti M, Quinn RJ (2014) Natural products as lead structures: chemical transformations to create lead-like libraries. Drug Discov Today 19:215
4. Cragg GM, Newman DJ (2005) International collaboration in drug discovery and development from natural sources. Pure Appl Chem 77:1923
5. Bergmann W, Burke DC (1955) Contribution to the study of marine products. XXXIX. The nucleosides of sponges. III. Spongothymidine and spongouridine. J Org Chem 20:1501
6. *MarinLit* (2015) http://pubs.rsc.org/marinlit. Royal Society of Chemistry, London
7. Sze P (1993) A biology of the algae. Wm. C Brown Publishers, Dubuque, IA
8. Darley WM (1982) Algal biology: a physiological approach. In: Wilkinson JF (ser ed) Basic microbiology, vol 9. Blackwell Scientific Publications, Oxford, UK
9. Blunt JW, Copp BR, Keyzers RA, Munro MHG, Prinsep MR (2015) Marine natural products. Nat Prod Rep 32:116, and previous reviews in this series
10. Erickson KL (1983) Constituents of *Laurencia*. In: Scheuer PJ (ed) Marine natural products: chemical and biological perspectives, vol V. Academic Press, New York, p 131
11. Kladi M, Vagias C, Roussis V (2004) Volatile halogenated metabolites from marine algae. Phytochem Rev 3:337
12. Wang BG, Gloer JB, Ji NY, Zhao JC (2013) Halogenated organic molecules of Rhodomelaceae origin: chemistry and biology. Chem Rev 113:3632
13. Ji NY, Wang BG (2014) Nonhalogenated organic molecules from *Laurencia* algae. Phytochem Rev 13:653

14. Gribble GW (2010) Naturally occurring organohalogen compounds – a comprehensive update. In: Kinghorn AD, Falk H, Kobayashi J (eds) Progress in the chemistry of organic natural products, vol 91. Springer, Wien, and previous reviews in this series
15. Guiry MD in Guiry MD, Guiry GM (2015) AlgaeBase. World-wide electronic publication, National University of Ireland, Galway. http://www.algaebase.org. August 2015
16. Cabrita MT, Vale C, Pilar A (2010) Halogenated compounds from marine algae. Mar Drugs 8:2301
17. Cassano V, Oliveira MC, Gil-Rodríguez MC, Sentíes A, Díaz-Larrea J, Fujii MT (2012) Molecular support for the establishment of the new genus *Laurenciella* within the *Laurencia* complex (Ceramiales, Rhodophyta). Bot Mar 55:349
18. Martin-Lescanne J, Rousseau F, De Reviers B, Payri C, Couloux A, Cruaud C, Le Gall L (2010) Phylogenetic analyses of the *Laurencia* complex (Rhodomelaceae, Ceramiales) support recognition of five genera: *Chondrophycus*, *Laurencia*, *Osmundea*, *Palisada* and *Yuzurua* stat. nov. Eur J Phycol 45:51
19. Saito Y (1967) Studies on Japanese species of *Laurencia*, with special reference to their comparative morphology. Mem Fac Fish Hokkaido Univ 15:1
20. Nam KW, Maggs CA, Garbary DJ (1994) Resurrection of the genus *Osmundea* with an emendation of the generic delineation of *Laurencia* (Ceramiales, Rhodophyta). Phycologia 33:384
21. Garbary DJ, Harper JT (1998) A phylogenetic analysis of the *Laurencia* complex (Rhodomelaceae) of the red algae. Cryptogamie Algol 19:185
22. Nam KW (1999) Morphology of *Chondrophycus undulata* and *C. parvipapillata* and its implications for the taxonomy of the *Laurencia* (Ceramiales, Rhodophyta) complex. Eur J Phycol 35:455
23. Nam KW (2006) Phylogenetic re-evaluation of the *Laurencia* complex (Rhodophyta) with a description of *L. succulenta* sp. nov. from Korea. J Appl Phycol 18:679
24. Kylin H (1956) Die Gattungen der Rhodophyceen. CWK Gleerups Forlag, Lund, Sweden
25. McDermid KJ (1988) Section V. *Laurencia* (Rhodophyta, Rhodomelaceae). In: Abbott IA (ed) Taxonomy of economic seaweeds, with reference to some Pacific and Caribbean species, vol 2. California Sea Grant College Program, La Jolla, CA
26. Harper JT, Garbary DJ (1997) Marine algae of Northern Senegal: the flora and its biogeography. Bot Mar 40:129
27. Serio D, Cormaci M, Furnari G (1999) *Osmundea maggsiana* sp. nov. (Ceramiales, Rhodophyta) from the Mediterranean Sea. Phycologia 38:277
28. Nam KW, Choi HG (2000) A detailed morphological study of the type species of *Osmundea* (Rhodomelaceae, Rhodophyta). Bot Mar 43:291
29. Yoneshigue-Valentin Y, Fujii MT, Gurgel CFD (2003) *Osmundea lata* (M. Howe & W.R. Taylor) comb. nov. (Ceramiales, Rhodophyta) from the Brazilian south-eastern continental shelf. Phycologia 42:301
30. Fenical W, Norris JN (1975) Chemotaxonomy in marine algae: chemical separation of some *Laurencia* species (Rhodophyta) from the Gulf of California. J Phycol 11:104
31. Masuda M, Abe T, Suzuki T, Suzuki M (1996) Morphological and chemotaxonomic studies on *Laurencia composita* and *L. okamurai* (Ceramiales, Rhodophyta). Phycologia 35:550
32. Barrow KD, Karsten U, King RJ, West JA (1995) Floridoside in the genus *Laurencia* (Rhodomelaceae: Ceramiales) – a chemosystematic study. Phycologia 34:279
33. Fujii MT, Cassano V, Stein ÉM, Carvalho LR (2011) Overview of the taxonomy and of the major secondary metabolites and their biological activities related to human health of the *Laurencia* complex (Ceramiales, Rhodophyta) from Brazil. Braz J Pharmacogn 21:268
34. Fujii MT, Cassano V, Sentíes A, Díaz-Larrea J, Machín-Sánchez M, Candelaria Gil-Rodríguez M (2012) Comparative analysis of the corps en cerise in several species of *Laurencia* (Ceramiales, Rhodophyta) from the Atlantic Ocean. Braz J Pharmacogn 22:795
35. Kuwano K, Matsuka S, Kono S, Ninomiya M, Onishi J, Saga N (1998) Growth and the content of laurinterol and debromolaurinterol in *Laurencia okamurai* (Ceramiales, Rhodophyta). J Appl Phycol 10:9

36. Bischoff-Bäsmann B, Bartsch I, Xia B, Wiencke C (1997) Temperature responses of macroalgae from the tropical island Hainan (P.R. China). Phycol Res 45:91
37. Horta PA, Vieira-Pinto T, Martins CDL, Sissini MN, Ramlov F, Lhullier C, Scherner F, Sanches PF, Farias JN, Bastos E, Bouzon JL, Munoz P, Valduga E, Arantes NP, Batista MB, Ruil P, Almeida RS, Paes E, Fonseca A, Schenkel EP, Rorig L, Bouzon Z, Barufi JB, Colepicolo P, Yokoya N, Copertino MS, de Oliveira EC (2012) Evaluation of impacts of climate change and local stressors on the biotechnological potential of marine macroalgae - a brief theoretical discussion of likely scenarios. Braz J Pharmacogn 22:768
38. Rodrigues CL, Caeiro S, Raikar SV (1997) Marine macrophyte communities on the reef flat at Agatti Atoll (Lakshadweep, India). Bot Mar 40:557
39. Jadeja RN, Tewari A (2011) Impact of discharge of soda ash industry effluent on abundance and community structure of rocky intertidal macroalgae of the Arabian Sea, Gujarat, India. Ind J Geo-Mar Sci 40:71
40. Turna II, Ertan ÖO, Cormaci M, Furnari G (2002) Seasonal variations in the biomass of macro-algal communities from the gulf of Antalya (north-eastern Mediterranean). Turk J Bot 26:19
41. Díez I, Secilla A, Santolaria A, Gorostiaga JM (1999) Phytobenthic intertidal community structure along an environmental pollution gradient. Mar Pollut Bull 38:463
42. Nonomura AM, West JA (1980) Ultrastructure of the parasite *Janczewskia morimotoi* and its host *Laurencia nipponica* (Ceramiales, Rhodophyta). J Ultrastruct Res 73:183
43. Hay ME (1996) Marine chemical ecology: what's known and what's next? J Exp Mar Biol Ecol 200:103
44. Kilar JA, Lou RM (1984) Ecological and behavioral studies of the decorator crab, *Microphrys bicornutus* Latreille (Decapoda: Brachyura): a test of optimum foraging theory. J Exp Mar Biol Ecol 74:157
45. Kilar JA, Lou RM (1986) The subtleties of camouflage and dietary preference of the decorator crab, *Microphrys bicornutus* Latreille (Decapoda: Brachyura). J Exp Mar Biol Ecol 101:143
46. Stoner AW, Waite JM (1991) Trophic biology of *Strombus gigas* in nursery habitats: diets and food sources in seagrass meadows. J Mollusc Stud 57:451
47. Boettcher AA, Targett NM (1996) Induction of metamorphosis in queen conch, *Strombus gigas* Linnaeus, larvae by cues associated with red algae from their nursery grounds. J Exp Mar Biol Ecol 196:29
48. Kuffner IB, Walters LJ, Becerro MA, Paul VJ, Ritson-William R, Beach KS (2006) Inhibition of coral recruitment by macroalgae and cyanobacteria. Mar Ecol Prog Ser 323:107
49. Suenaga K (2004) Bioorganic studies on marine natural products with bioactivity, such as antitumor activity and feeding attractance. Bull Chem Soc Jpn 77:443
50. Wessels M, König GM, Wright AD (2000) New natural product isolation and comparison of the secondary metabolite content of three distinct samples of the sea hare *Aplysia dactylomela* from Tenerife. J Nat Prod 63:920
51. Rogers CN, De Nys R, Charlton TS, Steinberg PD (2000) Dynamics of algal secondary metabolites in two species of sea hare. J Chem Ecol 26:721
52. Yamamura S, Hirata Y (1963) Structures of aplysin and aplysinol, naturally occurring bromo-compounds. Tetrahedron 19:1485
53. de Oliveira LS, Gregoracci GB, Silva GGZ, Salgado LT, Filho GA, Alves-Ferreira M, Pereira RC, Thomps FL (2012) Transcriptomic analysis of the red seaweed *Laurencia dendroidea* (Florideophyceae, Rhodophyta) and its microbiome. BMC Genomics 13:487
54. Faulkner DJ (1995) Chemical riches from the oceans. Chem Br 31:680
55. Butler A, Carter-Franklin JN (2004) The role of vanadium bromoperoxidase in the biosynthesis of halogenated marine natural products. Nat Prod Rep 21:180
56. Winter JM, Moore BS (2009) Exploring the chemistry and biology of vanadium-dependent haloperoxidases. J Biol Chem 284:18577
57. Carter-Franklin JN, Butler A (2004) Vanadium bromoperoxidase-catalyzed biosynthesis of halogenated marine natural products. J Am Chem Soc 126:15060

58. Suzuki M, Takahashi Y, Nakano S, Abe T, Masuda M, Ohnishi T, Noya Y, Seki K (2009) An experimental approach to study the biosynthesis of brominated metabolites by the red algal genus *Laurencia*. Phytochemistry 70:1410
59. Salgado LT, Viana NB, Andrade LR, Leal RN, da Gama BAP, Attias M, Pereira RC, Amado Filho GM (2008) Intra-cellular storage, transport and exocytosis of halogenated compounds in marine red alga *Laurencia obtusa*. J Struct Biol 162:345
60. Suzuki M, Kurosawa E (1979) Halogenated and non-halogenated aromatic sesquiterpenes from the red algae *Laurencia okamurai* Yamada. Bull Chem Soc Jpn 52:3352
61. Suzuki M, Kurosawa E, Kurata K (1987) (*E*)-2-Tridecyl-2-heptadecenal, an unusual metabolite from the red alga *Laurencia* species. Bull Chem Soc Jpn 60:3793
62. Vairappan CS, Suzuki M, Abe T, Masuda M (2001) Halogenated metabolites with antibacterial activity from the Okinawan *Laurencia* species. Phytochemistry 58:517
63. Reis VM, Oliveira LS, Passos RMF, Viana NB, Mermelstein C, Sant'Anna C, Pereira RC, Paradas WC, Thompson FL, Amado-Filho GM, Salgado LT (2013) Traffic of secondary metabolites to cell surface in the red alga *Laurencia dendroidea* depends on a two-step transport by the cytoskeleton. PloS One 8:e63929
64. Paradas WC, Salgado LT, Sudatti DB, Crapez MA, Fujii MT, Coutinho R, Pereira RC, Amado Filho GM (2010) Induction of halogenated vesicle transport in cells of the red seaweed *Laurencia obtusa*. Biofouling 26:277
65. de Nys R, Dworjanyn SA, Steinberg PD (1998) A new method for determining surface concentrations of marine natural products on seaweeds. Mar Ecol Prog Ser 162:79
66. Sudatti DB, Rodrigues SV, Pereira RC (2006) Quantitative GC-ECD analysis of halogenated metabolites: determination of surface and within-thallus elatol of *Laurencia obtusa*. J Chem Ecol 32:835
67. Sudatti DB, Rodrigues SV, Coutinho R, da Gama BAP, Salgado LT, Amado Filho GM, Pereira RC (2008) Transport and defensive role of elatol at the surface of the red seaweed *Laurencia obtusa* (Ceramiales, Rhodophyta). J Phycol 44:584
68. Howard BM, Nonomura AM, Fenical W (1980) Chemotaxonomy in marine algae: secondary metabolite synthesis by *Laurencia* in unialgal culture. Biochem Syst Ecol 8:329
69. Masuda M, Kawaguchi S, Abe T, Kawamoto T, Suzuki M (2002) Additional analysis of chemical diversity of the red algal genus *Laurencia* (Rhodomelaceae) from Japan. Phycol Res 50:135
70. Sudatti DB, Fujii MT, Rodrigues SV, Turra A, Pereira RC (2011) Effects of abiotic factors on growth and chemical defenses in cultivated clones of *Laurencia dendroidea* J. Agarth (Ceramiales, Rhodophyta). Mar Biol 158:1439
71. Oliveira AS, Sudatti DB, Fujii MT, Rodrigues SV, Pereira RC (2013) Inter- and intrapopulation variation in the defensive chemistry of the red seaweed *Laurencia dendroidea* (Ceramiales, Rhodophyta). Phycologia 52:130
72. de Carvalho LR, Farias JN, Riul P, Fujii MT (2015) An overview of global distribution of the diterpenes synthesized by the red algae *Laurencia* complex (Ceramiales, Rhodomelaceae). In: Kim SK, Chojnacka K (eds) Marine algae extracts: processes, products, and applications. Wiley, Weinheim
73. Kokkotou K, Ioannou E, Nomikou M, Pitterl F, Vonaparti A, Siapi E, Zervou M, Roussis V (2014) An integrated approach using UHPLC-PDA-HRMS and 2D HSQC NMR for the metabolic profiling of the red alga *Laurencia*: dereplication and tracing of natural products. Phytochemistry 108:208
74. Ji N-Y, Li X-M, Li K, Ding L-P, Gloer JB, Wang B-G (2007) Diterpenes, sesquiterpenes, and a $C_{15}$ acetogenin from the marine red alga *Laurencia mariannensis*. J Nat Prod 70:1901
75. Suzuki T, Furusaki A, Hashiba N, Kurosawa E (1977) Novel skeletal bromo ether from the marine alga *Laurencia nipponica* Yamada. Tetrahedron Lett 18:3731
76. Suzuki M, Furusaki A, Kurosawa E (1979) The absolute configurations of halogenated chamigrene derivatives from the marine alga *Laurencia glandulifera* Kützing. Tetrahedron 35:823

77. Suzuki M, Koizumi K, Kikuchi H, Suzuki T, Kurosawa E (1983) Epilaurallene, a new nonterpenoid $C_{15}$-bromoallene from the red alga *Laurencia nipponica* Yamada. Bull Chem Soc Jpn 56:715
78. Suzuki M, Segawa M, Kikuchi H, Suzuki T, Kurosawa E (1985) (5*S*,7*R*,10*R*)-Selin-4(14)-en-5$\alpha$-ol, a sesquiterpene alcohol from the red alga *Laurencia nipponica*. Phytochemistry 24:2011
79. Li XD, Ding W, Miao FP, Ji NY (2012) Halogenated chamigrane sesquiterpenes from *Laurencia okamurai*. Magn Reson Chem 50:174
80. Howard BM, Fenical W (1976) 10-Bromo-α-chamigrene. Tetrahedron Lett 17:2519
81. König GM, Wright AD (1997) *Laurencia rigida*: chemical investigations of its antifouling dichloromethane extract. J Nat Prod 60:967
82. Wolinsky LE, Faulkner DJ (1976) A biomimetic approach to the synthesis of *Laurencia* metabolites. Synthesis of 10-bromo-α-chamigrene. J Org Chem 41:597
83. Guella G, Öztunç A, Mancini I, Pietra F (1997) Stereochemical features of sesquiterpene metabolites as a distinctive trait of red seaweeds in the genus *Laurencia*. Tetrahedron Lett 38:8261
84. Li XD, Miao FP, Yin XL, Liu JL, Ji NY (2012) Sesquiterpenes from the marine red alga *Laurencia composita*. Fitoterapia 83:1191
85. Ji N-Y, Li X-M, Li K, Wang B-G (2009) Halogenated sesquiterpenes from the marine red alga *Laurencia saitoi* (Rhodomelaceae). Helv Chim Acta 92:1873
86. Dias T, Brito I, Moujir L, Paiz N, Darias J, Cueto M (2005) Cytotoxic sesquiterpenes from *Aplysia dactylomela*. J Nat Prod 68:1677
87. Rovirosa J, Soto H, Cueto M, Dárias J, Herrera J, San-Martín A (1999) Sesquiterpenes from *Laurencia claviformis*. Phytochemistry 50:745
88. Ji N, Li X, Li K, Gloer JB, Wang B (2008) Halogenated sesquiterpenes and non-halogenated linear $C_{15}$ acetogenins from the marine red alga *Laurencia composita* and their chemotaxonomic significance. Biochem Syst Ecol 36:938
89. Jongaramruong J, Blackman AJ, Skelton BW, White AH (2002) Chemical relationships between the sea hare *Aplysia parvula* and the red seaweed *Laurencia filiformis* from Tasmania. Aust J Chem 55:275
90. Elsworth JF, Thomson RH (1989) A new chamigrane from *Laurencia glomerata*. J Nat Prod 52:893
91. Wright AD, König GM, Sticher O (1991) New sesquiterpenes and $C_{15}$ acetogenins from the marine red alga *Laurencia implicata*. J Nat Prod 54:1025
92. Suzuki M, Matsuo Y, Masuda M (1993) Structures of laurenenyne A and B, novel halogenated acetogenins from a species of the red algal *Laurencia*. Tetrahedron 49:2033
93. Takahashi Y, Suzuki M, Abe T, Masuda M (1998) Anhydroaplysiadiol from *Laurencia japonensis*. Phytochemistry 48:987
94. Vairappan CS, Zanil II, Kamada T (2014) Structural diversity and geographical distribution of halogenated secondary metabolites in red algae, *Laurencia nangii* Masuda (Rhodomelaceae, Ceramiales), in the coastal waters of North Borneo Island. J Appl Phycol 26:1189
95. Kimura J, Kamada N, Tsujimoto Y (1999) Fourteen chamigrane derivatives from a red alga *Laurencia nidifica*. Bull Chem Soc Jpn 72:289
96. Suzuki T, Kikuchi H, Kurosawa E (1982) Six new sesquiterpenoids from the red alga *Laurencia nipponica* Yamada. Bull Chem Soc Jpn 55:1561
97. Lyakhova EG, Kalinovsky AI, Kolesnikova SA, Vaskovsky VE, Stonik VA (2004) Halogenated diterpenoids from the red alga *Laurencia nipponica*. Phytochemistry 65:2527
98. Alarif WM, Al-Lihaibi SS, Ayyad SEN, Abdel-Rhman MH, Badria FA (2012) Laurene-type sesquiterpenes from the Red Sea red alga *Laurencia obtusa* as potential antitumor-antimicrobial agents. Eur J Med Chem 55:462
99. Ojika M, Shizuri Y, Yamada K (1982) A halogenated chamigrane epoxide and six related halogen-containing sesquiterpenes from the red alga *Laurencia okamurai*. Phytochemistry 21:2410

100. Ji N-Y, Li X-M, Zhang Y, Wang B-G (2007) Two new halogenated chamigrane-type sesquiterpenes and other secondary metabolites from the marine red alga *Laurencia okamurai* and their chemotaxonomic significance. Biochem Syst Ecol 35:627
101. Abou-Elnaga ZS, Alarif WM, Al-Iihaibi SS (2011) New larvicidal acetogenin from the red alga *Laurencia papillosa*. CLEAN 39:787
102. Kennedy DJ, Selby IA, Thomson RH (1988) Chamigrane metabolites from *Laurencia obtusa* and *L. scoparia*. Phytochemistry 27:1761
103. König GM, Wright AD (1994) New $C_{15}$ acetogenins and sesquiterpenes from the red alga *Laurencia* sp. cf. *L. gracilis*. J Nat Prod 57:477–485
104. Howard BM, Fenical W (1975) Structures and chemistry of two new halogen-containing chamigrene derivatives from *Laurencia*. Tetrahedron Lett 16:1687
105. Suzuki M, Kawamoto T, Vairappan CS, Ishii T, Abe T, Masuda M (2005) Halogenated metabolites from Japanese *Laurencia* spp. Phytochemistry 66:2787
106. Suzuki M, Nakano S, Takahashi Y, Abe T, Masuda M, Takahashi H, Kobayashi K (2002) Brominated labdane-type diterpenoids from an Okinawan *Laurencia* sp. J Nat Prod 65:801
107. San-Martín A, Darias J, Soto H, Contreras C, Herrera JS, Rovirosa J (1997) A new $C_{15}$ acetogenin from the marine alga *Laurencia claviformis*. Nat Prod Lett 10:303
108. Suzuki M, Kurosawa E, Furusaki A (1988) The structure and absolute stereochemistry of a halogenated chamigrene derivative from the red alga *Laurencia* species. Bull Chem Soc Jpn 61:3371
109. Liang Y, Li XM, Cui CM, Li CS, Sun H, Wang BG (2012) Sesquiterpene and acetogenin derivatives from the marine red alga *Laurencia okamurai*. Mar Drugs 10:2817
110. Fenical W (1976) Chemical variation in a new bromochamigrene derivative from the red seaweed *Laurencia pacifica*. Phytochemistry 15:511
111. Suzuki M, Segawa M, Suzuki T, Kurosawa E (1983) Structure of halogenated chamigrene derivatives, minor constituents from the red alga *Laurencia nipponica* Yamada. Bull Chem Soc Jpn 56:3824
112. Suzuki M, Kurosawa E, Irie T (1974) Glanduliferol, a new halogenated sesquiterpenoid from *Laurencia glandulifera* Kützing. Tetrahedron Lett 15:1807
113. Suzuki M, Kurosawa E, Irie T (1974) Three new sesquiterpenoids containing bromine, minor constituents of *Laurencia glandulifera* Kützing. Tetrahedron Lett 15:821
114. Wright AD, Coll JC, Price IR (1990) Tropical marine algae, VII. The chemical composition of marine algae from North Queensland waters. J Nat Prod 53:845
115. Rovirosa J, Astudillo L, Ramirez ME, San-Martin A (1991) Chemical relationship between *Aplysia dactylomela* and *Laurencia claviformis* Borgesen from Easter Island. Bol Soc Chil Quim 36:153
116. Suzuki M, Kurosawa E (1978) Two new halogenated sesquiterpenes from the red alga *Laurencia majuscula* Harvey. Tetrahedron Lett 19:4805
117. Suzuki M, Furusaki A, Hashiba N, Kurosawa E (1979) The structures and absolute stereochemistry of two halogenated chamigrenes from the red alga *Laurencia majuscula* Harvey. Tetrahedron Lett 20:879
118. Palaniveloo K, Vairappan CS (2014) Chemical relationship between red algae genus *Laurencia* and sea hare (*Aplysia dactylomela* Rang) in the North Borneo Island. J Appl Phycol 26:1199
119. Niwa H, Yoshida Y, Hasegawa T, Yamada K (1985) Total synthesis of (±)-(Z)-9-(bromomethylene)-1,5,5-trimethyl*spiro*[5.5]undeca-1,7-dien-3-one, a brominated sesquiterpene of chamigrane type. Chem Lett 14:1687
120. Niwa H, Yoshida Y, Hasegawa T, Yamada K (1991) Total synthesis of (±)-(Z)-9-(bromomethylene)-1,5,5-trimethyl*spiro*[5.5]undeca-1,7-dien-3-one, a brominated sesquiterpene of the chamigrane type. Tetrahedron 47:2155
121. Zhu JL, Huang PW, You RY, Lee FY, Tsao SW, Chen IC (2011) Total syntheses of (±)-(Z)- and (±)-(*E*)-9-(bromomethylene)-1,5,5-trimethyl*spiro*[5.5]undeca-1,7-dien-3-one and (±)-majusculone. Synthesis 43:715

122. Suzuki M, Kurosawa E, Kurata K (1987) Majusculone, a novel norchamigrane-type metabolite from the red alga *Laurencia majuscula* Harvey. Bull Chem Soc Jpn 60:3795
123. Srikrishna A, Vasantha Lakshmi B, Mathews M (2006) Construction of *spiro*[5.5]undecanes containing a quaternary carbon atom adjacent to a *spiro*centre via an Ireland ester Claisen rearrangement and RCM reaction sequence. Total syntheses of (±)-α-chamigrene, (±)-β-chamigrene and (±)-laurencenone C. Tetrahedron Lett 47:2103
124. Brennan MR, Erickson KL, Minott DA, Pascoe KO (1987) Chamigrane metabolites from a Jamaican variety of *Laurencia obtusa*. Phytochemistry 26:1053
125. Kaiser CR, Pitombo LF, Pinto AC (2000) NMR analysis of a complex spin system from a *spiro*-chamigrene. Spectrosc Lett 33:457
126. Machado FLD, Pacienza-Lima W, Rossi-Bergmann B, Gestinari LMD, Fujii MT, de Paula JC, Costa SS, Lopes NP, Kaiser CR, Soares AR (2011) Antileishmanial sesquiterpenes from the Brazilian red alga *Laurencia dendroidea*. Planta Med 77:733
127. Capon RJ, Ghisalberti EL, Mori TA, Jefferies PR (1988) Sesquiterpenes from *Laurencia* spp. J Nat Prod 51:1302
128. Díaz-Marrero AR, Brito I, de la Rosa JM, D'Croz L, Fabelo O, Ruiz-Perez C, Darias J, Cueto M (2009) Novel lactone chamigrene-derived metabolites from *Laurencia majuscula*. Eur J Org Chem 2009:1407
129. González AG, Darias J, Díaz A, Fourneron JD, Martín JD, Pérez C (1976) Evidence for the biogenesis of halogenated chamigrenes from the red alga *Laurencia obtusa*. Tetrahedron Lett 17:3051
130. González AG, Martín JD, Martín VS, Martínez-Ripoll M, Fayos J (1979) X-Ray study of sesquiterpene constituents of the alga *L. obtusa* leads to structure revision. Tetrahedron Lett 20:2717
131. Gerwick WH, Lopez A, Davila R, Albors R (1987) Two new chamigrene sesquiterpenoids from the tropical red alga *Laurencia obtusa*. J Nat Prod 50:1131
132. Martín JD, Caballero P, Fernández JJ, Norte M, Pérez R, Rodríguez ML (1989) Metabolites from *Laurencia obtusa*. Phytochemistry 28:3365
133. Dorta E, Díaz-Marrero AR, Cueto M, D'Croz L, Maté JL, Darias J (2004) Chamigrenelactone, a polyoxygenated sesquiterpene with a novel structural type and devoid of halogen from *Laurencia obtusa*. Tetrahedron Lett 45:7065
134. Davyt D, Fernández R, Suescun L, Mombrú AW, Saldaña J, Domínguez L, Coll J, Fujii MT, Manta E (2001) New sesquiterpene derivatives from the red alga *Laurencia scoparia*. Isolation, structure determination and anthelmintic activity. J Nat Prod 64:1552
135. Díaz-Marrero AR, de la Rosa JM, Brito I, Darias J, Cueto M (2012) Dactylomelatriol, a biogenetically intriguing omphalane-derived marine sesquiterpene. J Nat Prod 75:115
136. Perales A, Martínez-Ripoll M, Fayos J (1979) Structure of obtusol acetate, a halogenated chamigrene-type sesquiterpene. Acta Crystallogr B35:2771
137. Waraszkiewicz SM, Erickson KL (1974) Halogenated sesquiterpenoids from the Hawaiian marine alga *Laurencia nidifica*: nidificene and nidifidiene. Tetrahedron Lett 15:2003
138. Kikuchi H, Suzuki T, Kurosawa E, Suzuki M (1991) The structure of notoryne, a halogenated $C_{15}$ nonterpenoid with a novel carbon skeleton from the red alga *Laurencia nipponica* Yamada. Bull Chem Soc Jpn 64:1763
139. Brito I, Cueto M, Díaz-Marrero AR, Darias J, San Martín A (2002) Oxachamigrenes, new halogenated sesquiterpenes from *Laurencia obtusa*. J Nat Prod 65:946
140. McPhail KL, Davies-Coleman MT, Copley RCB, Eggleston DS (1999) New halogenated sesquiterpenes from South African specimens of the circumtropical sea hare *Aplysia dactylomela*. J Nat Prod 62:1618
141. Ali MS (2004) Algal halo-chamigranes. J Chem Soc Pak 26:310
142. Juagdan EG, Kalidindi R, Scheuer P (1997) Two new chamigranes from an Hawaiian red alga *Laurencia cartilaginea*. Tetrahedron 53:521
143. Francisco MEY, Turnbull MM, Erickson KL (1998) Cartilagineol, the fourth lineage of *Laurencia*-derived polyhalogenated chamigrene. Tetrahedron Lett 39:5289

144. Shubina LK, Fedorov SN, Kalinovskiy AI, Dmitrenok AS, Jin JO, Song MG, Kwak JY, Stonik VA (2007) Four new chamigrane sesquiterpenoids from the opistobranch mollusk *Aplysia dactylomela*. Russ Chem Bull 56:2109
145. Ayyad SEN, Dawidar AAM, Dias HW, Howie RA, Jakupovic J, Thomson RH (1990) Three halogenated metabolites from *Laurencia obtusa*. Phytochemistry 29:3193
146. Da Silva Machado FL, Ventura TLB, de Souza Gestinari LM, Cassano V, Resende JALC, Kaiser CR, Lasunskaia EB, Muzitano MF, Soares AR (2014) Sesquiterpenes from the Brazilian red alga *Laurencia dendroidea* J. Agarth. Molecules 19:3181
147. Suescun L, Mombrú AW, Mariezcurrena RA, Davyt D, Fernández R, Manta E (2001) Two natural products from the algae *Laurencia scoparia*. Acta Crystallogr C57:286
148. Francisco MEY, Erickson KL (2001) Ma'iliohydrin, a cytotoxic chamigrene dibromohydrin from a Philippine *Laurencia* species. J Nat Prod 64:790
149. de Nys R, König G, Wright A, Sticher O (1993) Two metabolites from the red alga *Laurencia flexilis*. Phytochemistry 34:725
150. Tan KL, Matsunaga S, Vairappan CS (2011) Halogenated chamigranes of red alga *Laurencia snackeyi* (Weber-van Bosse) Masuda from Sulu-Sulawesi Sea. Biochem Syst Ecol 39:213
151. Fedorov SN, Reshetnyak MV, Shchedrin AP, Il'in GS, Struchkov YT, Stonik VA, Elyakov GB (1989) New halogenated chamigrane sesquiterpenoid from the mollusc *Aplysia* sp. Structure and absolute configuration. Dokl Akad Nauk SSSR 305:877
152. Li XD, Miao FP, Li K, Ji NY (2012) Sesquiterpenes and acetogenins from the marine red alga *Laurencia okamurai*. Fitoterapia 83:518
153. de Nys R, Coll JC, Bowden BF (1992) Tropical marine algae. VIII. The structural determination of novel sesquiterpenoid metabolites from the red alga *Laurencia majuscula*. Aust J Chem 45:1611
154. Rashid MA, Gustafson KR, Cardellina JH II, Boyd MR (1995) Brominated chamigrane sesquiterpenes produce a novel profile of differential cytotoxicity in the NCI in vitro screen. Nat Prod Lett 6:255
155. Vairappan CS, Anangdan SP, Lee KT, Matsunaga S (2010) Role of secondary metabolites as defense chemicals against ice-ice disease bacteria in biofouler at carrageenophyte farms. J Appl Phycol 22:305
156. Schmitz FJ, Michaud DP, Schmidt PG (1982) Marine natural products: parguerol, deoxyparguerol and isoparguerol. New brominated diterpenes with modified pimarane skeletons from the sea hare *Aplysia dactylomela*. J Am Chem Soc 104:6415
157. Bansemir A, Just N, Michalik M, Lindequist U, Lalk M (2004) Extracts and sesquiterpene derivatives from the red alga *Laurencia chondrioides* with antibacterial activity against fish and human pathogenic bacteria. Chem Biodivers 1:463
158. Coll JC, Wright AD (1989) Tropical marine algae. III. New sesquiterpenes from *Laurencia majuscula* (Rhodophyta, Rhodophyceae, Ceramiales, Rhodomelaceae). Aust J Chem 42:1591
159. Furusaki A, Matsumoto T, Kurata K, Suzuki T, Suzuki M, Kurosawa E (1983) X-ray structure determination of (–)-obtusane, a new sesquiterpene from the red alga *Laurencia nipponica* Yamada. Bull Chem Soc Jpn 56:3501
160. Ji N-Y, Li X-M, Cui C-M, Wang B-G (2007) Terpenes and polybromoindoles from the marine red alga *Laurencia decumbens* (Rhodomelaceae). Helv Chim Acta 90:1731
161. González AG, Martín JD, Martín VS, Norte M (1979) Carbon-13 NMR application to *Laurencia* polyhalogenated sesquiterpenes. Tetrahedron Lett 20:2719
162. Vairappan CS, Phang S (2005) Morphology and halochamigrene metabolite content of *Laurencia majuscula* (Rhodomelaceae, Ceramiales) from the Spratly Islands. Malay J Sci 24:29
163. Guella G, Mancini I, Pietra F (1992) $C_{15}$ acetogenins and terpenes of the sponge *Spongia zimocca* of Il Rogiolo: a case of seaweed-metabolite transfer to, and elaboration within, a sponge? Comp Biochem Physiol B 103:1019
164. Guella G, Chiasera G, Mancini I, Pietra F (1991) Conformational analysis of marine polyhalogenated β-chamigrenes through temperature-dependent NMR spectra. Helv Chim Acta 74:774

165. Vairappan CS, Daitoh M, Suzuki M, Abe T, Masuda M (2001) Antibacterial halogenated metabolites from the Malaysian *Laurencia* species. Phytochemistry 58:291
166. Vairappan CS, Anangdan SP, Matsunaga S (2009) Diet-derived halogenated metabolite from the sea hare *Aplysia parvula*. Malay J Sci 28:269
167. Hegazy MEF, Moustfa AY, Mohamed AEHH, Alhammady MA, Elbehairi SEIE, Ohta S, Paré PW (2014) New cytotoxic halogenated sesquiterpenes from the Egyptian sea hare *Aplysia oculifera*. Tetrahedron Lett 55:1711
168. Martín JD, Pérez C, Ravelo JL (1986) Enantioselective ring construction: synthesis of halogenated marine natural *spiro*[5.5]undecane sesquiterpenes. J Am Chem Soc 108:7801
169. González AG, Martín JD, Martín VS, Norte M, Fayos J, Martínez-Ripoll M (1978) A new polyhalogenated sesquiterpene from *Laurencia obtusa*. Tetrahedron Lett 19:2035
170. König GM, Wright AD (1997) Pulsed field gradient spectroscopy (PFGS): application to the structure elucidation of (+)-(10*S*)-10-bromo-β-chamigrene. Phytochem Anal 8:167
171. Martín JD, Palazón JM, Pérez C, Ravelo JL (1986) Syntheses of marine molecules. Pure Appl Chem 58:395
172. dos Santos AO, Veiga-Santos P, Ueda-Nakamura T, Dias BP, Sudatti DB, Bianco EM, Pereira RC, Nakamura CV (2010) Effect of elatol, isolated from red seaweed *Laurencia dendroidea*, on *Leishmania amazonensis*. Mar Drugs 8:2733
173. Born FS, Bianco EM, da Camara CAG (2012) Acaricidal and repellent activity of terpenoids from seaweeds collected in Pernambuco, Brazil. Nat Prod Commun 7:463
174. Sims JJ, Lin GHY, Wing RM (1974) Marine natural products X. Elatol, a halogenated sesquiterpene alcohol from the red alga *Laurencia elata*. Tetrahedron Lett 15:3487
175. Lhullier C, Donnangelo A, Caro M, Palermo JA, Horta PA, Falkenberg M, Schenkel EP (2009) Isolation of elatol from *Laurencia microcladia* and its palatability to the sea urchin *Echinometra lucunter*. Biochem Syst Ecol 37:254
176. Ji N, Li X, Ding L, Wang B (2007) Aristolane sesquiterpenes and highly brominated indoles from the marine red alga *Laurencia similis* (Rhodomelaceae). Helv Chim Acta 90:385
177. Rose AF, Sims JJ (1977) Marine natural products XIV. 1-*S*-Bromo-4-*R*-hydroxyselin-7-ene, a metabolite of the marine alga *Laurencia* sp. Tetrahedron Lett 18:2935
178. Jiménez-Romero C, Mayer AMS, Rodríguez AD (2014) Dactyloditerpenol acetate, a new prenylbisabolane-type diterpene from *Aplysia dactylomela* with significant in vitro anti-neuroinflammatory activity. Bioorg Med Chem Lett 24:344
179. Suzuki M, Segawa M, Suzuki T, Kurosawa E (1985) Structures of two new halochamigrene derivatives from the red alga *Laurencia nipponica* Yamada. Bull Chem Soc Jpn 58:2435
180. Waraszkiewicz SM, Erickson KL (1975) Halogenated sesquiterpenoids from the Hawaiian marine alga *Laurencia nidifica*. II. Nidifidienol. Tetrahedron Lett 16:281
181. Masuda M, Itoh T, Matsuo Y, Suzuki M (1997) Sesquiterpenoids of *Laurencia majuscula* (Ceramiales, Rhodophyta) from the Ryukyu Islands, Japan. Phycol Res 45:59
182. Sakai R, Higa T, Jefford CW, Bernardinelli G (1986) The absolute configurations and biogenesis of some new halogenated chamigrenes from the sea hare *Aplysia dactylomela*. Helv Chim Acta 69:91
183. McMillan JA, Paul IC, White RH, Hager LP (1974) Molecular structure of acetoxyintricatol: a new bromo compound from *Laurencia intricata*. Tetrahedron Lett 15:2039
184. Cox PJ, Howie RA (1989) Structure of 2,10-dibromo-3-chloro-7*R*,8*S*-epoxychamigrene. Z Krist 188:1
185. Bano S, Ali MS, Ahmad VU (1987) Marine natural products. VI. A halogenated chamigrene epoxide from the red alga *Laurencia pinnatifida*. Planta Med 53:508
186. Furusaki A, Katayama C, Matsumoto T, Suzuki M, Suzuki T, Kikuchi H, Kurosawa E (1982) The crystal and molecular structure of 7,8-epoxyhalochamigrene. Bull Chem Soc Jpn 55:3398
187. Vairappan CS, Ishii T, Lee TK, Suzuki M, Zhaoqi Z (2010) Antibacterial activities of a new brominated diterpene from Borneon *Laurencia* spp. Mar Drugs 8:1743

188. Bittner ML, Silva M, Paul VJ, Fenical W (1985) A rearranged chamigrene derivative and its potential biogenetic precursor from a new species of the marine red algal genus *Laurencia* (Rhodomelaceae). Phytochemistry 24:987
189. Kaiser CR, Pitombo LF, Pinto AC (1998) C-13 and H-1 NMR assignments of the chamigrenes prepacifenol and dehydroxyprepacifenol epoxides. Spectrosc Lett 31:573
190. Faulkner DJ, Stallard MO, Ireland C (1974) Prepacifenol epoxide, a halogenated sesquiterpene diepoxide. Tetrahedron Lett 15:3571
191. Ireland C, Stallard MO, Faulkner DJ, Finer J, Clardy J (1976) Some chemical constituents of the digestive gland of the sea hare *Aplysia californica*. J Org Chem 41:2461
192. Suzuki M, Kurosawa E (1985) A C-15 non-terpenoid from the red alga *Laurencia okamurai*. Phytochemistry 24:1999
193. Vairappan CS, Suzuki M, Ishii T, Okino T, Abe T, Masuda M (2008) Antibacterial activity of halogenated sesquiterpenes from Malaysian *Laurencia* spp. Phytochemistry 69:2490
194. Pitombo LF, Kaiser CR, Pinto AC (1996) Occurrence of chamigrenes in *Aplysia dactylomela* from Brazilian waters. Bol Soc Chil Quim 41:433
195. Masuda M, Kawaguchi S, Takahashi Y, Matsuo Y, Suzuki M (1997) A taxonomic study of the genus *Laurencia* (Ceramiales, Rhodophyta) from Vietnam. I. *Laurencia caduciramulosa* Masuda et Kawaguchi, sp. nov. Crypt Algol 18:71
196. Hall JG, Reiss JA (1986) Elatenyne - a pyrano[3,2-*b*]pyranyl vinyl acetylene from the red alga *Laurencia elata*. Aust J Chem 39:1401
197. Fronczek FR, Caccamese S (1989) Redetermination of the absolute configuration of deoxyprepacifenol from the Mediterranean red alga *Laurencia majuscula*. Acta Crystallogr C45:1102
198. de Nys R, Coll JC, Bowden BF (1993) Tropical marine algae. IX. A new sesquiterpenoid metabolite from the red alga *Laurencia marianensis*. Aust J Chem 46:933
199. Kikuchi H, Suzuki T, Suzuki M, Kurosawa E (1985) A new chamigrane-type bromo diether from the red alga *Laurencia nipponica* Yamada. Bull Chem Soc Jpn 58:2437
200. Watanabe K, Umeda K, Miyakado M (1989) Isolation and identification of three insecticidal principles from the red alga *Laurencia nipponica* Yamada. Agric Biol Chem 53:2513
201. Mao SC, Guo YW (2010) Sesquiterpenes from Chinese red alga *Laurencia okamurai*. Chin J Nat Med 8:321
202. Sims JJ, Fenical W, Wing RM, Radlick P (1973) Marine natural products. IV. Prepacifenol, a halogenated epoxy sesquiterpene and precursor to pacifenol from the red alga *Laurencia filiformis*. J Am Chem Soc 95:972
203. Kurata K, Furusaki A, Katayama C, Kikuchi H, Suzuki T (1981) A new labile sesquiterpene diol having bromine from the marine red alga *Laurencia nipponica* Yamada. Chem Lett 10:773
204. Kigoshi H, Shizuri Y, Niwa H, Yamada K (1981) Laurencenyne, a plausible precursor of various nonterpenoid $C_{15}$-compounds and neolaurencenyne from the red alga *Laurencia okamurai*. Tetrahedron Lett 22:4729
205. Kurata K, Suzuki T, Suzuki M, Kurosawa E, Furusaki A, Suehiro K, Matsumoto T, Katayama C (1983) Structures of two new halogenated chamigrane-type sesquiterpenoids from the red alga *Laurencia nipponica* Yamada. Chem Lett 12:561
206. Caccamese S, Compagnini A, Toscano RM, Nicolo F, Chapuis G (1987) A new labile bromoterpenoid from the red alga *Laurencia majuscula*: dehydrochloroprepacifenol. Tetrahedron 43:5393
207. González AG, Martín JD, Norte M, Pérez R, Weyler V, Perales A, Fayos J (1983) New halogenated constituents of the digestive gland of the sea hare *Aplysia dactylomela*. Tetrahedron Lett 24:847
208. Ahmad VU, Ali MS (1991) Pinnatifinone, a new halogenated chamigrene from the red alga *Laurencia pinnatifida*. Sci Pharm 59:243
209. White RH, Hager LP (1975) A biogenetic sequence of halogenated sesquiterpenes from *Laurencia intricata*. Phys Chem Sci Res Rep 1:633

210. Stallard MO, Faulkner DJ (1974) Chemical constituents of the digestive gland of the sea hare *Aplysia californica* - II. Chemical transformations. Comp Biochem Physiol 49B:37
211. Selover SJ, Crews P (1980) Kylinone, a new sesquiterpene skeleton from the marine alga *Laurencia pacifica*. J Org Chem 45:69
212. Stallard MO, Faulkner DJ (1974) Chemical constituents of the digestive gland of the sea hare *Aplysia californica* - I. Importance of diet. Comp Biochem Physiol 49B:25
213. Kaiser CR, Pitombo LF, Pinto AC (2001) Complete $^{1}$H and $^{13}$C NMR assignments of chamigrenes from *Aplysia dactylomela*. Magn Reson Chem 39:147
214. Ji NY, Li XM, Wang BG (2010) Sesquiterpenes and other metabolites from the marine red alga *Laurencia composita* (Rhodomelaceae). Helv Chim Acta 93:2281
215. Waraszkiewicz SM, Erickson KL (1976) Halogenated sesquiterpenoids from the Hawaiian marine alga *Laurencia nidifica*. IV. Nidifocene. Tetrahedron Lett 17:1443
216. Waraszkiewicz SM, Erickson KL, Finer J, Clardy J (1977) Nidifocene: a reassignment of structure. Tetrahedron Lett 18:2311
217. Iwata C, Akiyama T, Miyashita K (1988) An approach to the stereoselective synthesis of nidifocene: regio- and stereoselective synthesis of *vic-trans*-bromochlorocyclohexane ring system. Chem Pharm Bull 36:2878
218. Miyashita K, Yoneda K, Akiyama T, Koga Y, Tanaka M, Yoneyama T, Iwata C (1993) An approach to the stereoselective synthesis of nidifocene. III. Total syntheses of stereoisomers of (±)-nidifocene from (±)-dehalogenonidifocene. Chem. Pharm Bull 41:465
219. Ahmad VU, Ali MS (1991) Terpenoids from marine red alga *Laurencia pinnatifida*. Phytochemistry 30:4172
220. Cassano V, De-Paula JC, Fujii MT, Da Gama BAP, Teixeira VL (2008) Sesquiterpenes from the introduced red seaweed *Laurencia caduciramulosa* (Rhodomelaceae, Ceramiales). Biochem Syst Ecol 36:223
221. Dias DA, Urban S (2011) Phytochemical studies of the southern Australian marine alga *Laurencia elata*. Phytochemistry 72:2081
222. Caccamese S, Compagnini A, Toscano RM (1986) Pacifenol from the Mediterranean red alga *Laurencia majuscula*. J Nat Prod 49:173
223. Fronczek FR, Caccamese S (1986) Redetermination of pacifenol, a halogenated sesquiterpene from the Mediterranean red alga *Laurencia majuscula*. Acta Crystallogr C42:1649
224. Suzuki T (1980) Two new sesquiterpene alcohols containing bromine from the marine alga *Laurencia nipponica* Yamada. Chem Lett 9:541
225. Sims JJ, Fenical W, Wing RM, Radlick P (1971) Marine natural products. I. Pacifenol, a rare sesquiterpene containing bromine and chlorine from the red alga *Laurencia pacifica*. J Am Chem Soc 93:3774
226. Argandona VH, San-Martín A, Rovirosa J (1993) Halogenated sesquiterpenes pacifenol and pacifenol derivatives on the aphid *Schizaphis graminum*. Phytochemistry 32:1159
227. Rovirosa J, Darias J, Manriquez V, Brito I, Lara N, Argandona V, San Martin A (1994) Structure and insecticidal activities of chamigrene derivatives. Bol Soc Chil Quim 39:193
228. Rao CB, Satyanarayana C, Rao DV (1994) A new chamigrane derivative from *Aplysia dactylomela* of the Indian Ocean. In: Thompson MF, Nagabhushanam R, Sarojini R, Fingerman M (eds) Recent developments in biofouling control. Indo-United States conference, Bangalore, July 1992. AA Balkema, Rotterdam
229. Li XD, Miao FP, Liang XR, Wang BG, Ji NY (2013) Two halosesquiterpenes from *Laurencia composita*. RSC Adv 3:1953
230. Sims JJ, Fenical W, Wing RM, Radlick P (1972) Marine natural products III. Johnstonol, an unusual halogenated epoxide from the red alga *Laurencia johnstonii*. Tetrahedron Lett 13:195
231. Irie T, Suzuki M, Hayakawa Y (1969) Isolation of aplysin, debromoaplysin and aplysinol from *Laurencia okamurai* Yamada. Bull Chem Soc Jpn 42:843
232. Sun J, Shi DY, Ma M, Li SA, Wang SJ, Han LJ, Yang YC, Fan X, Shi JG, He L (2005) Sesquiterpenes from the red alga *Laurencia tristicha*. J Nat Prod 68:915

233. Atta-ur-Rahman, Ahmad VU, Bano S, Abbas SA, Alvi KA, Ali MS, Lu HSM, Clardy J (1988) Pinnatazane, a bridged cyclic ether sesquiterpene from *Laurencia pinnatifida*. Phytochemistry 27:3879
234. Aknin M, Ahond A, Chiaroni A, Poupat C, Riche C, Kornprobst JM (1989) Isolation, détermination structurale et configuration absolue de l'almadioxyde. Tetrahedron Lett 30:559
235. Bano S, Ali MS, Ahmad VU (1988) Marine natural products, IX. A new halogenated sesquiterpene pinnatifidone from the red alga *Laurencia pinnatifida*. Z Naturforsch 43B:1347
236. Taber DF, Sikkander IMI, Storck PH (2007) Enantioselective synthesis of (+)-majusculone. J Org Chem 72:4098
237. Iwata C, Akiyama T, Miyashita K (1988) Synthesis of four possible isomers of 9-(bromomethylene)-1,2,5-trimethy*spiro*[5.5]undeca-1,7-dien-3-one: structure elucidation of a brominated rearranged chamigrane-type sesquiterpene. Chem Pharm Bull 36:2872
238. Suzuki M, Daitoh M, Vairappan CS, Abe T, Masuda M (2001) Novel halogenated metabolites from the Malaysian *Laurencia pannosa*. J Nat Prod 64:597
239. Liang Y, Li XM, Cui CM, Li CS, Wang BG (2009) A new rearranged chamigrane sesquiterpene from *Laurencia okamurai*. Chin Chem Lett 20:190
240. Fedorov SN, Shubina LK, Kalinovsky AI, Lyakhova EG, Stonik VA (2000) Structure and absolute configuration of a new rearranged chamigrane-type sesquiterpenoid from the sea hare *Aplysia* sp. Tetrahedron Lett 41:1979
241. Atta-ur-Rahman (1989) Isolation and structural studies on new natural products of potential biological importance. Pure Appl Chem 61:453
242. Fukuzawa A, Shea CM, Masamune T, Furusaki A, Katayama C, Matsumoto T (1981) Spironippol, new sesquiterpene from the marine alga *Laurencia nipponica* Yamada. Tetrahedron Lett 22:4087
243. Kazlauskas R, Murphy PT, Quinn RJ, Wells RT (1976) New laurene derivatives from *Laurencia filiformis*. Aust J Chem 29:2533
244. Irie T, Yasunari Y, Suzuki T, Imai N, Kurosawa E, Masamune T (1965) A new sesquiterpene hydrocarbon from *Laurencia glandulifera*. Tetrahedron Lett 6:3619
245. Irie T, Suzuki T, Ito S, Kurosawa E (1967) The absolute configuration of laurene and α-cuparenone. Tetrahedron Lett 8:3187
246. Irie T, Suzuki T, Yasunari Y, Kurosawa E, Masamune T (1969) Laurene, a sesquiterpene hydrocarbon from *Laurencia* species. Tetrahedron 25:459
247. Suzuki M, Kurosawa E (1978) New aromatic sesquiterpenoids from the red alga *Laurencia okamurai* Yamada. Tetrahedron Lett 19:2503
248. Wratten SJ, Faulkner DJ (1977) Metabolites of the red alga *Laurencia subopposita*. J Org Chem 42:3343
249. Findlay JA, Li GQ (2002) Novel terpenoids from the sea hare *Aplysia punctata*. Can J Chem 80:1697
250. Srikrishna A, Sundarababu G (1990) A radical cyclisation based strategy to cuparenoids: synthesis of (±)-α-cuparenone, (±)-epilaurene and laurenes. Tetrahedron 46:3601
251. Bailey WF, Jiang XL, McLeod CE (1995) Conformational control in the cyclization of an unsaturated vinyllithium: synthesis of (±)-laurene. J Org Chem 60:7791
252. Kulkarni MG, Pendharkar DS (1997) An efficient total synthesis of (±)-laurene. J Chem Soc Perkin Trans 1:3127
253. Oh CH, Han JW, Kim JS, Um SY, Jung HH, Jang WH, Won HS (2000) A short synthesis of (±)-laurene: mechanistic reinvestigation in palladium-catalyzed cycloreductions of 1,6-enynes. Tetrahedron Lett 41:8365
254. Nemoto H, Nagamochi M, Fukumoto K (1993) Chiral cyclobutanones as versatile synthons: the first enantioselective total synthesis of (+)-laurene. J Chem Soc Perkin Trans 1:2329
255. Kladi M, Xenaki H, Vagias C, Papazafiri P, Roussis V (2006) New cytotoxic sesquiterpenes from the red algae *Laurencia obtusa* and *Laurencia microcladia*. Tetrahedron 62:182
256. Izac RR, Sims JJ (1979) Marine natural products. 18. Iodinated sesquiterpenes from the red algal genus *Laurencia*. J Am Chem Soc 101:6136

257. Dias DA, White JM, Urban S (2009) *Laurencia filiformis*: phytochemical profiling by conventional and HPLC-NMR approaches. Nat Prod Commun 4:157
258. Suzuki M, Kurosawa E (1979) Halogenated sesquiterpene phenols and ethers from the red alga *Laurencia glandulifera* Kützing. Bull Chem Soc Jpn 52:3349
259. König GM, Wright AD (1997) Sesquiterpene content of the antibacterial dichloromethane extract of the marine red alga *Laurencia obtusa*. Planta Med 63:186
260. Appleton DR, Babcock RC, Copp BR (2001) Novel tryptophan-derived dipeptides and bioactive metabolites from the sea hare *Aplysia dactylomela*. Tetrahedron 57:10181
261. Gewali MB, Ronald RC (1982) Synthesis of allolaurinterol. J Org Chem 47:2792
262. Kladi M, Vagias C, Papazafiri P, Furnari G, Serio D, Roussis V (2007) New sesquiterpenes from the red alga *Laurencia microcladia*. Tetrahedron 63:7606
263. Caccamese S, Hager LP, Rinehart KL, Setzer RB (1979) Characterization of *Laurencia* species by GC-MS. Bot Mar 22:41
264. Blunt JW, Lake RJ, Munro MHG (1984) Sesquiterpenes from the marine red alga *Laurencia distichophylla*. Phytochemistry 23:1951
265. Irie T, Suzuki M, Kurosawa E, Masamune T (1970) Laurinterol, debromolaurinterol and isolaurinterol, constituents of *Laurencia intermedia* Yamada. Tetrahedron 26:3271
266. Harrowven DC, Lucas MC, Howes PD (2001) The synthesis of a natural product family: from debromoisolaurinterol to the aplysins. Tetrahedron 57:791
267. Mao S, Guo Y (2005) Cuparene-derived sesquiterpenes from the Chinese red alga *Laurencia okamurai* Yamada. Helv Chim Acta 88:1034
268. Ryu G, Park SH, Choi BW, Lee NH, Hwang HJ, Ryu SY, Lee BH (2002) Cytotoxic activities of brominated sesquiterpenes from the red alga *Laurencia okamurai*. Nat Prod Sci 8:103
269. Ji N-Y, Li X-M, Li K, Ding L-P, Wang B-G (2008) Laurane-derived sesquiterpenes from the marine red alga *Laurencia tristicha* (Rhodomelaceae). Nat Prod Res 22:715
270. Schmitz FJ, Gopichand Y, Michaud D, Prasad RS, Remaley S, Hossain MB, Rahman A, Sengupta PK, van der Helm D (1981) Recent developments in research on metabolites from Caribbean marine invertebrates. Pure Appl Chem 51:853
271. Angawi RF, Alarif WM, Hamza RI, Badria FA, Ayyad SEN (2014) New cytotoxic laurene-, cuparene- and laurokamurene-type sesquiterpenes from the red alga *Laurencia obtusa*. Helv Chim Acta 97:1388
272. González AG, Arteaga JM, Fernández JJ, Martín JD, Norte M, Ruano JZ (1984) Terpenoids of the red alga *Laurencia pinnatifida*. Tetrahedron 40:2751
273. Irie T, Fukuzawa A, Izawa M, Kurosawa E (1969) Laurenisol, a new sesquiterpenoid containing bromine from *Laurencia nipponica* Yamada. Tetrahedron Lett 10:1343
274. Kladi M, Vagias C, Furnari G, Moreau D, Roussakis C, Roussis V (2005) Cytotoxic cuparene sesquiterpenes from *Laurencia microcladia*. Tetrahedron Lett 46:5723
275. Yu XQ, He WF, Liu DQ, Feng MT, Fang Y, Wang B, Feng LH, Guo YW, Mao SC (2014) A *seco*-laurane sesquiterpene and related laurane derivatives from the red alga *Laurencia okamurai* Yamada. Phytochemistry 103:162
276. Ahmad VU, Ali MS, Bano S (1990) Marine natural products. Part 12. Laurol, a new metabolite from the red alga *Laurencia pinnatifida*. Sci Pharm 58:299
277. Rao CB, Satyanarayana C, Rao DV, Fahy E, Faulkner DJ (1989) Metabolites of *Aplysia dactylomela* from the Indian Ocean. Indian J Chem B28:322
278. Shizuri Y, Yamada A, Yamada K (1984) Laurequinone, a cyclolaurane sesquiterpene from the red alga *Laurencia nidifica*. Phytochemistry 23:2672
279. Mao S, Guo Y (2006) A laurane sesquiterpene and rearranged derivatives from the Chinese red alga *Laurencia okamurai* Yamada. J Nat Prod 69:1209
280. Yamada K, Yazawa H, Uemura D, Toda M, Hirata Y (1969) Total synthesis of (±)-aplysin and (±)-debromoaplysin. Tetrahedron 25:3509
281. Feutrill GI, Mirrington RN, Nichols RJ (1973) The total synthesis of (±)-laurinterol and related compounds. Aust J Chem 26:345
282. Ronald RC (1976) A new stereoselective synthesis of (±)-debromoaplysin and (±)-aplysin. Tetrahedron Lett 17:4413

283. Biswas S, Ghosh A, Venkateswaran RV (1990) Stereocontrolled synthesis of (±)-debromoaplysin, (±)-aplysin, (±)-debromoaplysinol, (±)-aplysinol and (±)-isoaplysin. J Org Chem 55:3498
284. Laronze JY, Boukili RE, Patigny D, Dridi S, Cartier D, Levy J (1991) The rearrangement of some cyclopentanone-aryloximes: synthesis of (±)-aplysin, (±)-filiformin and of their debromo analogues. Tetrahedron 47:10003
285. Ronald RC, Gewali MB, Ronald BP (1980) Total synthesis of (–)-aplysin and (–)-debromoaplysin. J Org Chem 45:2224
286. Takano S, Moriya M, Ogasawara K (1992) Enantiocontrolled syntheses of the cuparene sesquiterpenes, (–)-herbertene, (+)-β-cuparenone, (–)-debromoaplysin and (–)-aplysin. Tetrahedron Lett 33:329
287. Nemoto H, Nagamochi M, Ishibashi H, Fukumoto K (1994) A remarkable substituent effect on the enantioselectivity of tandem asymmetric epoxidation and enantiospecific ring expansion of cyclopropylidene alcohols: a new enantiocontrolled synthesis of (–)-debromoaplysin and (–)-aplysin. J Org Chem 59:74
288. Srikrishna A, Chandrasekhar Babu N (2001) An enantiospecific formal total sunthesis of (–)-aplysin and (–)-debromoaplysin. Tetrahedron Lett 42:4913
289. Fletcher CJ, Blair DJ, Wheelhouse KMP, Aggarwal VK (2012) The total synthesis of (–)-aplysin via a lithiation-borylation-propenylation sequence. Tetrahedron 68:7598
290. Sun J, Shi D-Y, Li S, Wang S-J, Han L-J, Fan Z, Yang Y-C, Shi J-G (2007) Chemical constituents of the red alga *Laurencia tristicha*. J Asian Nat Prod Res 9:725
291. Miyamoto T, Ebisawa Y, Higuchi R (1995) Aplyparvunin, a bioactive acetogenin from the sea hare *Aplysia parvula*. Tetrahedron Lett 36:6073
292. Cameron AF, Ferguson G, Robertson JM (1967) The crystal structure and absolute stereochemistry of laurinterol. The absolute stereochemistry of aplysin. J Chem Soc Chem Commun 271
293. Cameron AF, Ferguson G, Robertson JM (1969) *Laurencia* natural products II. Crystal structure and absolute stereochemistry of laurinterol acetate, a bicyclo[3.1.0]hexane derivative. J Chem Soc B:692
294. Sun J, Han LJ, Shi DY, Fan X, Wang SJ, Li S, Yang YC, Shi JG (2005) Sesquiterpenes from red alga *Laurencia tristicha*. Chin Chem Lett 16:1611
295. Sun J, Han LJ, Shi DY, Fan X, Wang SJ, Li S, Yang YC, Shi JG (2006) Sesquiterpene components of *Laurencia tristicha*. Chin Trad Herb Drugs 37:329
296. Suzuki M, Kurata K, Kurosawa E (1986) The structure of isoaplysin, a brominated rearranged cuparane-type sesquiterpenoid from the red alga *Laurencia okamurai* Yamada. Bull Chem Soc Jpn 59:3981
297. McMillan JA, Paul IC, Caccamese S, Rinehart KL (1976) Aplysinol from *Laurencia decidua*: crystal structure and absolute stereochemistry. Tetrahedron Lett 17:4219
298. Wu Z (1989) Chemical constituents of marine algae *Laurencia okamurai*. Chin J Mar Drugs 8:1
299. Copley RCB, Davies-Coleman MT, Edmonds DR, Faulkner DJ, McPhail KL (2002) Absolute stereochemistry of ibhayinol from a South African sea hare. J Nat Prod 65:580
300. Nemoto H, Hakamata H, Nagamochi M, Fukumoto K (1994) An efficient route to chiral benzooxabicyclo[3.2.1]octane ring system-the first enantiocontrolled total synthesis of (–)-filiformin. Heterocycles 39:467
301. Suzuki M, Kurosawa E (1976) New bromo-compounds from *Laurencia glandulifera* Kützing. Tetrahedron Lett 17:4817
302. Yoo S, Suh JH, Yi KY (1998) Total synthesis of (±)-filiforminol and (±)-bromoether A. Synthesis 30:771
303. Su S, Sun WS, Wang B, Cheng W, Liang H, Zhao YY, Zhang QY, Wu J (2010) A novel brominated cuparene-derived sesquiterpene ether from the red alga *Laurencia* sp. J Asian Nat Prod Res 12:916
304. Izac RR, Drage JS, Sims JJ (1981) Caraibical, a new aromatic sesquiterpene from the marine alga *Laurencia caraibica*. Tetrahedron Lett 22:1799

305. Ichiba T, Higa T (1986) New cuparene-derived sesquiterpenes with unprecedented oxygenation patterns from the sea hare *Aplysia dactylomela*. J Org Chem 51:3364
306. Srikrishna A, Krishnan K (1992) Total syntheses of (±)-cyclolaurene, (±)-epicyclolaurene and (±)-β-cuparenones. Tetrahedron 48:3429
307. Irie T, Suzuki M, Kurosawa E, Masamune T (1966) Laurinterol and debromolaurinterol, constituents from *Laurencia intermedia*. Tetrahedron Lett 7:1837
308. Okamoto Y, Nitanda N, Ojika M, Sakagami Y (2001) Aplysiallene, a new bromoallene as an $Na^+$, $K^+$-ATPase inhibitor from the sea hare *Aplysia kurodai*. Biosci Biotechnol Biochem 65:474
309. Tsukamoto S, Yamashita Y, Ohta T (2005) New cytotoxic and antibacterial compounds isolated from the sea hare *Aplysia kurodai*. Mar Drugs 3:22
310. Masuda M, Abe T, Kogame K, Kawaguchi S, Phang SM, Daitoh M, Sakai T, Takahashi Y, Suzuki M (2002) Taxonomic notes on marine algae from Malaysia. VIII. Three species of *Laurencia* (Rhodophyceae). Bot Mar 45:571
311. Takahashi H, Tonoi Y, Matsumoto K, Minami H, Fukuyama Y (1998) Total synthesis of (–)-laurequinone. Chem Lett 27:485
312. Shizuri Y, Yamada K (1985) Laurebiphenyl, a dimeric sesquiterpene of the cyclolaurane-type from the red alga *Laurencia nidifica*. Phytochemistry 24:1385
313. Srikrishna A, Khan IA, Babu RR, Sajjanshetty A (2007) The first total synthesis of (±)-laurokamurene B. Tetrahedron 63:12616
314. Srikrishna A, Beeralah B, Babu RR (2008) Enantioselective total synthesis and assignment of the absolute configuration of (+)-laurokamurene B. Tetrahedron Asymm 19:624
315. Sun J, Shi D, Ma M, Li S, Wang S, Han L, Yang Y, Fan X, Shi J, He L (2008) Addition and correction to "Sesquiterpenes from the red alga *Laurencia tristicha*". J Nat Prod 71:296
316. Chen P, Wang J, Liu K, Li C (2008) Synthesis and structural revision of (±)-laurentristich-4-ol. J Org Chem 73:339
317. Howard BM, Fenical W (1976) α- and β-Snyderol: new bromo-monocyclic sesquiterpenes from the seaweed *Laurencia*. Tetrahedron Lett 16:41
318. González AG, Martín JD, Norte M, Rivera P, Ruano JZ (1984) Two new $C_{15}$ acetylenes from the marine red alga *Laurencia obtusa*. Tetrahedron 40:3443
319. Topcu G, Aydoğmuş Z, Imre S, Gšren AC, Pezzuto JM, Clement JA, Kingston DGI (2003) Brominated sesquiterpenes from the red alga *Laurencia obtusa*. J Nat Prod 66:1505
320. Howard BM, Fenical W (1978) Obtusadiol, a unique bromoditerpenoid from the marine red alga *Laurencia obtusa*. Tetrahedron Lett 18:2453
321. Imre S, Aydoğmuş Z (1997) Secondary metabolites from the red alga *Laurencia obtusa*. Pharmazie 52:883
322. González AG, Martín JD, Pérez C, Ramírez MA (1976) Bromonium ion-induced cyclization of methyl farnesate: application to the synthesis of snyderol. Tetrahedron Lett 17:137
323. Kuniyoshi M, Marma MS, Higa T, Bernardinelli G, Jefford CW (2001) New bromoterpenes from the red alga *Laurencia luzonensis*. J Nat Prod 64:696
324. Kuniyoshi M, Wahome PG, Miono T, Hashimoto T, Yokoyama M, Shrestha KL, Higa T (2005) Terpenoids from *Laurencia luzonensis*. J Nat Prod 68:1314
325. Su H, Shi D-Y, Li J, Guo S-J, Li L-L, Yuan Z-H, Zhu X-B (2009) Sesquiterpenes from *Laurencia similis*. Molecules 14:1889
326. Ioannou E, Nappo M, Avila C, Vagias C, Roussis V (2009) Metabolites from the sea hare *Aplysia fasciata*. J Nat Prod 72:1716
327. Norte M, González R, Padilla A, Fernández JJ, Vázquez JT (1991) New halogenated sesquiterpenes from the red alga *Laurencia caespitosa*. Can J Chem 69:518
328. Paul VJ, Fenical W (1980) Palisadins A, B and related monocyclofarnesol-derived sesquiterpenoids from *Laurencia* cf. *palisada*. Tetrahedron Lett 21:2787
329. Masuda M, Takahashi Y, Okamoto K, Matsuo Y, Suzuki M (1997) Morphology and halogenated secondary metabolites of *Laurencia snackeyi* (Weber-van Bosse) stat. nov. (Ceramiales, Rhodophyra). Eur J Phycol 32:293

330. Vairappan CS, Kamada T, Lee WW, Jeon YJ (2013) Anti-inflammatory activity of halogenated secondary metabolites of *Laurencia snackeyi* (Weber-van Bosse) Masuda in LPS-stimulated RAW 264.7 macrophages. J Appl Phycol 25:1805
331. de Nys R, Wright AD, König GM, Sticher O, Alino PM (1993) Five new sesquiterpenes from the red alga *Laurencia flexilis*. J Nat Prod 56:877
332. König GM, Wright AD (1994) X-ray crystal structure of 3,4-epoxypalisadin A. J Nat Prod 57:151
333. Su J, Zhong Y, Zeng L, Wu H, Ma K (1995) Terpenoids from *Laurencia karlae*. Phytochemistry 40:195
334. Makhanu DS, Yokoyama M, Miono T, Maesato T, Maedomari M, Wisespongpand P, Kuniyoshi M (2006) New sesquiterpenes from the Okinawan red alga *Laurencia luzonensis*. Bull Fac Sci Univ Ryukuyus 81:115
335. Su H, Yuan Z-H, Li J, Guo S-J, Deng L-P, Han L-J, Zhu X-B, Shi D-Y (2009) Sesquiterpenes from the marine red alga *Laurencia saitoi*. Helv Chim Acta 92:1291
336. Huang Y, Hu S, Zhong Y, Su J (1994) Structure of palisadin B, $C_{15}H_{24}Br_2O$. Chin J Struct Chem 13:48
337. Vairappan CS, Anangdan SP, Lee TK (2007) Additional halogenated secondary metabolites from the sea hare *Aplysia dactylomela*. Malay J Sci 26:57
338. Couladouros EA, Vidali VP (2004) Novel stereocontrolled approach to *syn*- and *anti*-oxepene-cyclogeranyl *trans*-fused polycyclic systems: asymmetric total synthesis of (–)-aplysistatin, (+)-palisadin A, (+)-palisadin B, (+)-12-hydroxypalisadin B and the AB ring system of adociasulfate-2 and toxicol A. Chem Eur J 10:3822
339. Vairappan CS, Lee TK (2005) Halogenated secondary metabolites from sea hare *Aplysia dactylomela*. Malay J Sci 24:17
340. Tanaka J, Kuniyoshi M, Tanaka C, Issa HH, Balansa W, Otsuka M, Githige WR, Higa T (2005) Diverse metabolites of coral reef organisms. Pure Appl Chem 77:83
341. Capon R, Ghisalberti EL, Jefferies PR, Skelton BW, White AH (1981) Sesquiterpene metabolites from *Laurencia filiformis*. Tetrahedron 37:1613
342. Sun LL, Wang CY, Dai HF, Shao CL, Mei WL, Tao-Liu MZD (2011) Chemical constituents of *Chondrophycus papillosus* and their cytotoxicity in vitro. Chem Nat Comp 47:650
343. Pettit GR, Herald CL, Allen MS, Von Dreele RB, Vanell LD, Kao JPY, Blake W (1977) The isolation and structure of aplysistatin. J Am Chem Soc 99:262
344. White JD, Nishiguchi T, Skeean RW (1982) Stereoselective, biogenetically patterned synthesis of (±)-aplysistatin. J Am Chem Soc 104:3923
345. Hoye TR, Caruso AJ, Dellaria JF, Kurth MJ (1982) Two syntheses of *dl*-aplysistatin. J Am Chem Soc 104:6704
346. Gosselin P, Rouessac F (1983) Polycyclisations cationiques de polyenes via leurs bromohydrines — II. Synthese de la (±) aplysistatine. Tetrahedron Lett 24:5515
347. Shieh H-M, Prestwich GD (1982) Chiral, biomimetic total synthesis of (–)-aplysistatin. Tetrahedron Lett 23:4643
348. Faulkner DJ (1976) 3$\beta$-Bromo-8-epicaparrapi oxide, the major metabolite of *Laurencia obtusa*. Phytochemistry 15:1993
349. Suzuki M, Takahashi Y, Matsuo Y, Guiry MD, Masuda M (1997) Scanlonenyne, a novel halogenated $C_{15}$ acetogenin from the red alga *Laurencia obtusa* in Irish waters. Tetrahedron 53:4271
350. Recsei C, Chan B, McErlean CSP (2014) Synthesis of (+)-luzofuran and (−)-ancistrofuran. J Org Chem 79:880
351. Horsley SB, Cardellina JH, Meinwald J (1981) Secondary metabolites from a red alga (*Laurencia intricata*): sesquiterpene alcohols. J Org Chem 46:5033
352. Schmitz FJ, McDonald FJ, Vanderah DJ (1978) Marine natural products: sesquiterpene alcohols and ethers from the sea hare *Aplysia dactylomela*. J Org Chem 43:4220
353. Sun HH, Waraszkiewicz SM, Erickson KL (1976) Sesquiterpenoid alcohols from the Hawaiian marine alga *Laurencia nidifica*. III. Tetrahedron Lett 17:585

354. Howard BM, Fenical W, Finer J, Hirotsu K, Clardy J (1977) Neoconcinndiol hydroperoxide, a novel marine diterpenoid from the red alga *Laurencia*. J Am Chem Soc 99:6440
355. Oppolzer W, Briner PH, Snowden RL (1980) A short synthesis of 3-methyl-5-(2,3,6-trimethylphenyl)-1-penten-3-ol, a sesquiterpene isolated from *Laurencia nidifica*. Helv Chim Acta 63:967
356. Schmitz FJ, McDonald FJ (1974) Marine natural products: dactyloxene-B, a sesquiterpene ether from the sea hare *Aplysia dactylomela*. Tetrahedron Lett 15:2541
357. Maurer B, Hauser A, Thommen W, Schulte-Elte KH, Ohloff G (1980) Synthesis and configuration of the eight diastereoisomeric racemates of dactyloxene-B. The relative configuration of dactyloxene-B and -C. Helv Chim Acta 63:293
358. Maurer B, Hauser A, Ohloff G (1980) Synthesis and absolute configuration of naturally occurring dactyloxene-B and -C. Helv Chim Acta 63:2503
359. Paquette LA, Lord MD, Negri JT (1993) Enantioselective synthesis of natural (+)-dactyloxene B and C by actuation of oxonium ion-initiated pinacol rearrangement. Tetrahedron Lett 34:5693
360. Lord MD, Negri JT, Paquette LA (1995) Oxonium ion-initiated pinacolic ring expansion reactions. Application to the enantioselective synthesis of the *spiro*cyclic sesquiterpene ethers dactyloxene-B and C. J Org Chem 60:191
361. Chattopadhyay SK, Karmakar S, Sarkar K (2005) Short new route to the chiral *spiro*-tetrahydrofuran subunit common to some terpenoids. Synth Commun 35:2125
362. Ayyad SN, Jakupovic J, Abdel-Mogib M (1994) A sesquiterpene ether from *Laurencia obtusa*. Phytochemistry 36:1077
363. Suzuki T, Kikuchi H, Kurosawa E (1980) (*E*)-γ-Bisabolen-8,9-epoxide and isocycloeudesmol, two new sesquiterpenoids from *Laurencia nipponica* Yamada. Chem Lett 9:1267
364. Martín JD, Pérez C, Ravelo JL (1985) Stereocontrolled syntheses of (*E*)- and (*Z*)-γ-bisabolene 8,9-epoxide. J Am Chem Soc 107:516
365. Vazquez JT, Chang M, Nakanishi K, Martin JD, Martin VS, Perez R (1988) Puertitols: novel sesquiterpenes from *Laurencia obtusa*. Structure elucidation and absolute configuration and conformation based on circular dichroism. J Nat Prod 51:1257
366. Davyt D, Fernandez R, Suescun L, Mombrú AW, Saldaña J, Domínguez L, Fujii MT, Manta E (2006) Bisabolanes from the red alga *Laurencia scoparia*. J Nat Prod 69:1113
367. Norte M, Fernández JJ, Padilla A (1992) Bisabolane halogenated sesquiterpenes from *Laurencia*. Phytochemistry 31:326
368. Brito I, Dias T, Díaz-Marrero AR, Darias J, Cueto M (2006) Aplysiadiol from *Aplysia dactylomela* suggested a key intermediate for a unified biogenesis of regular and irregular marine algal bisabolene-type metabolites. Tetrahedron 62:9655
369. Chang M, Vazquez JT, Nakanishi K, Cataldo F, Estrada DM, Fernandez J, Gallardo A, Martin JD, Norte M, Perez R (1989) Regular and irregular sesquiterpenes containing a halogenated hydropyran from *Laurencia caespitosa*. Phytochemistry 28:1417
370. Lhullier C, Falkenberg M, Ioannou E, Quesada A, Papazafiri P, Horta PA, Schenkel EP, Vagias C, Roussis V (2010) Cytotoxic halogenated metabolites from the Brazilian red alga *Laurencia catarinensis*. J Nat Prod 73:27
371. Hollenbeak KH, Schmitz FJ, Hossain MB, van der Helm D (1979) Marine natural products. Deodactol, antineoplastic sesquiterpenoid from the sea hare *Aplysia dactylomela*. Tetrahedron 35:541
372. Gopichand Y, Schmitz FJ, Shelly J, Rahman A, Van Der Helm D (1981) Halogenated acetylenic ethers from the sea hare *Aplysia dactylomela*. J Org Chem 46:5192
373. De Carvalho LR, Fujii MT, Roque NF, Lago JHG (2006) Aldingenin derivatives from the red alga *Laurencia aldingensis*. Phytochemistry 67:1331
374. Takahashi S, Yasuda M, Nakamura T, Hatano K, Matsuoka K, Koshino H (2014) Synthesis and structural revision of a brominated sesquiterpenoid, aldingenin C. J Org Chem 79:9373
375. González AG, Darias J, Martín JD (1973) Caespitol, a new halogenated sesquiterpene from *Laurencia caespitosa*. Tetrahedron Lett 14:2381

376. González AG, Darias J, Martín JD, Pérez C (1974) Revised structure of caespitol and its correlation with isocaespitol. Tetrahedron Lett 15:1249
377. Masuda M, Kogame K, Arisawa S, Suzuki M (1998) Morphology and halogenated secondary metabolites of three Gran Canarian species of *Laurencia* (Ceramiales, Rhodophyra). Bot Mar 41:265
378. de Carvalho LR, Fujii MT, Roque NF, Kato MJ, Lago JHG (2003) Aldingenin A, new brominated sesquiterpene from red algae *Laurencia aldingensis*. Tetrahedron Lett 44:2637
379. Mukhina OA, Koshino H, Crimmins MT, Kutateladze AG (2015) Computationally driven reassignment of the structures of aldingenins A and B. Tetrahedron Lett 56:4900
380. Schmitz FJ, Michaud DP, Hollenbeak KH (1980) Marine natural products: dihydroxydeodactol monoacetate, a halogenated sesquiterpene ether from the sea hare *Aplysia dactylomela*. J Org Chem 45:1525
381. González AG, Martín JD, Pérez C, Ramírez MA, Ravelo F (1980) Total synthesis of 8-desoxy-isocaespitol, a new polyhalogenated sesquiterpene from *Laurencia caespitosa*. Tetrahedron Lett 21:187
382. González AG, Darias J, Martín JD, Pérez C, Sims JJ, Lin GHY, Wing RM (1975) Isocaespitol, a new halogenated sesquiterpene from *Laurencia caespitosa*. Tetrahedron 31:2449
383. González AG, Martín JD, Melián MA (1976) Synthesis of marine terpenoids III. Synthesis of ($\pm$)-isocaespitol. Tetrahedron Lett 17:2279
384. González AG, Darias V, Estévez E (1982) Chemotherapeutic activity of polyhalogenated terpenes from Spanish algae. Planta Med 44:44
385. Yang J, Tummatorn J, Slegeris R, Tlais SF, Dudley GB (2011) Synthesis of the tricyclic core of aldingenin B by oxidative cyclo-ketalization of an alkyne-diol. Org Lett 13:2065
386. Crimmins MT, Hughes CO (2012) Total synthesis of the proposed structure of aldingenin B. Org Lett 14:2168
387. Estrada DM, Martín JD, Pérez R, Rivera P, Rodríguez ML, Ruano JZ (1987) Furocaespitane and related $C_{12}$ metabolites from *Laurencia caespitosa*. Tetrahedron Lett 28:687
388. González AG, Darias J, Martín JD (1973) Furocaespitane, a new furan from *Laurencia caespitosa*. Tetrahedron Lett 14:3625
389. González AG, Martín JD, Norte M, Pérez R, Rivera P, Ruano JZ, Rodríguez ML, Fayos J, Perales A (1983) X-Ray structure determination of new brominated metabolites isolated from the red seaweed *Laurencia obtusa*. Tetrahedron Lett 24:4143
390. Erickson KL, Beutler JA, Gray GN, Cardellina JH II, Boyd MR (1995) Majapolene A, a cytotoxic peroxide, and related sesquiterpenes from the red alga *Laurencia majuscula*. J Nat Prod 58:1848
391. Monde K, Taniguchi T, Miura N, Vairappan CS, Suzuki M (2006) Absolute configurations of brominated sesquiterpenes determined by vibrational circular dichroism. Chirality 18:335
392. Brito I, Dias T, Díaz-Marrero AR, Darias J, Cueto M (2007) Corrigendum to "Aplysiadiol from *Aplysia dactylomela* suggested a key intermediate for a unified biogenesis of regular and irregular marine algal bisabolene-type metabolites". Tetrahedron 63:3908
393. Iliopoulou D, Roussis V, Pannecouque C, De Clercq E, Vagias C (2002) Halogenated sesquiterpenes from the red alga *Laurencia obtusa*. Tetrahedron 58:6749
394. González AG, Darias J, Martín JD (1977) Biomimetic interconversions of two new types of metabolite from *Laurencia perforata*. Tetrahedron Lett 18:3375
395. Howard BM, Fenical W (1979) Guadalupol and epiguadalupol, rearranged sesquiterpene alcohols from *Laurencia snyderiae* var. *guadalupensis*. Phytochemistry 18:1224
396. González AG, Darias J, Martín JD, Melián MA (1978) Total synthesis of racemic perforenone and 3-debromo-perforatone. Tetrahedron Lett 19:481
397. Majetich G, Ringold C (1987) A stereospecific synthesis of ($\pm$)-perforenone. Heterocycles 25:271
398. González AG, Aguiar JM, Martín JD, Norte M (1975) Three new sesquiterpenoids from the marine alga *Laurencia perforata*. Tetrahedron Lett 16:2499

399. Wright AD, Goclik E, König GM (2003) Three new sesquiterpenes from the red alga *Laurencia perforata*. J Nat Prod 66:435
400. González AG, Aguiar JM, Darias J, González E, Martín JD, Martín VS, Pérez C, Fayos J, Martínez-Ripoll M (1978) Perforenol, a new polyhalogenated sesquiterpene from *Laurencia perforata*. Tetrahedron Lett 19:3931
401. González AG, Aguiar JM, Martín JD, Rodríguez ML (1976) Perforene, a new halogenated sesquiterpene from the red alga *Laurencia perforata*. Tetrahedron Lett 17:205
402. Coll JC, Skelton BW, White AH, Wright AD (1989) Tropical marine algae V. The structure determination of two novel sesquiterpenes from the red alga *Laurencia tenera* (Rhodophyceae, Ceramiales, Rhodomelaceae). Aust J Chem 42:1695
403. Wright AD, König GM, Sticher O, Rüegger H (1992) The application of two-dimensional nuclear magnetic resonance methodologies to the structure solution of the new natural product tenerol acetate from *Laurencia tenera*. Phytochem Anal 3:263
404. Caccamese S, Amico V, Neri P (1990) Two new rearranged sesquiterpenoids from the red alga *Laurencia obtusa*. J Nat Prod 53:1287
405. Iliopoulou D, Vagias C, Galanakis D, Argyropoulos D, Roussis V (2002) Brasilane-type sesquiterpenoids from *Laurencia obtusa*. Org Lett 4:3263
406. Stallard MO, Fenical W, Kittredge JS (1978) The brasilenols, rearranged sesquiterpene alcohols isolated from the marine opisthobranch *Aplysia brasiliana*. Tetrahedron 34:2077
407. Manzo E, Ciavatta ML, Gavagnin M, Puliti R, Mollo E, Guo YW, Mattia CA, Mazzarella L, Cimino G (2005) Structure and absolute stereochemistry of novel $C_{15}$-halogenated acetogenins from the anaspidean mollusc *Aplysia dactylomela*. Tetrahedron 61:7456
408. Greene AE, Serra AA, Barreiro EJ, Costa PRR (1987) Expeditious, stereocontrolled syntheses of racemic and natural brasilenol through intramolecular asymmetry transfer. Absolute stereochemistry of brasilenol. J Org Chem 52:1169
409. Greene AE, Coelho F, Barreiro EJ, Costa PRR (1986) A selective synthesis of brasilenol, a novel sesquiterpene from the sea hare *Aplysia brasiliana* and the red alga *Laurencia obtusa*. J Org Chem 51:4250
410. Amico V, Caccamese S, Neri P, Russo G, Foti M (1991) Brasilane-type sesquiterpenoids from the Mediterranean red alga *Laurencia obtusa*. Phytochemistry 30:1921
411. Mihopoulos N, Vagias C, Scoullos M, Roussis V (1999) Laurencienyne B, a new acetylenic cyclic ether from the red alga *Laurencia obtusa*. Nat Prod Lett 13:151
412. Tori M, Nakashima K, Seike M, Asakawa Y, Wright AD, König GM, Sticher O (1994) Revised structure of a brasilane-type sesquiterpene isolated from the red alga *Laurencia implicata* and its absolute configuration. Tetrahedron Lett 35:3105
413. Aydoğmuş Z, Imre S, Ersoy L, Wray V (2004) Halogenated secondary metabolites from *Laurencia obtusa*. Nat Prod Res 18:43
414. Suzuki T, Suzuki M, Kurosawa E (1975) α-Bromocuparene and α-isobromocuparene, new bromo compounds from *Laurencia* species. Tetrahedron Lett 16:3057
415. Coll JC, Wright AD (1989) Tropical marine algae IV. Novel metabolites from the red alga *Laurencia implicata* (Rhodophyta, Rhodophyceae, Ceramiales, Rhodomelaceae). Aust J Chem 42:1685
416. Wright AD, König GM, de Nys R, Sticher O (1993) Seven new metabolites from the marine red alga *Laurencia majuscula*. J Nat Prod 56:394
417. Sun HH, Erickson KL (1978) Sesquiterpenoids from the Hawaiian marine alga *Laurencia nidifica*. 7. (+)-Selin-4,7(11)-diene. J Org Chem 43:1613
418. Fukuzawa A, Aye M, Takaya Y, Masamune T, Murai A (1990) A sesquiterpene alcohol from the red alga *Laurencia nipponica*. Phytochemistry 29:2337
419. Dieter RK, Kinnel R, Meinwald J, Eisner T (1979) Brasudol and isobrasudol, two bromosesquiterpenes from a sea hare (*Aplysia brasiliana*). Tetrahedron Lett 20:1645
420. Brennan MR, Erickson KL (1982) Austradiol acetate and austradiol diacetate, 4,6-dihydroxy-(+)-selinane derivatives from an Australian *Laurencia* sp. J Org Chem 47:3917
421. Rochfort SJ, Capon RJ (1996) Parguerenes revisited: new brominated diterpenes from the southern Australian marine red alga *Laurencia filiformis*. Aust J Chem 49:19

422. Suzuki M, Takahashi Y, Mitome Y, Itoh T, Abe T, Masuda M (2002) Brominated metabolites from an Okinawan *Laurencia intricata*. Phytochemistry 60:861
423. Alarif WM, Al-Footy KO, Zubair MS, Halid PHM, Ghandourah MA, Basaif SA, Al-Lihaibi SS, Ayyad SEN, Badria FA (2016) The role of new eudesmane-type sesquiterpenoid and known eudesmane derivatives from the red alga *Laurencia obtusa* as potential antifungal-antitumour agents. Nat Prod Res 30:1150
424. Howard BM, Fenical W (1977) Structure, chemistry and absolute configuration of (–)-(1*S*,4*R*)-bromo-hydoxy-selin-7-ene from a marine red alga *Laurencia* sp. J Org Chem 42:2518
425. Rose AF, Sims JJ, Wing RM, Wiger GM (1978) Marine natural products. XVII. The structure of (1*S*,4*R*,7*R*)-1-bromo-4-hydroxy-7-chloroselinane, a metabolite of the marine alga *Laurencia* sp. Tetrahedron Lett 19:2533
426. Kazlauskas R, Murphy PT, Wells RJ, Daly JJ, Oberhänsli WE (1977) Heterocladol, a halogenated selinane sesquiterpene of biosynthetic significance from *Laurencia filiformis*: its isolation, crystal structure and absolute configuration. Aust J Chem 30:2679
427. Baker B, Ratnapala L, Mahindaratne MPD, de Silva ED, Tillekeratne LMV, Jeong JH, Scheuer PJ, Seff K (1988) Lankalapuol A and B: two *cis*-eudesmanes from the sea hare *Aplysia dactylomela*. Tetrahedron 44:4695
428. Suzuki T, Furusaki A, Kikuchi H, Kurosawa E, Katayama C (1981) The absolute configuration of cycloeudesmol from the red alga *Laurencia nipponica* Yamada. Tetrahedron Lett 22:3423
429. Fenical W, Sims JJ (1974) Cycloeudesmol, an antibiotic cyclopropane containing sesquiterpene from the marine alga *Chondria oppositiclada* Dawson. Tetrahedron Lett 15:1137
430. Guella G, Skropeta D, Mancini I, Pietra F (2002) The first 6,8-cycloeudesmane sesquiterpene from a marine organism: the red seaweed *Laurencia microcladia* from the Baia di Calenzana, Elba Island. Z Naturforsch 57B:1147
431. Li C, Li X, Cui C, Wang B (2010) Brominated metabolites from the marine red alga *Laurencia similis*. Z Naturforsch 65B:87
432. Kamada T, Vairappan CS (2013) New bioactive secondary metabolites from Bornean red alga *Laurencia similis* (Ceramiales). Nat Prod Commun 8:287
433. Rahelivao MP, Gruner M, Andriamanantoanina H, Andriamihaja B, Bauer I, Knölker HJ (2015) Red algae (Rhodophyta) from the coast of Madagascar: preliminary bioactivity studies and isolation of natural products. Mar Drugs 13:4197
434. Ji NY, Li XM, Ding LP, Wang BG (2007) Two new aristolane sesquiterpenes from *Laurencia similis*. Chin Chem Lett 18:178
435. Imre S, Islimyeli S, Öztunc A, Thomson RH (1981) Obtusenol, a sesquiterpene from *Laurencia obtusa*. Phytochemistry 20:833
436. González AG, Martín JD, Pérez C, Ramírez MA, Ravelo F (1981) Total synthesis of obtusenol. Tetrahedron Lett 22:5071
437. Takeda S, Iimura Y, Tanaka K, Kurosawa E, Suzuki T (1990) A new naturally occurring racemic compound from the marine red alga *Laurencia obtusa* (Hudson) Lamouroux. Chem Lett 19:155
438. Suzuki T, Kurosawa E (1979) New bromo-acetal from the marine alga *Laurencia nipponica* Yamada. Chem Lett 8:301
439. Kurata K, Suzuki T, Suzuki M, Kurosawa E, Furusaki A, Matsumoto T (1983) Laureacetal-D and -E, two new secochamigrane derivatives from the red alga *Laurencia nipponica* Yamada. Chem Lett 12:557
440. Kurata K, Suzuki T, Suzuki M, Kurosawa E (1983) Laureacetal-C, an unusual secochamigrane sesquiterpene from the red alga *Laurencia nipponica* Yamada. Chem Lett 12:29
441. Fenical W, Schulte GR, Finer J, Clardy J (1978) Poitediol, a new nonisoprenoid sesquiterpene diol from the marine alga *Laurencia poitei*. J Org Chem 43:3628
442. Schmitz FJ, Hollenbeak KH, Vanderah DJ (1978) Marine natural products: dactylol, a new sesquiterpene alcohol from a sea hare. Tetrahedron 34:2719

443. Gadwood RC (1985) Stereoelectronic effects in cyclo-octanes: synthesis of (±)-dactylol and (±)-isodactylol. J Chem Soc Chem Commun:123
444. Feldman KS, Wu MJ, Rotella DP (1990) Total synthesis of (±)-dactylol and related studies. J Am Chem Soc 112:8490
445. Molander GA, Eastwood PR (1995) Total synthesis of (+)-dactylol via a novel [3+5] annulation approach. J Org Chem 60:4559
446. Fürstner A, Langemann K (1996) A concise total synthesis of dactylol via ring closing metathesis. J Org Chem 61:8746
447. Gadwood RC, Lett RM, Wissinger JE (1984) Total synthesis of (±)-poitediol and (±)-4-epipoitediol. J Am Chem Soc 106:3869
448. Hall SS, Faulkner DJ, Fayos J, Clardy J (1973) Oppositol, a brominated sesquiterpene alcohol of a new skeletal class from the red alga *Laurencia subopposita*. J Am Chem Soc 95:7187
449. Fukuzawa A, Sato H, Masamune T (1987) Synthesis of (±)-prepinnaterpene, a bromoditerpene from the red alga *Laurencia pinnata* Yamada. Tetrahedron Lett 28:4303
450. Kim D, Kim IH (1997) A stereoselective total synthesis of (±)-oppositol by a doubly diastereoselective intramolecular ester enolate alkylation. Tetrahedron Lett 38:415
451. Wijesinghe WAJP, Kang MC, Lee WW, Lee HS, Kamada T, Vairappan CS, Jeon YJ (2014) 5β-Hydroxypalisadin B isolated from red alga *Laurencia snackeyi* attenuates inflammatory response in lipopolysaccharide-stimulated RAW 264.7 macrophages. Algae 29:333
452. Kurata K, Suzuki T, Suzuki M, Kurosawa E, Furusaki A, Matsumoto T (1983) Laurencial, a novel sesquiterpene α,β-unsaturated aldehyde from the red alga *Laurencia nipponica* Yamada. Chem Lett 12:299
453. Miyashita K, Tanaka A, Shintaku H, Iwata C (1998) Regiospecific bromination-cyclization of *spiro*cyclic cyclohexanones and its application to the synthesis of (±)-laurencial. Tetrahedron 54:1395
454. Suzuki M, Kurosawa E, Irie T (1970) Spirolaurenone, a new sesquiterpenoid containing bromine from *Laurencia glandulifera* Kützing. Tetrahedron Lett 11:4995
455. Suzuki M, Kowata N, Kurosawa E (1980) The structure of *spiro*laurenone, a halogenated sesquiterpenoid from the red alga *Laurencia glandulifera* Kützing. Tetrahedron 36:1551
456. Murai A, Kato K, Masamune T (1982) Total synthesis of (±)-*spiro*laurenone. Tetrahedron Lett 23:2887
457. Fukuzawa A, Matsue H, Masamune T, Furusaki A, Katayama C, Matsumoto T (1984) Laurenones A and B, new sesquiterpenes from the red alga *Laurencia nipponica* Yamada. Chem Lett 13:1349
458. Gressler V, Stein EM, Dörr F, Fujii MT, Colepicolo P, Pinto E (2011) Sesquiterpenes from the essential oil of *Laurencia dendroidea* (Ceramiales, Rhodophyta): isolation, biological activities and distribution among seaweeds. Braz J Pharmacogn 21:248
459. Norte M, Fernández JJ, Souto ML (1994) Viridianol, a rearranged sesquiterpene with a novel carbon skeleton from *Laurencia viridis*. Tetrahedron Lett 35:4607
460. Xu X, Zeng L, Su J (1997) Tricyclic sesquiterpene from *Laurencia majuscula*. Chem Res Chin Univ 13:176
461. Brito I, Cueto M, Dorta E, Darias J (2002) Bromocyclococanol, a halogenated sesquiterpene with a novel carbon skeleton from the red alga *Laurencia obtusa*. Tetrahedron Lett 43:2551
462. Guella G, Skropeta D, Breuils S, Mancini I, Pietra F (2001) Calenzanol, the first member of a new class of sesquiterpene with a novel skeleton isolated from the red seaweed *Laurencia microladia* from the Bay of Calenzana, Elba Island. Tetrahedron Lett 42:723
463. Guella G, Skropeta D, Mancini I, Pietra F (2003) Calenzanane sesquiterpenes from the red seaweed *Laurencia microcladia* from the Bay of Calenzana, Elba Island: acid-catalyzed stereospecific conversion of calenzanol into indene- and guaiazulene-type sesquiterpenes. Chem Eur J 9:5770
464. Caccamese S, Amico V, Neri P, Foti M (1991) The structure of laurobtusol, a new rearranged sesquiterpenoid from the Mediterranean red alga *Laurencia obtusa*. Tetrahedron 47:10101

465. Blanchfield JT, Chow S, Bernhardt PV, Kennard CHL, Kitching W (2004) Concerning the proposed structure of (+)-laurobtusol: spectral discrepancies with synthetic, racemic stereoisomers. Aust J Chem 57:673
466. González AG, Martín JD, Martín VS, Pérez R, Drexler SA, Clardy J (1984) Structure of güimarediol, a new rearranged sesquiterpenoid from the red alga *Laurencia* sp. Chem Lett 13:1865
467. Díaz-Marrero AR, Brito I, de la Rosa JM, Darias J, Cueto M (2008) Gomerones A-C, halogenated sesquiterpenoids with a novel carbon skeleton from *Laurencia majuscula*. Tetrahedron 64:10821
468. Huwyler N, Carreira EM (2012) Total synthesis and stereochemical revision of the chlorinated sesquiterpene (±)-gomerone C. Angew Chem Int Ed 51:13066
469. González AG, Martín JD, Martín VS, Pérez R, Tagle B, Clardy J (1985) Rhodolaureol and rhodolauradiol, two new halogenated tricyclic sesquiterpenes from a marine alga. J Chem Soc Chem Commun:260
470. González AG, Martín JD, Martín VS, Norte M, Pérez R (1982) Bioimetic approach to the syntheses of rhodolaureol and rhodolauradiol. Tetrahedron Lett 23:2395
471. Fedorov SN, Radchenko OS, Shubina LK, Kalinovsky AI, Gerasimenko AV, Popov DY, Stonik VA (2001) Aplydactone, a new sesquiterpenoid with an unprecedented carbon skeleton from the sea hare *Aplysia dactylomela*, and its Cargill-like rearrangement. J Am Chem Soc 123:504
472. Guella G, Chiasera G, Pietra F (1992) Conformational studies of marine polyhalogenated α-chamigrenes using temperature-dependent NMR spectra. Cyclohexene-ring flipping and rigid-chair cyclohexane ring in the presence of equatorial halogen atoms at C(8) and C(9). Helv Chim Acta 75:2012
473. Guella G, Chiasera G, Pietra F (1992) Conformational studies of marine polyhalogenated α-chamigrenes using temperature-dependent NMR spectra. Inverted-chair and twist-boat cyclohexane moieties in the presence of an axial halogen atom at C(8). Helv Chim Acta 75:2026
474. González AG, Darias J, Martín JD, Martín VS, Norte M, Pérez C, Perales A, Fayos J (1980) *Laurencia* sesquiterpene biogenetic-type interconversions. Tetrahedron Lett 21:1151
475. Takeda S, Kurosawa E, Komiyama K, Suzuki T (1990) The structures of cytotoxic diterpenes containing bromine from the marine red alga *Laurencia obtusa* (Hudson) Lamouroux. Bull Chem Soc Jpn 63:3066
476. Bian WT, You ZJ, Wang CY, Shao CL (2014) Brominated pimarane diterpenoids from the sea hare *Aplysia pulmonica* from the South China Sea. Chem Nat Comp 50:557
477. Kurata K, Taniguchi K, Agatsuma Y, Suzuki M (1998) Diterpenoid feeding-deterrents from *Laurencia saitoi*. Phytochemistry 47:363
478. Tsukamoto S, Yamashita Y, Yoshida T, Ohta T (2004) Parguerol and isoparguerol isolated from the sea hare *Aplysia kurodai* induce neurite outgrowth in PC-12 cells. Mar Drugs 2:170
479. Suzuki T, Takeda S, Hayama N, Tanaka I, Komiyama K (1989) The structure of brominated diterpene from the marine red alga *Laurencia obtusa* (Hudson) Lamouroux. Chem Lett 18:969
480. Ji NY, Li XM, Wang BG (2008) Halogenated terpenes and a $C_{15}$ acetogenin from the marine red alga *Laurencia saitoi*. Molecules 13:2894
481. Masuda M, Takahashi Y, Matsuo Y, Suzuki M (1997) A taxonomic study of the genus *Laurencia* (Ceramiales, Rhodophyta) from Vietnam. II. *Laurencia lageniformis* sp. nov. Crypt Algol 18:163
482. Higgs MD, Faulkner DJ (1982) A diterpene from *Laurencia obtusa*. Phytochemistry 21:789
483. Takeda S, Matsumoto T, Komiyama K, Kurosawa E, Suzuki T (1990) A new cytotoxic diterpene from the marine red alga *Laurencia obtusa* (Hudson) Lamouroux. Chem Lett 19:277
484. Suzuki M, Kurosawa E, Kurata K (1988) Venustanol, a brominated labdane diterpene from the red alga *Laurencia venusta*. Phytochemistry 27:1209

485. Yamamura S, Hirata Y (1971) A naturally-occurring bromo-compound, aplysin-20 from *Aplysia kurodai*. Bull Chem Soc Jpn 44:2560
486. Matsuda H, Tomiie Y, Yamamura S, Hirata Y (1967) The structure of aplysin-20. Chem Commun:898
487. Yamaguchi Y, Uyehara T, Kato T (1985) Biogenetic type synthesis of (±)-concinndiol and (±)-aplysin 20. Tetrahedron Lett 26:343
488. Murai A, Abiko A, Masamune T (1984) Total synthesis of (±)-aplysin-20. Tetrahedron Lett 25:4955
489. Ojika M, Kigoshi H, Yoshikawa K, Nakayama Y, Tamada K (1992) A new bromo diterpene, *epi*-aplysin-20, and *ent*-isoconcinndiol from the marine mollusc *Aplysia kurodai*. Bull Chem Soc Jpn 65:2300
490. Sims JJ, Lin GHY, Wing RM, Fenical W (1973) Marine natural products. Concinndiol, a bromo-diterpene alcohol from the red alga *Laurencia concinna*. J Chem Soc Chem Commun:470
491. Howard BM, Fenical W (1980) Isoconcinndiol, a brominated diterpenoid from *Laurencia snyderae* var. *guadalupensis*. Phytochemistry 19:2774
492. Rodríguez ML, Martín JD, Estrada D (1989) The absolute configuration of (+)-isoconcinndiol. Acta Crystallogr C45:306
493. Fujiwara S, Takeda K, Uyehara T, Kato T (1986) Structural revision of isoconcinndiol by its synthesis. Chem Lett 15:1763
494. Fukuzawa A, Miyamoto M, Kumagai Y, Abiko A, Takaya Y, Masamune T (1985) Structure of new bromoditerpenes, pinnatols, from the marine red alga *Laurencia pinnata* Yamada. Chem Lett 14:1259
495. González AG, Ciccio JF, Rivera AP, Martín JD (1985) New halogenated diterpenes from the red alga *Laurencia perforata*. J Org Chem 50:1261
496. Yamamura S, Terada Y (1977) Isoaplysin-20, a natural bromine-containing diterpene, from *Aplysia kurodai*. Tetrahedron Lett 18:2171
497. Imamura PM, Rúveda EA (1980) The C-13 configuration of the bromine-containing diterpene isoaplysin-20. Synthesis of debromoisoaplysin-20 and its C-13 epimer. J Org Chem 45:510
498. Nishizawa M, Takenaka H, Hirotsu K, Higuchi T, Hayashi Y (1984) Synthesis and structure determination of isoaplysin-20. J Am Chem Soc 106:4290
499. Nishizawa M, Takenaka H, Hayashi Y (1986) Chemical simulation of polycyclic diterpenoid biosynthesis using mercury (II) triflate/*N*, *N*-dimethylaniline complex: mechanistic aspects of a biomimetic olefin cyclization. J Org Chem 51:806
500. Briand A, Kornprobst JM, Aleasa HS, Rizk AFM, Toupet L (1997) (–)-Paniculatol, a new *ent*-labdane bromoditerpene from *Laurencia paniculata*. Tetrahedron Lett 38:3399
501. Iliopoulou D, Mihopoulos N, Roussis V, Vagias C (2003) New brominated labdane diterpenes from the red alga *Laurencia obtusa*. J Nat Prod 66:1225
502. Kuniyoshi M, Marma MS, Higa T, Bernardinelli G, Jefford CW (2000) 3-Bromobarekoxide, an unusual diterpene from *Laurencia luzonensis*. Chem Commun:1155
503. Justicia J, Oller-Lopez JL, Campana AG, Oltra JE, Cuerva JM, Bunuel E, Cardenas DJ (2005) 7-*endo* radical cyclizations catalyzed by titanocene(III). Straightforward synthesis of terpenoids with seven-membered carbocycles. J Am Chem Soc 127:14911
504. Howard BM, Fenical W (1978) Structures of the irieols, new dibromoditerpenoids of a unique skeletal class from *Laurencia irieii*. J Org Chem 43:4401
505. Fukuzawa A, Takaya Y, Matsue H, Masamune T (1985) Structure of a new bromoditerpene, prepinnaterpene, from the marine red alga *Laurencia pinnata* Yamada. Chem Lett 14:1263
506. Fenical W, Howard B, Gifkins KB, Clardy J (1975) Irieol A and iriediol, dibromoditerpenes of a new skeletal class from *Laurencia*. Tetrahedron Lett 16:3983
507. Howard BM, Fenical W, Donovan SF, Clardy J (1982) Neoirieone, a diterpenoid of a new skeletal class from the red marine alga *Laurencia* cf. *irieii*. Tetrahedron Lett 23:3847

508. Fukuzawa A, Kumagai Y, Masamune T, Furusaki A, Matsumoto T, Katayama C (1982) Pinnaterpenes A, B and C, new dibromoditerpenes from the red alga *Laurencia pinnata* Yamada. Chem Lett 11:1389
509. Ji NY, Li XM, Cui CM, Wang BG (2007) Two new brominated diterpenes from *Laurencia decumbens*. Chin Chem Lett 18:957
510. Pettit GR, Herald CL, Einck JJ, Vanell LD, Brown P, Gust D (1978) Isolation and structure of angasiol. J Org Chem 43:4685
511. Atta-ur-Rahman, Alvi KA, Abbas SA, Sultana T, Shameel M, Choudhary MI, Clardy JC (1991) A diterpenoid lactone from *Aplysia juliana*. J Nat Prod 54:886
512. Takahashi H, Takahashi Y, Suzuki M, Abe T, Masuda M (2010) Neoirietriol. Acta Crystallogr E66:o1795
513. Takahashi Y, Daitoh M, Suzuki M, Abe T, Masuda M (2002) Halogenated metabolites from the new Okinawan red alga *Laurencia yonaguniensis*. J Nat Prod 65:395
514. Takahashi H, Takahashi Y, Suzuki M, Abe T, Masuda M (2007) Crystal structure and absolute stereochemistry of neoirietetraol. Anal Sci 23:x103
515. Petraki A, Ioannou E, Papazafiri P, Roussis V (2015) Dactylomelane diterpenes from the sea hare *Aplysia depilans*. J Nat Prod 78:462
516. Fernández JJ, Souto ML, Gil LV, Norte M (2005) Isolation of naturally occurring dactylomelane metabolites as *Laurencia* constituents. Tetrahedron 61:8910
517. Estrada DM, Ravelo JL, Ruiz-Pérez C, Martín JD (1989) Dactylomelol, a new class of diterpene from the sea hare *Aplysia dactylomela*. Tetrahedron Lett 30:6219
518. Guella G, Marchetti F, Pietra F (1997) Rogioldiol A, a new obtusane diterpene, and rogiolal, a degraded derivative, of the red seaweed *Laurencia microcladia* from Il Rogiolo along the coast of Tuscany: a synergism in structural elucidation. Helv Chim Acta 80:684
519. Schmitz FJ, Hollenbeak KH, Carter DC, Hossain MB, van der Helm D (1979) Marine natural products: 14-bromoobtus-1-ene-3,11-diol, a new diterpenoid from the sea hare *Aplysia dactylomela*. J Org Chem 44:2445
520. Guella G, Pietra F (1998) Antipodal pathways to secondary metabolites in the same eukaryotic organism. Chem Eur J 4:1692
521. Mohammed KA, Hossain CF, Zhang L, Bruick RK, Zhou YD, Nagle DG (2004) Laurenditerpenol, a new diterpene from the tropical marine alga *Laurencia intricata* that potently inhibits HIF-1 mediated hypoxic signaling in breast tumor cells. J Nat Prod 67:2002
522. Jung ME, Im G-YJ (2009) Total synthesis of racemic laurenditerpenol, an HIF-1 inhibitor. J Org Chem 74:8739
523. Jung ME, Im G-YJ (2008) Convergent total synthesis of the racemic HIF-1 inhibitor laurenditerpenol. Tetrahedron Lett 49:4962
524. Chittiboyina AG, Kumar GM, Carvalho PB, Liu Y, Zhou Y-D, Nagle DG, Avery MA (2007) Total synthesis and absolute configuration of laurenditerpenol: a hypoxia inducible factor-1 activation inhibitor. J Med Chem 50:6299
525. Mukherjee S, Scopton AP, Corey EJ (2010) Enantioselective pathway for the synthesis of laurenditerpenol. Org Lett 12:1836
526. Pitsinos EN, Athinaios N, Vidali VP (2012) Enantioselective total synthesis of (-)-laurenditerpenol. Org Lett 14:4666
527. Mihopoulos N, Vagias C, Mikros E, Scoullos M, Roussis V (2001) Prevezols A and B: new brominated diterpenes from the red alga *Laurencia obtusa*. Tetrahedron Lett 42:3749
528. Iliopoulou D, Mihopoulos N, Vagias C, Papazafiri P, Roussis V (2003) Novel cytotoxic brominated diterpenes from the red alga *Laurencia obtusa*. J Org Chem 68:7667
529. Leung AE, Rubbiani R, Gasser G, Tuck KL (2014) Enantioselective total syntheses of the proposed structures of prevezol B and evaluation of anti-cancer activity. Org Biomol Chem 12:8239
530. Leung AE, Blair M, Forsyth CM, Tuck KL (2013) Synthesis of the proposed structures of prevezol C. Org Lett 15:2198
531. Blair M, Forsyth CM, Tuck KL (2010) Towards the synthesis of prevezol C: total enantioselective synthesis of (–)-2-*epi*-prevezol C. Tetrahedron Lett 51:4808

532. Guella G, Pietra F (2000) A new-skeleton diterpenoid, new prenylbisabolanes, and their putative biogenetic precursor, from the red seaweed *Laurencia microcladia* from Il Rogiolo: assigning the absolute configuration when two chiral halves are connected by single bonds. Helv Chim Acta 83:2946
533. Chatter R, Kladi M, Tarhouni S, Maatoug R, Kharrat R, Vagias C, Roussis V (2009) Neorogioltriol: a brominated diterpene with analgesic activity from *Laurencia glandulifera*. Phytochem Lett 2:25
534. Norte M, Souto ML, Fernández JJ (1996) Viridiols, two new diterpenes from *Laurencia viridis*. Nat Prod Lett 8:263
535. Caccamese S, Toscano RM, Cerrini S, Gavuzzo E (1982) Laurencianol, a new halogenated diterpenoid from marine alga *Laurencia obtusa*. Tetrahedron Lett 23:3415
536. Kladi M, Ntountaniotis D, Zervou M, Vagias C, Ioannou E, Roussis V (2014) Glandulaurencianols A-C, brominated diterpenes from the red alga *Laurencia glandulifera* and the sea hare *Aplysia punctata*. Tetrahedron Lett 55:2835
537. Brennan MR, Kim IK, Erickson KL (1993) Kahukuenes, new diterpenoids from the marine alga *Laurencia majuscula*. J Nat Prod 56:76
538. Chatter R, Cenac N, Roussis V, Kharrat R, Vergnolle N (2012) Inhibition of sensory afferents activation and visceral pain by a brominated diterpene. Neurogastroenterol Motil 24, e336
539. Ojika M, Yoshida Y, Okumura M, Ieda S, Yamada K (1990) Aplysiadiol, a new brominated diterpene from the marine mollusc *Aplysia kurodai*. J Nat Prod 53:1619
540. Niwa H, Ieda S, Inagaki H, Yamada K (1990) A biogenetic-type synthesis of (±)-aplysiadiol, a brominated diterpene isolated from the marine mollusc *Aplysia kurodai*. Tetrahedron Lett 31:7157
541. Sun J, Han LJ, Yang RY, Shi DY, Uan ZH, Shi JG (2007) Studies on chemical constituents of *Laurencia tristicha*. Chin J Chin Mat Med 32:2610
542. Mahdi F, Falkenberg M, Ioannou E, Roussis V, Zhou YD, Nagle DG (2011) Thyrsiferol inhibits mitochondrial respiration and HIF-1 activation. Phytochem Lett 4:75
543. Ji NY, Li XM, Xie H, Ding J, Li K, Ding LP, Wang BG (2008) Highly oxygenated triterpenoids from the marine red alga *Laurencia mariannensis* (Rhodomelaceae). Helv Chim Acta 91:1940
544. Suzuki T, Suzuki M, Furusaki A, Matsumoto T, Kato A, Imanaka Y, Kurosawa E (1985) Teurilene and thyrsiferyl 23-acetate, *meso* and remarkably cytotoxic compounds from the marine red alga *Laurencia obtusa* (Hudson) Lamouroux. Tetrahedron Lett 26:1329
545. Blunt JW, Hartshorn MP, McLennan TJ, Munro MHG, Robinson WT, Yorke SC (1978) Thyrsiferol, a squalene-derived metabolite of *Laurencia thyrsifera*. Tetrahedron Lett 19:69
546. Blunt JW, McCombs JD, Munro MHG, Thomas FN (1989) Complete assignment of the $^{13}C$ and $^{1}H$ nmr spectra of thyrsiferyl acetate. Magn Reson Chem 27:792
547. Sakemi S, Higa T, Jefford CW, Bernardinelli G (1986) Venustatriol, a new antiviral triterpene tetracyclic ether from *Laurencia venusta*. Tetrahedron Lett 27:4287
548. Fernández JJ, Souto ML, Norte M (1998) Evaluation of the cytotoxic activity of polyethers isolated from *Laurencia*. Bioorg Med Chem 6:2237
549. Manzo E, Gavagnin M, Bifulco G, Cimino P, Di Micco S, Ciavatta ML, Guo YW, Cimino G (2007) Aplysiols A and B, squalene-derived polyethers from the mantle of the sea hare *Aplysia dactylomela*. Tetrahedron 63:9970
550. Broka CA, Hu L, Lee WJ, Shen T (1987) Synthetic studies on thyrsiferol. Tetrahedron Lett 28:4993
551. Broka CA, Lin Y (1988) Synthetic studies on thyrsiferol. Elaboration of the bromotetrahydropyran ring. J Org Chem 53:5876
552. Hashimoto M, Kan T, Nozaki K, Yanagiya M, Shirahama H, Matsumoto T (1990) Total syntheses of (+)-thyrsiferol, (+)-thyrsiferyl 23-acetate, and (+)-venustatriol. J Org Chem 55:5088
553. González IC, Forsyth CJ (2000) Total synthesis of thyrsiferyl 23-acetate, a specific inhibitor of protein phosphatase 2A and an anti-leukemic inducer of apoptosis. J Am Chem Soc 122:9099

554. Norte M, Fernández JJ, Souto ML (1997) New polyether squalene derivatives from *Laurencia*. Tetrahedron 53:4649
555. Cen-Pacheco F, Mollinedo F, Villa-Pulgarín JA, Norte M, Fernández JJ, Daranas AH (2012) Saiyacenols A and B: the key to solve the controversy about the configuration of aplysiols. Tetrahedron 68:7275
556. Suzuki T, Hasegawa M, Hirayama Y, Takahashi Y, Matsuo Y (1995) The structure of squalene-derived polyether, 15(28)-anhydrothyrsiferyl 23-acetate isolated from the marine red alga *Laurencia obtusa* (Hudson) Lamouroux. J Hokkaido Univ Ed Sect II A 46:57
557. Suzuki T, Takeda S, Suzuki M, Kurosawa E, Kato A, Imanaka Y (1987) Cytotoxic squalene-derived polyethers from the marine red alga *Laurencia obtusa* (Hudson) Lamouroux. Chem Lett 16:361
558. Manríquez CP, Souto ML, Gavín JA, Norte M, Fernández JJ (2001) Several new squalene-derived triterpenes from *Laurencia*. Tetrahedron 57:3117
559. Souto ML, Manríquez CP, Norte M, Fernández JJ (2002) Novel marine polyethers. Tetrahedron 58:8119
560. Norte M, Fernández JJ, Souto ML, García-Grávalos MD (1996) Two new antitumoral polyether squalene derivatives. Tetrahedron Lett 37:2671
561. Matsuo Y, Suzuki M, Masuda M, Iwai T, Morimoto Y (2008) Squalene-derived triterpene polyethers from the red alga *Laurencia omaezakiana*. Helv Chim Acta 91:1261
562. Corey EJ, Ha D-C (1988) Total synthesis of venustatriol. Tetrahedron Lett 29:3171
563. Cen-Pacheco F, Villa-Pulgarin JA, Mollinedo F, Martín MN, Fernández JJ, Hernádez Daranas AH (2011) New polyether triterpenoids from *Laurencia viridis* and their biological evaluation. Mar Drugs 9:2220
564. Cen-Pacheco F, Santiago-Benítez AJ, García C, Álvarez-Méndez SJ, Martín-Rodríguez AJ, Norte M, Martín VS, Gavín JA, Fernández JJ, Daranas AH (2015) Oxasqualenoids from *Laurencia viridis*: combined spectroscopic-computational analysis and antifouling potential. J Nat Prod 78:712
565. Ola ARB, Babey AM, Motti C, Bowden BF (2010) Aplysiols C-E, brominated triterpene polyethers from the marine alga *Chondria armata* and a revision of the structure of aplysiol B. Aust J Chem 63:907
566. Suzuki M, Matsuo Y, Takahashi Y, Masuda M (1995) Callicladol, a novel cytotoxic bromotriterpene polyether from a Vietnamese species of the red algal genus *Laurencia*. Chem Lett 24:1045
567. Vera B, Rodríguez AD, Avilés E, Ishikawa Y (2009) Aplysqualenols A and B: squalene-derived polyethers with antitumoral and antiviral activity from the Caribbean sea slug *Aplysia dactylomela*. Eur J Org Chem 2009:5327
568. Norte M, Fernández JJ, Souto ML, Gavin JA, García-Grávalos MD (1997) Thyrsenols A and B, two unusual polyether squalene derivatives. Tetrahedron 53:3173
569. Cen-Pacheco F, Villa-Pulgarin JA, Mollinedo F, Norte M, Daranas AH, Fernández JJ (2011) Cytotoxic oxasqualenoids from the red alga *Laurencia viridis*. Eur J Med Chem 46:3302
570. Cen-Pacheco F, Nordström L, Souto ML, Martín MN, Fernández JJ, Daranas AH (2010) Studies on polyethers produced by red algae. Mar Drugs 8:1178
571. Hioki H, Motosue M, Mizutani Y, Noda A, Shimoda T, Kubo M, Harada K, Fukuyama Y, Kodama M (2009) Total synthesis of pseudodehydrothyrsiferol. Org Lett 11:579
572. Clausen DJ, Wan S, Floreancig PE (2011) Total synthesis of the protein phosphatase 2A inhibitor lactodehydrothyrsiferol. Angew Chem Int Ed 50:5178
573. Suenaga K, Shibata T, Takada N, Kigoshi H, Yamada K (1998) Aurilol, a cytotoxic bromoditerpene isolated from the sea hare *Dolabella auricularia*. J Nat Prod 61:515
574. Morimoto Y, Nishikawa Y, Takaishi M (2005) Total synthesis and complete assignment of the stereostructure of a cytotoxic bromoditerpene polyether (+)-aurilol. J Am Chem Soc 127:5806
575. Matsuo Y, Suzuki M, Masuda M (1995) Enshuol, a novel squalene-derived pentacyclic triterpene alcohol from a new species of the red algal genus *Laurencia*. Chem Lett 24:1043

576. Morimoto Y, Yata H, Nishikawa Y (2007) Assignment of the absolute configuration of the marine pentacyclic polyether (+)-enshuol by total synthesis. Angew Chem Int Ed 46:6481
577. Hashimoto M, Yanagiya M, Shirahama H (1988) Total synthesis of *meso*-triterpene ether, teurilene. Chem Lett 17:645
578. Hashimoto M, Harigaya H, Yanagiya M, Shirahama H (1991) Total synthesis of the *meso*-triterpene polyether teurilene. J Org Chem 56:2299
579. Iwai T, Kinoshita T, Morimoto Y (1998) Highly efficient total synthesis of cytotoxic *meso* polyether teurilene featuring diastereoselective method for construction of tetrahydrofuran rings by means of rhenium (VII) oxide. Symp Chem Nat Prod 40:277
580. Morimoto Y, Iwai T, Kinoshita T (1999) Effective combination of two-directional synthesis and rhenium(VII) chemistry: total synthesis of *meso* polyether teurilene. J Am Chem Soc 121:6792
581. Morimoto Y, Iwai T, Kinoshita T (2002) Total synthesis of highly symmetric squalene-derived cytotoxic polyethers. J Synth Org Chem Jpn 60:1112
582. Rodríguez-López J, Crisóstomo FP, Ortega N, Rodríguez-López M, Martín VS, Martín T (2013) Epoxide-opening cascades triggered by a Nicholas reaction: total synthesis of teurilene. Angew Chem Int Ed 52:3659
583. Suzuki M, Matsuo Y, Takeda S, Suzuki T (1993) Intricatetraol, a halogenated triterpene alcohol from the red alga *Laurencia intricata*. Phytochemistry 33:651
584. Umezawa T, Oguri Y, Matsuura H, Yamazaki S, Suzuki M, Yoshimura E, Furuta T, Nogata Y, Serisawa Y, Matsuyama-Serisawa K, Abe T, Matsuda F, Suzuki M, Okino T (2014) Omaezallene from red alga *Laurencia* sp.: structure elucidation, total synthesis and antifouling activity. Angew Chem Int Ed 53:3909
585. Morimoto Y, Okita T, Takaishi M, Tanaka T (2007) Total synthesis and determination of the absolute configuration of (+)-intricatetraol. Angew Chem Int Ed 46:1132
586. Authors' unpublished data
587. Kigoshi H, Ojika M, Shizuri Y, Niwa H, Yamada K (1986) Isolation of (10*R*,11*R*)-(+)-squalene-10,11-epoxide from the red alga *Laurencia okamurai* and its enantioselective synthesis. Tetrahedron 42:3789
588. Kigoshi H, Ojika M, Shizuri Y, Niwa H, Yamada K (1982) (10*R*,11*R*)-(+)-squalene-10,11-epoxide: isolation from *Laurencia okamurai* and the asymmetric synthesis. Tetrahedron Lett 23:5413
589. Kigoshi H, Itoh T, Ogawa T, Ochi K, Okada M, Suenaga K, Yamada K (2001) Auriculol, a cytotoxic oxygenated squalene from the Japanese sea hare *Dolabella auricularia*: isolation, stereostructure and synthesis. Tetrahedron Lett 42:7461
590. Xiong Z, Busch R, Corey EJ (2010) A short total synthesis of (+)-omaezakianol via an epoxide-initiated cationic cascade reaction. Org Lett 12:1512
591. Fernández JJ, Souto ML, Norte M (2000) Marine polyether triterpenes. Nat Prod Rep 17:235
592. Norte M, Fernández JJ, Ruano JZ, Matías L, Rodríguez ML, Pérez R (1988) Graciosin and graciosallene, two bromoethers from *Laurencia obtusa*. Phytochemistry 27:3537
593. Norte M, Fernández JJ, Runao JZ (1989) Three new bromo ethers from the red alga *Laurencia obtusa*. Tetrahedron 45:5987
594. Kamada T, Vairappan CS (2012) A new bromoallene-producing chemical type of the red alga *Laurencia nangii* Masuda. Molecules 17:2119
595. Suzuki T, Koizumi K, Suzuki M, Kurosawa E (1983) Kumausynes and deacetylkumausynes, four new halogenated C-15 acetylenes from the red alga *Laurencia nipponica* Yamada. Chem Lett 12:1643
596. Martin T, Soler MA, Betancort JM, Martin VS (1997) Biomimetic-type synthesis of halogenated tetrahydrofurans from *Laurencia*. Total synthesis of (+)-*trans*-deacetylkumausyne. J Org Chem 62:1570
597. Brown MJ, Harisson T, Overman LE (1991) General approach to halogenated tetrahydrofuran natural products from the red algae of the genus *Laurencia*. Synthesis of (±)-*trans*-kumausyne and demonstration of an asymmetric synthesis strategy. J Am Chem Soc 113:5378

598. Osumi K, Sugimura H (1995) Total synthesis of (–)-*trans*-kumausyne. Tetrahedron Lett 36:5789
599. Andrey O, Glanzmann C, Landais Y, Parra-Rapado L (1997) 1,3-Asymmetric induction in electrophilic addition onto homoallylsilanes. An approach towards the total synthesis of (±)-kumausyne. Tetrahedron 53:2835
600. Lee E, Yoo SK, Cho YS, Cheon HS, Chong YH (1997) Radical cyclisation of β-alkoxyacrylates: stereoselective synthesis of (–)-*trans*-kumausyne. Tetrahedron Lett 38:7757
601. Mereyala HB, Gadikota RR (2000) A general strategy for the formal synthesis of (–)-*trans*-kumausyne and total synthesis of (5*R*)-Hagen's gland lactones from diacetone-D-glucose. Tetrahedron Asymm 11:743
602. García C, Martín T, Martín VS (2001) β-Hydroxy-γ-lactones as chiral building blocks for the enantioselective synthesis of marine natural products. J Org Chem 66:1420
603. Gadikota RR, Callam CS, Lowary TL (2001) Total synthesis of (2*S*,3*S*,5*S*,10*S*)-6,9-epoxynonadec-18-ene-7,10-diol and formal total synthesis of (+)-*trans*-kumausyne from D-arabinose. J Org Chem 66:9046
604. Chandler CL, Phillips AJ (2005) A total synthesis of (±)-*trans*-kumausyne. Org Lett 7:3493
605. Gutiérrez-Cepeda A, Daranas AH, Fernández JJ, Norte M, Souto ML (2014) Stereochemical determination of five-membered cyclic ether acetogenins using a spin-spin coupling constant approach and DFT calculations. Mar Drugs 12:4031
606. Fukuzawa A, Aye M, Takaya Y, Fukui TM, Murai A, Masamune T (1989) Laureoxolane a new bromo ether from *Laurencia nipponica*. Tetrahedron Lett 30:3665
607. Kladi M, Vagias C, Papazarifi P, Brogi S, Tafi A, Roussis V (2009) Tetrahydrofuran acetogenins from *Laurencia glandulifera*. J Nat Prod 72:190
608. Suzuki M, Nakano S, Takahashi Y, Abe T, Masuda M (1999) Bisezakyne-A and -B, halogenated $C_{15}$ acetogenins from a Japanese *Laurencia* species. Phytochemistry 51:657
609. Fukuzawa A, Kurosawa E, Tobetsu I (1980) Laureepoxide, new bromo ether from the marine red alga *Laurencia nipponica* Yamada. Tetrahedron Lett 21:1471
610. Imre S, Aydoğmuş Z, Güner H, Lotter H, Wagner H (1995) Polybrominated non-terpenoid $C_{15}$ compounds from *Laurencia paniculata* and *Laurencia obtusa*. Z Naturforsch 50C:743
611. Ji N-Y, Li X-M, Li K, Wang B-G (2007) Laurendecumallenes A-B and laurendecumenynes A-B, halogenated nonterpenoid $C_{15}$ acetogenins from the marine red alga *Laurencia decumbens*. J Nat Prod 70:1499
612. Ji NY, Li XM, Li K, Wang BG (2010) Erratum to "Laurendecumallenes A-B and laurendecumenynes A-B, halogenated nonterpenoid $C_{15}$ acetogenins from the marine red alga *Laurencia decumbens*". J Nat Prod 73:1192
613. Dyson BS, Burton JW, Sohn TI, Kim B, Bae H, Kim D (2012) Total synthesis and structure confirmation of elatenyne: success of computational methods for NMR prediction with highly flexible diastereomers. J Am Chem Soc 134:11781
614. Kim K, Brennan MR, Erickson KL (1989) Lauroxolanes from the marine alga *Laurencia majuscula*. Tetrahedron Lett 30:1757
615. Sheldrake HM, Jamieson C, Burton JW (2006) The changing faces of halogenated marine natural products: total synthesis of the reported structures of elatenyne and an enyne from *Laurencia majuscula*. Angew Chem Int Ed 45:7199
616. Sheldrake HM, Jamieson C, Pascu SI, Burton JW (2009) Synthesis of the originally proposed structures of elatenyne and an enyne from *Laurencia majuscula*. Org Biomol Chem 7:238
617. Smith SG, Paton RS, Burton JW, Goodman JM (2008) Stereostructure assignment of flexible five-membered rings by GIAO $^{13}C$ NMR calculations: prediction of the stereochemistry of elatenyne. J Org Chem 73:4053
618. Brkljaca R, Urban S (2013) Relative configuration of the marine natural product elatenyne using NMR spectroscopic and chemical derivatization methodologies. Nat Prod Commun 8:729
619. Abdel-Mageed WM, Ebel R, Valeriote FA, Jaspars M (2010) Laurefurenynes A-F, new cyclic ether acetogenins from a marine red alga *Laurencia* sp. Tetrahedron 66:2855

620. Shepherd DJ, Broadwith PA, Dyson BS, Paton RS, Burton JW (2013) Structure reassignment of laurefurenynes A and B by computation and total synthesis. Chem Eur J 19:12644
621. Holmes MT, Britton R (2013) Total synthesis and structural revision of laurefurenynes A and B. Chem Eur J 19:12649
622. Suzuki T, Koizumi K, Suzuki M, Kurosawa E (1983) Kumausallene, a new bromoallene from the marine red alga *Laurencia nipponica* Yamada. Chem Lett 12:1639
623. Grese TA, Hutchinson KD, Overman LE (1993) General approach to halogenated tetrahydrofuran natural products from red algae of genus *Laurencia*. Total synthesis of (±)-kumausallene and (±)-1-*epi*-kumausallene. J Org Chem 58:2468
624. Lee E, Yoo SK, Choo H, Song HY (1998) Radical cyclization of β-alkoxyacrylates: a formal synthesis of (–)-kumausallene. Tetrahedron Lett 39:317
625. Evans PA, Murthy VS, Roseman JD, Rheingold AL (1999) Enantioselective total synthesis of the nonisoprenoid sesquiterpene (–)-kumausallene. Angew Chem Int Ed 38:3175
626. Werness JB, Tang W (2011) Stereoselective total synthesis of (–)-kumausallene. Org Lett 13:3664
627. Nesbitt CL, McErlean CSP (2009) An expedient synthesis of 2,5-disubstituted-3-oxygenated tetrahydrofurans. Tetrahedron Lett 50:6318
628. Okamoto Y, Nitanda N, Ojika M, Sakagami Y (2003) Aplysiallene, a new bromoallene as an $Na^+$,$K^+$-ATPase inhibitor from the sea hare *Aplysia kurodai* (Erratum). Biosci Biotechnol Biochem 67:460
629. Wang J, Pagenkopf BL (2007) First total synthesis and structural reassignment of (–)-aplysiallene. Org Lett 9:3703
630. Suzuki M, Kurosawa E (1981) Okamurallene, a novel halogenated $C_{15}$ metabolite from the red alga *Laurencia okamurai* Yamada. Tetrahedron Lett 22:3853
631. Suzuki M, Sasage Y, Ikura M, Hikichi K, Kurosawa E (1989) Structure revision of okamurallene and structure elucidation of further $C_{15}$ non-terpenoid bromoallenes from *Laurencia intricata*. Phytochemistry 28:2145
632. Suzuki M, Kondo H, Tanaka I (1991) The absolute stereochemistry of okamurallene and its congeners, halogenated $C_{15}$ nonterpenoids from the red alga *Laurencia intricata*. Chem Lett 20:33
633. Suzuki M, Kurosawa E (1982) Deoxyokamurallene and isookamurallene, new halogenated nonterpenoid $C_{15}$-compounds from the red alga *Laurencia okamurai* Yamada. Chem Lett 11:289
634. Kinnel R, Duggan AJ, Eisner T, Meinwald J, Miura I (1977) Panacene, an aromatic bromoallene from a sea hare (*Aplysia brasiliana*). Tetrahedron Lett 18:3913
635. Feldman KS (1982) Biomimetic synthesis of (±)-panacene. Tetrahedron Lett 23:3031
636. Feldman KS, Mechem CC, Nader L (1982) Total synthesis of (±)-panacene. J Am Chem Soc 104:4011
637. Sabot C, Bérard D, Canesi S (2008) Expeditious total syntheses of natural allenic products via aromatic ring umpolung. Org Lett 10:4629
638. Howard BM, Fenical W, Arnold EV, Clardy J (1979) Obtusin, a unique bromine-containing polycyclic ketal from the red marine alga *Laurencia obtusa*. Tetrahedron Lett 20:2841
639. Caccamese S, Toscano RM (1986) Neoobtusin, a new brominated ketal from the marine red alga *Laurencia obtusa*. Gazz Chim Ital 116:177
640. Sugimura H, Hasegawa Y, Osumi K (2000) Studies relating to the synthesis of laurenenynes: construction of the alkylidene side chain via [2,3]-Wittig-Still rearrangement at the anomeric center of a furanoside derivative. Heterocycles 52:99
641. Liu X, Li XM, Li CS, Ji NY, Wang BG (2010) Laurenidificin, a new brominated $C_{15}$ acetogenin from the marine red alga *Laurencia nidifica*. Chin Chem Lett 21:1213
642. Schulte GR, Chung MCH, Scheuer PJ (1981) Two bicyclic $C_{15}$ enynes from the sea hare *Aplysia oculifera*. J Org Chem 46:3870
643. Waraszkiewicz SM, Sun HH, Erickson KL (1976) $C_{15}$-halogenated compounds from the Hawaiian marine alga *Laurencia nidifica*. V. The maneonenes. Tetrahedron Lett 17:3021

644. Waraszkiewicz SM, Sun HH, Erickson KL, Finer J, Clardy J (1978) $C_{15}$ halogenated compounds from the Hawaiian marine alga *Laurencia nidifica*. Maneonenes and isomaneonenes. J Org Chem 43:3194
645. Ayyad SEN, Al-Footy KO, Alarif WM, Sobahi TR, Bassaif SA, Makki MS, Asiri AM, Al Halwani AY, Badria AF, Badria FA (2011) Bioactive $C_{15}$ acetogenins from the red alga *Laurencia obtusa*. Chem Pharm Bull 59:1294
646. Holmes AB, Jennings-White CLD, Kendrick DA (1983) Total synthesis of *cis*-maneonenes-A and B. J Chem Soc Chem Commun:415
647. Sun HH, Waraszkiewicz SM, Erickson KL (1976) $C_{15}$-halogenated compounds from the Hawaiian marine alga *Laurencia nidifica*. VI. The isomaneonenes. Tetrahedron Lett 17:4227
648. Holmes AB, Jennings-White CLD, Kendrick DA (1984) Total synthesis of (±)-*trans*-maneonene-B. J Chem Soc Chem Commun:1594
649. Vanderah DJ, Schmitz FJ (1976) Marine natural products; Isodactylyne, a halogenated acetylenic ether from the sea hare *Aplysia dactylomela*. J Org Chem 41:3480
650. Gao L, Murai A (1992) Total synthesis of (–)-dactylyne and (–)-isodactylyne. Tetrahedron Lett 33:4349
651. Gao L, Murai A (1996) Total synthesis of (–)-dactylynes. Heterocycles 42:745
652. McDonald FJ, Campbell DC, Vanderah DJ, Schmitz FJ, Washecheck DM, Burks JE, Van Der Helm D (1975) Marine natural products. Dactylyne an acetylenic dibromochloro ether from the sea hare *Aplysia dactylomela*. J Org Chem 40:665
653. Lee H, Kim KW, Park J, Kim H, Kim S, Kim D, Hu X, Yang W, Hong J (2008) A general strategy for construction of both 2,6-*cis* and 2,6-*trans*-disubstituted tetrahydropyrans: substrate-controlled asymmetric total synthesis of (+)-scanlonenyne. Angew Chem Int Ed 47:4200
654. de Silva ED, Schwartz RE, Scheuer PJ, Shoolery JN (1983) Srilankenyne, a new metabolite from the sea hare *Aplysia oculifera*. J Org Chem 48:395
655. Takahashi Y, Suzuki M, Abe T, Masuda M (1999) Japonenynes, halogenated $C_{15}$ acetogenins from *Laurencia japonensis*. Phytochemistry 50:799
656. Imre S, Öztunç A, Islimyeli S (1987) Chemical investigation of some marine organisms from Turkish waters. Turk Kim Derg 11:119
657. Kozikowski AP, Lee J (1990) A synthesis approach to the *cis*-fused marine pyranopyrans (3*E*)- and (3*Z*)-dactomelyne. X-ray structure of a rare organomercurial. J Org Chem 55:863
658. Lee E, Park CM, Yun JS (1995) Total synthesis of dactomelynes. J Am Chem Soc 117:8017
659. Lee E (1996) Oxacycle synthesis via radical cyclization of β-alkoxyacrylates. Pure Appl Chem 68:631
660. Fukuzawa A, Masamune T (1981) Laurepinnacin and isolaurepinnacin, new acetylenic cyclic ethers from the marine red alga *Laurencia pinnata* Yamada. Tetrahedron Lett 22:4081
661. Kotsuki H, Ushio Y, Kadota I, Ochi M (1989) Stereoselective reduction of bicyclic ketals. A new, enantioselective synthesis of isolaurepinnacin and lauthisan skeletons. J Org Chem 54:5153
662. Berger D, Overman LE, Renhowe PA (1993) Enantioselective total synthesis of (+)-isolaurepinnacin. J Am Chem Soc 115:9305
663. Berger D, Overman LE, Renhowe PA (1997) Total synthesis of (+)-isolaurepinnacin. Use of acetal-alkene cyclizations to prepare highly functionalized seven-membered cyclic ethers. J Am Chem Soc 119:2446
664. Suzuki T, Matsumura R, Nagai Y, Sato K, Sekiguchi H, Hagiwara H, Ando M (1997) Stereospecific construction of medium-sized cyclic ethers and its applications to synthesis of marine natural products. Symp Chem Nat Prod 39:91
665. Suzuki T, Matsumura R, Oku KI, Taguchi K, Hagiwara H, Hoshi T, Ando M (2001) Formal synthesis of (+)-isolaurepinnacin. Tetrahedron Lett 42:65
666. Rodríguez-López J, Ortega N, Martín VS, Martín T (2014) β-Hydroxy-γ-lactones as nucleophiles in the Nicholas reaction for the synthesis of oxepene rings. Enantioselective formal synthesis of (−)-isolaurepinnacin and (+)-rogioloxepane A. Chem Commun 50:3685

667. Guella G, Mancini I, Chiasera G, Pietra F (1992) On the unusual propensity by the red seaweed *Laurencia microcladia* of Il Rogiolo to form $C_{15}$ oxepanes: isolation of rogioloxepane A, B, C, and their likely biogenetic acyclic precursor, prerogioloxepane. Helv Chim Acta 75:310
668. Matsumura R, Suzuki T, Hagiwara H, Hoshi T, Ando M (2001) The first total synthesis of (+)-rogioloxepane A. Tetrahedron Lett 42:1543
669. Crimmins MT, DeBaillie AC (2003) Enantioselective total synthesis of (+)-rogioloxepane A. Org Lett 5:3009
670. Lyakhova EG, Kalinovsky AI, Dmitrenok AS, Kolesnikova SA, Fedorov SN, Vaskovsky VE, Stonik VA (2006) Structures and absolute stereochemistry of nipponallene and neonipponallene, new brominated allenes from the red alga *Laurencia nipponica*. Tetrahedron Lett 47:6549
671. Guella G, Mancici I, Chiasera G, Pietra F (1992) Rogiolenyne D, the likely immediate precursor of rogiolenyne A and B, branched $C_{15}$ acetogenins isolated from the red seaweed *Laurencia microcladia* of Il Rogiolo. Conformation and absolute configuration in the whole series. Helv Chim Acta 75:303
672. Guella G, Pietra F (1991) Rogiolenyne A, B and C: the first branched marine $C_{15}$ acetogenins. Isolation from the red seaweed *Laurencia microcladia* or the sponge *Spongia zimocca* of Il Rogiolo. Helv Chim Acta 74:47
673. Kurosawa E, Fukuzawa A, Irie T (1973) Isoprelaurefucin, new bromo compound from *Laurencia nipponica* Yamada. Tetrahedron Lett 14:4135
674. Suzuki M, Kurata K, Suzuki T, Kurosawa E (1986) The absolute configuration of isoprelaurefucin. Bull Chem Soc Jpn 59:2953
675. Lee H, Kim Y, Yoon T, Kim B, Kim S, Kim H-D, Kim D (2005) Novel "protecting group-dependent" alkylation-RCM strategy to medium-sized oxacycles: first total synthesis of (-)-isoprelaurefucin. J Org Chem 70:8723
676. Suzuki M, Mizuno Y, Matsuo Y, Masuda M (1996) Neoisoprelaurefucin, a halogenated $C_{15}$ non-terpenoid compound from *Laurencia nipponica*. Phytochemistry 43:121
677. Lee H, Kim H, Baek S, Kim S, Kim D (2003) Total synthesis and determination of the absolute configuration of (+)-neoisoprelaurefucin. Tetrahedron Lett 44:6609
678. Falshaw CP, King TJ, Imre S, Islimyeli S, Thomson RH (1980) Laurenyne, a new acetylene from *Laurencia obtusa*, crystal structure and absolute configuration. Tetrahedron Lett 21:4951
679. Öztunç A, Imre S, Lotter H, Wagner H (1991) Two $C_{15}$ bromoallenes from the red alga *Laurencia obtusa*. Phytochemistry 30:255
680. Overman LE, Thompson AS (1988) Total synthesis of (–)-laurenyne. Use of acetyl-initiated cyclizations to prepare functionalized eight-membered cyclic ethers. J Am Chem Soc 110:2248
681. Boeckman RK, Zhang J, Reeder MR (2002) Synthetic and mechanistic studies of the retro-Claisen rearrangement 4. An application to the total synthesis of (+)-laurenyne. Org Lett 4:3891
682. Clark JS, Freeman RP, Cacho M, Thomas AW, Swallow S, Wilson C (2004) Stereoselective synthesis of the cyclic ether core of (+)-laurenyne. Tetrahedron Lett 45:8639
683. Suzuki M, Kurosawa E, Furusaki A, Matsumoto T (1983) The structures of (3*Z*)-epoxyvenustin, (3*Z*)-venustin, and (3*Z*)-venustinene, new halogenated $C_{15}$-nonterpenoids from the red alga *Laurencia venusta* Yamada. Chem Lett 12:779
684. Howard BM, Fenical W, Hirotsu K, Solheim B, Clardy J (1980) The rhodophytin and chondriol natural products, structures of several new acetylenes from *Laurencia* and a reassignment of structure of *cis*-rhodophytin. Tetrahedron 36:171
685. Fenical W (1974) Rhodophytin, a halogenated vinyl peroxide of marine origin. J Am Chem Soc 96:5580
686. Fenical W, Sims JJ, Radlick P (1973) Chondriol, a halogenated acetylene from the marine alga *Chondria oppositiclada*. Tetrahedron Lett 14:313

687. Fenical W, Gifkins KB, Clardy J (1974) X-ray determination of chondriol, a re-assignment of structure. Tetrahedron Lett 15:1507
688. Suzuki M, Kurosawa E (1980) Venustin A and B, new halogenated $C_{15}$ metabolites from the red alga *Laurencia venusta* Yamada. Chem Lett 9:1177
689. Norte M, González AG, Cataldo F, Rodríguez ML, Brito I (1991) New examples of acyclic and cyclic C-15 acetogenins from *Laurencia pinnatifida*. Reassignment of the absolute configuration for *E* and *Z* pinnatifidienyne. Tetrahedron 47:9411
690. Notre M, Fernández JJ, Cataldo F, González AG (1989) *E*-Dihydrorhodophytin, a $C_{15}$ acetogenin from the red alga *Laurencia pinnatifida*. Phytochemistry 28:647
691. Vairappan CS, Lee TK (2009) C-15 halogenated acetogenin with antibacterial activity against food pathogens. Malay J Sci 28:263
692. Kinnel RB, Dieter RK, Meinwald J, Van Engen D, Clardy J, Eisner T, Stallard MO, Fenical W (1979) Brasilenyne and *cis*-dihydrorhodophytin, antifeedant medium-ring haloethers from a sea hare (*Aplysia brasiliana*). Proc Natl Acad Sci USA 76:3576
693. Fukuzawa A, Takasugi Y, Murai A (1991) Prelaureatin, a new biogenetic key intermediate isolated from *Laurencia nipponica*. Tetrahedron Lett 32:5597
694. Fukuzawa A, Aye M, Murai A (1990) A direct enzymatic synthesis of laurencin from laurediol. Chem Lett 19:1579
695. Fukuzawa A, Takasugi Y, Murai A, Nakamura M, Tamura M (1992) Enzymatic single-step formation of laureatin and its key intermediate, prelaureatin, from (3*Z*,6*S*,7*S*)-laurediol. Tetrahedron Lett 33:2017
696. Fukuzawa A, Aye M, Takasugi Y, Nakamura M, Tamura M, Murai A (1994) Enzymic bromo-ether cyclization of laurediols with bromoperoxidase. Chem Lett 13:2307
697. Ishihara J, Kanoh N, Murai A (1995) Enzymatic reaction of (3*E*,6*S*,7*S*)-laurediol and the molecular modeling studies on the cyclization of laurediols. Tetrahedron Lett 36:737
698. Crimmins MT, Tabet EA (2000) Total synthesis of (+)-prelaureatin and (+)-laurallene. J Am Chem Soc 122:5473
699. Fujiwara K, Souma SI, Mishima H, Murai A (2002) Total synthesis of prelaureatin. Synlett 13:1493
700. Sasaki M, Oyamada K, Takeda K (2010) Formal total syntheses of (+)-prelaureatin and (+)-laurallene by diastereoselective Brook rearrangement-mediated [3+4] annulation. J Org Chem 75:3941
701. Li J, Suh JM, Chin E (2010) Expedient enantioselective synthesis of the $\Delta^4$-oxocene cores of (+)-laurencin and (+)-prelaureatin. Org Lett 12:4712
702. Iliopoulou D, Vagias C, Harvala C, Roussis V (2002) $C_{15}$ acetogenins from the red alga *Laurencia obtusa*. Phytochemistry 59:111
703. González AG, Martín JD, Martín VS, Norte M, Pérez R, Ruano JZ, Drexler SA, Clardy J (1982) Non-terpenoid C-15 metabolites from the red seaweed *Laurencia pinnatifida*. Tetrahedron 38:1009
704. Kim H, Choi WJ, Jung J, Kim S, Kim D (2003) Construction of eight-membered ether rings by olefin geometry-dependent internal alkylation: first asymmetric total syntheses of (+)-3-(*E*)- and (+)-3-(*Z*)-pinnatifidenyne. J Am Chem Soc 125:10238
705. Snyder SA, Brucks AP, Treitler DS, Moga I (2012) Concise synthetic approaches for the *Laurencia* family: formal total syntheses of (±)- laurefucin and (±)-*E*- and (±)-*Z*-pinnatifidenyne. J Am Chem Soc 134:17714
706. Imre S, Lotter H, Wagner H, Thomson RH (1987) Epoxy-*trans*-isodihydrorhodophytin, a new metabolite from *Laurencia obtusa*. Z Naturforsch 42C:507
707. Caccamese S, Azzolina R, Duesler EN, Paul IC, Rinehart KL (1980) Laurencienyne; a new acetylenic cyclic ether from the marine red alga *Laurencia obtusa*. Tetrahedron Lett 21:2299
708. Rinehart KL Jr, Shaw PD, Shield LS, Gloer JB, Harbour GC, Koker MES, Samain D, Schwartz RE, Tymiak AA, Weller DL, Carter GT, Munro MHG, Hughes RG Jr, Renis HE, Swynenberg EB, Stringfellow DA, Vavra JJ, Coats JH, Zurenko GE, Kuentzel SL, Li LH, Bakus GJ, Brusca RC, Craft LL, Young DN, Connor JL (1981) Marine natural products as sources of antiviral, antimicrobial, and antineoplastic agents. Pure Appl Chem 53:795

709. Aydoğmuş Z, Imre S (1999) A new halogenated $C_{15}$ non-terpenoid compound from the marine red alga *Laurencia obtusa*. Acta Pharm Turc 41:93
710. Ojika M, Nemoto T, Yamada K (1993) Doliculols A and B, the non-halogenated $C_{15}$ acetogenins with cylic ether from the sea hare *Dolabella auricularia*. Tetrahedron Lett 34:3461
711. Gutiérrez-Cepeda A, Fernández JJ, Gil LV, López-Rodríguez M, Norte M, Souto ML (2011) Nonterpenoid $C_{15}$ acetogenins from *Laurencia marilzae*. J Nat Prod 74:441
712. Irie T, Suzuki M, Masamune T (1965) Laurencin, a constituent from *Laurencia* species. Tetrahedron Lett 6:1091
713. Irie T, Suzuki M, Masamune T (1968) Laurencin, a constituent of *Laurencia glandulifera* Kützing. Tetrahedron 24:4193
714. Forbes Cameron A, Cheung KK, Ferguson G, Monteath Robertson J (1969) *Laurencia* natural products: crystal structure and absolute stereochemistry of laurencin. J Chem Soc B:559
715. Robinson RA, Clark JS, Holmes AB (1993) Synthesis of (+)-laurencin. J Am Chem Soc 115:10400
716. Bratz M, Bullock WH, Overman LE, Takemoto T (1995) Total synthesis of (+)-laurencin. Use of acetal-vinyl sulfide cyclizations for forming highly functionalized eight-membered cyclic ethers. J Am Chem Soc 117:5958
717. Burton JW, Clark JS, Derrer S, Stork TC, Bendall JG, Holmes AB (1997) Synthesis of medium ring ethers. 5. The synthesis of (+)-laurencin. J Am Chem Soc 119:7483
718. Crimmins MT, Choy AL (1999) An asymmetric aldol-ring-closing metathesis strategy for the enantioselective synthesis of (+)-laurencin. J Am Chem Soc 121:5653
719. Crimmins MT, Emmitte KA (1999) Total synthesis of (+)-laurencin: an asymmetric alkylation-ring-closing metathesis approach to medium ring ethers. Org Lett 1:2029
720. Baek S, Jo H, Kim H, Kim H, Kim S, Kim D (2005) Highly stereoselective and efficient total synthesis of (+)-laurencin. Org Lett 7:75
721. Fujiwara K, Yoshimoto S, Takizawa A, Souma S, Mishima H, Murai A, Kawai H, Suzuki T (2005) Synthesis of (+)-laurencin via ring expansion of a C-glycoside derivative. Tetrahedron Lett 46:6819
722. Adsool VA, Pansare SV (2008) An enantioselective approach to (+)-laurencin. Org Biomol Chem 6:2011
723. Ortega N, Martin VS, Martin T (2010) An approach to lauroxanes by literative use of $Co_2(CO)_6$-acetylenic complexes. A formal synthesis of (+)-laurencin. J Org Chem 75:6660
724. White RH, Hager LP (1978) Intricenyne and related halogenated compounds from *Laurencia intricata*. Phytochemistry 17:939
725. Blunt JW, Lake RJ, Munro MHG (1984) Metabolites of the marine red alga *Laurencia thyrsifera*. III. Aust J Chem 37:1545
726. Cardellina JH II, Horsley SB, Clardy J, Leftow SR, Meinwald J (1982) Secondary metabolites from the red alga *Laurencia intricata*: halogenated enynes. Can J Chem 60:2675
727. Kim G, Sohn TI, Kim D, Paton R (2014) Asymmetric total synthesis of (+)-bermudenynol, a $C_{15}$ *Laurencia* metabolite with a vinyl chloride containing oxocene skeleton, through intramolecular amide enolate alkylation. Angew Chem Int Ed 53:272
728. Kladi M, Vagias C, Stavri M, Rahman MM, Gibbons S, Roussis V (2008) $C_{15}$ acetogenins with antistaphylococcal activity from the red alga *Laurencia glandulifera*. Phytochem Lett 1:31
729. Blunt JW, Lake RJ, Munro MHG, Yorke SC (1981) A new vinyl acetylene from the red alga *Laurencia thyrsifera*. Aust J Chem 34:2393
730. Howard BM, Schulte GR, Fenical W, Solheim B, Clardy J (1980) Three new vinyl acetylenes from the marine red alga *Laurencia*. Tetrahedron 36:1747
731. Fukuzawa A, Kurosawa E, Tobetsu I (1979) Laurallene, new bromoallene from the marine red alga *Laurencia nipponica* Yamada. Tetrahedron Lett 20:2797

732. Ishihara J, Shimada Y, Kanoh N, Takasugi Y, Fukuzawa A, Murai A (1997) Conversion of prelaureatin into laurallene, a bromo-allene compound, by enzymatic and chemical bromoetherification reactions. Tetrahedron 53:8371
733. Saitoh T, Suzuki T, Sugimoto M, Hagiwara H, Hoshi T (2003) Total synthesis of (+)-laurallene. Tetrahedron Lett 44:3175
734. Kim MJ, Sohn TI, Kim D, Paton RS (2012) Concise substrate-controlled asymmetric total syntheses of dioxabicyclic marine natural products with 2,10-dioxabicyclo-[7.3.0]dodecene and 2,9-dioxabicyclo[6.3.0]undecene skeletons. J Am Chem Soc 134:20178
735. Suzuki M, Takahashi Y, Matsuo Y, Masuda M (1996) Pannosallene, a brominated $C_{15}$ nonterpenoid from *Laurencia pannosa*. Phytochemistry 41:1101
736. Suzuki M, Kurosawa E (1987) (3*E*)-Laureatin and (3*E*)-isolaureatin, halogenated C-15 non-terpenoid compounds from the red alga *Laurencia nipponica* Yamada. Bull Chem Soc Jpn 60:3791
737. Yoshimitsu Y, Inuki S, Oishi S, Fujii N, Ohno H (2013) Palladium-catalyzed medium-ring formation for construction of the core structure of *Laurencia* oxacycles: synthetic study of laurendecumallene B. Org Lett 15:3046
738. Kurosawa E, Fukuzawa A, Irie T (1973) *trans*- and *cis*-Laurediol, unsaturated glycols from *Laurencia nipponica* Yamada. Tetrahedron Lett 14:2121
739. Irie T, Izawa M, Kurosawa E (1968) Laureatin, a constituent from *Laurencia nipponica* Yamada. Tetrahedron Lett 9:2091
740. Irie T, Izawa M, Kurosawa E (1970) Laureatin and isolaureatin, constituents of *Laurencia nipponica* Yamada. Tetrahedron 26:851
741. Kurosawa E, Furusaki A, Izawa M, Fukuzawa A, Irie T (1973) The absolute configurations of laureatin and isolaureatin. Tetrahedron Lett 14:3857
742. Sugimoto M, Suzuki T, Hagiwara H, Hoshi T (2007) The first total synthesis of (+)-(*Z*)-laureatin. Tetrahedron Lett 48:1109
743. Kim H, Lee H, Lee D, Kim S, Kim D (2007) Asymmetric total syntheses of (+)-(3*Z*)-laureatin and (+)-(3*Z*)-isolaureatin by "lone pair-lone pair interaction-controlled" isomerization. J Am Chem Soc 129:2269
744. Keshipeddy S, Martínez I, Castillo BF, Morton MD, Howell AR (2012) Toward a formal synthesis of laureatin: unexpected rearrangements involving cyclic ether nucleophiles. J Org Chem 77:7883
745. Irie T, Izawa M, Kurosawa E (1968) Isolaureatin, a constituent from *Laurencia nipponica* Yamada. Tetrahedron Lett 9:2735
746. Fukuzawa A, Kurosawa E, Irie T (1972) Laurefucin and acetyllaurefucin, new bromo compounds from *Laurencia nipponica* Yamada. Tetrahedron Lett 13:3
747. Furusaki A, Kurosawa E, Fukuzawa A, Irie T (1973) The revised structure and absolute configuration of laurefucin from *Laurencia nipponica* Yamada. Tetrahedron Lett 14:4579
748. McPhail KL, Davies-Coleman MT (2005) (3Z)-Bromofucin from a South African sea hare. Nat Prod Res 19:449
749. Fukuzawa A, Aye M, Nakamura M, Tamura M, Murai A (1990) Structure elucidation of laureoxanyne, a new nonisoprenoid $C_{15}$ enyne, using lactoperoxidase. Tetrahedron Lett 31:4895
750. Kennedy DJ, Selby IA, Cowe HJ, Cox PJ, Thomson RH (1984) Bromoallenes from the alga *Laurencia microcladia*. J Chem Soc Chem Commun:153
751. Park JH, Kim BS, Kim HS, Kim SH, Kim DJ (2007) Substrate-controlled asymmetric total synthesis of (+)-microcladallene B with bromination strategy based on a nucleophile-assisting leaving group. Angew Chem Int Ed 46:4726
752. Denmark SE, Yang S-M (2002) Intramolecular silicon-assisted cross-coupling: total synthesis of (+)-brasilenyne. J Am Chem Soc 124:15196
753. Denmark SE, Yang SM (2004) Total synthesis of (+)-brasilenyne. Application of an intramolecular silicon-assisted cross-coupling reaction. J Am Chem Soc 126:12432

754. King TJ, Imre S, Öztunc A, Thomson RH (1979) Obtusenyne, a new acetylenic nine-membered cyclic ether from *Laurencia obtusa*. Tetrahedron Lett 20:1453
755. Curtis NR, Holmes AB, Looney MG (1992) Studies towards the synthesis of obtusenyne. Synthesis of the hexahydrooxonin nucleus. Tetrahedron Lett 33:671
756. Curtis NR, Holmes AB, Looney MG (1991) Studies towards the synthesis of obtusenyne. A Claisen rearrangement approach to unsaturated nine-membered lactones. Tetrahedron 47:7171
757. Fujiwara K, Awakura D, Tsunashima M, Nakamura A, Honma T, Murai A (1999) Total synthesis of (+)-obtusenyne. J Org Chem 64:2616
758. Crimmins MT, Powell MT (2003) Enantioselective total synthesis of (+)-obtusenyne. J Am Chem Soc 125:7592
759. Urmura T, Suzuki T, Onodera N, Hagiwara H, Hoshi T (2007) Total synthesis of (+)-obtusenyne. Tetrahedron Lett 48:715
760. Frankie Mak SY, Curtis NR, Payne AN, Congreve MS, Wildsmith AJ, Francis CL, Davies JE, Pascu SI, Burton JW, Holmes AB (2008) Synthesis of (+)-obtusenyne. Chem Eur J 14:2867
761. Awakura D, Fujiwara K, Murai A (1999) Determination of the absolute configuration of Norte's obtusenynes by total synthesis of (12*R*,13*R*)-(–)- and (12*S*,13*R*)-(+)-obtusenynes. Chem Lett 28:461
762. Kurata K, Furusaki A, Suehiro K, Katayama C, Suzuki T (1982) Isolaurallene, a new nonterpenoid $C_{15}$-bromoallene, from the red alga *Laurencia nipponica* Yamada. Chem Lett 11:1031
763. Furusaki A, Katsuragi S-I, Suehiro K, Matsumoto T (1985) The conformations of (*Z*)-2,3,4,7,8,9-hexahydrooxonin and (*Z*)-cyclononene. X-ray structure determinations of isolaurallene and neolaurallene, and force-field calculations. Bull Chem Soc Jpn 58:803
764. Crimmins MT, Emmitte KA (2001) Asymmetric total synthesis of (–)-isolaurallene. J Am Chem Soc 123:1533
765. Crimmins MT, Emmitte KA, Choy AL (2002) Ring closing metathesis for the formation of medium ring ethers: the total synthesis of (–)-isolaurallene. Tetrahedron 58:1817
766. Suzuki M, Kurosawa E, Furusaki A, Katsuragi S-I, Matsumoto T (1984) Neolaurallene, a new halogenated C-15 nonterpenoid from the red alga *Laurencia okamurai* Yamada. Chem Lett 13:1033
767. Jeong W, Kim MJ, Kim H, Kim S, Kim D, Shin KJ (2010) Substrate-controlled asymmetric total synthesis and structure revision of (+)-itomanallene A. Angew Chem Int Ed 49:752
768. Braddock DC, Bhuva R, Millan DS, Pérez-Fuertes Y, Roberts CA, Sheppard RN, Solanki S, Stokes ESE, White AJP (2007) A biosynthetically-inspired synthesis of the tetrahydrofuran core of obtusallenes II and IV. Org Lett 9:445
769. Guella G, Chiasera G, Mancini I, Öztunç A, Pietra F (1997) Twelve-membered *O*-bridged cyclic ethers of red seaweeds in the genus *Laurencia* exist in solution as slowly interconverting conformers. Chem Eur J 3:1223
770. Guella G, Mancini I, Öztunç A, Pietra F (2000) Conformational bias in macrocyclic ethers and observation of high solvolytic reactivity at a masked furfuryl (=2-furylmethyl) C-atom. Helv Chim Acta 83:336
771. Ciavatta ML, Gavagnin M, Puliti R, Cimino G, Martínez E, Ortea J, Mattia CA (1997) Dactyllalene: a novel dietary $C_{15}$ bromoallene from the Atlantic anaspidean mollusc *Aplysia dactylomela*. Tetrahedron 53:17343
772. Braddock DC, Millan DS, Pérez-Fuertes Y, Pouwer RH, Sheppard RN, Solanki S, White AJP (2009) Bromonium ion induced transannular oxonium ion formation-fragmentation in model obtusallene systems and structural reassignment of obtusallenes V-VII. J Org Chem 74:1835
773. Öztunç A, Imre S, Wagner H, Norte M, Fernández JJ, González R (1991) A new and highly oxygenated bromoallene from a marine source. Tetrahedron Lett 32:4377
774. Cox PJ, Imre S, Islimyeli S, Thomson RH (1982) Obtusallene I, a new halogenated allene from *Laurencia obtusa*. Tetrahedron Lett 23:579
775. Cox PJ, Howie RA (1982) X-ray structure analysis of obtusallene. Acta Crystallogr B38:1386

776. Öztunç A, Imre S, Wagner H, Norte M, Fernández JJ, González R (1991) A new haloether from *Laurencia* possessing a lauroxacyclododecane ring. Structural and conformational studies. Tetrahedron 47:2273
777. Gutiérrez-Cepeda G, Fernández JJ, Norte M, Souto ML (2011) New bicyclotridecane $C_{15}$ nonterpenoid bromoallenes from *Laurencia marilzae*. Org Lett 13:2690
778. Wright AE, Wing RM, Sims JJ (1983) Poitediene a new metabolite from the marine red alga *Laurencia poitei*. Tetrahedron Lett 24:4649
779. Kigoshi H, Shizuri Y, Niwa H, Yamada K (1982) Isolation and structures of *trans*-laurencenyne, a possible precursor of the $C_{15}$ halogenated cyclic ethers, and *trans*-neolaurencenyne from *Laurencia okamurai*. Tetrahedron Lett 23:1475
780. Kigoshi H, Shizuri Y, Niwa H, Yamada K (1986) Four new $C_{15}$ acetylenic polyenes of biogenetic significance from the red alga *Laurencia okamurai*: structures and synthesis. Tetrahedron 42:3781
781. Holmeide AK, Skattebol I, Sydnes M (2001) The syntheses of three highly unsaturated marine lipid hydrocarbons. J Chem Soc Perkin Trans 1:1942
782. Fukuzawa A, Honma T, Takasugi Y, Murai A (1993) Biogenetic intermediates, (3*E* and 3*Z*,12*Z*)-laurediols and (3*E* and 3*Z*)-12,13-dihydrolaurediols, isolated from *Laurencia nipponica*. Phytochemistry 32:1435
783. Palazón JM, Martín VS (1988) Enantioselective total synthesis of 6(*R*),7(*R*)-3-*cis*-9-*cis*-12-*cis*, 6-acetoxy-7-chloropentadeca-3,9,12-trien-1-yne and its 3-*trans*-isomer. Tetrahedron Lett 29:681
784. Fukuzawa A, Sato H, Miyamoto M, Masamune T (1986) Synthesis of (6*S*,7*S*)-*trans*-laurediol and its [9,10-$^2H_2$]-analogue. Tetrahedron Lett 27:2901
785. Añorbe B, Martín VS, Palazón JM, Trujillo JM (1986) Enantiomeric syntheses of 6*R*,7*R* and 6*S*,7*S* *trans*- and *cis*-laurediol. Tetrahedron Lett 27:4991
786. Martin T, Martin VS (2000) A short synthesis of *trans*-(+)-laurediol. Tetrahedron Lett 41:2503
787. Gadikota RR, Keller AI, Callam CS, Lowary TL (2003) Efficient syntheses of *trans*-(+)-laurediol from carbohydrate precursors. Tetrahedron Asymm 14:737
788. Ishihara J, Kanoh N, Fukuzawa A, Murai A (1994) Isomerization of the (*Z*)-enyne unit to the (*E*)-enyne unit. Conversion of laureatin to (*E*)-isolaureatin. Chem Lett 13:1563
789. Braddock DC (2006) A hypothesis concerning the biosynthesis of the obtusallene family of marine natural products via electrophilic bromination. Org Lett 8:6055
790. Braddock DC, Rzepa HS (2008) Structural reassignment of obtusallenes V, VI and VII by GIAO-based density functional prediction. J Nat Prod 71:728
791. Su H, Yuan ZH, Li J, Guo SJ, Deng LP, Han LJ, Zhu XB, Shi DY (2009) Two new bromoindoles from red alga *Laurencia similis*. Chin Chem Lett 20:456
792. Sun WS, Su S, Zhu RX, Tu GZ, Cheng W, Liang H, Guo XY, Zhao YY, Zhang QY (2013) A pair of unprecedented *spiro*-trisindole enantiomers fused through a five-member ring from *Laurencia similis*. Tetrahedron Lett 54:3617
793. Carter GT, Rinehart KL, Li LH, Kuentzel S, Connor JL (1978) Brominated indoles from *Laurencia brongniartii*. Tetrahedron Lett 19:4479
794. Su H, Yuan Z, Li J, Guo S, Han L, Zhu X, Shi D (2009) Studies on chemical constituents of *Laurencia saitoi*. Chin J Chin Mat Med 34:871
795. Masuda M, Kawaguchi S, Takahashi Y, Okamoto K, Suzuki M (1999) Halogenated secondary metabolites of *Laurencia similis* (Rhodomelaceae, Rhodophyta). Bot Mar 42:199
796. Vairappan CS, Yen AM, Yi OC, Moi PS (2004) Biologically active polybrominated indoles in the red alga *Laurencia similis* from the coastal waters of Sabah (Rhodomelaceae, Ceramiales). Malay J Sci 23:119
797. Suárez-Castillo OR, Beiza-Granados L, Meléndez-Rodríguez M, Alvarez-Hernández A, Morales-Ríos MS, Joseph-Nathan P (2006) Synthesis of bromoindole alkaloids from *Laurencia brogniartii*. J Nat Prod 69:1596

798. Tanaka J, Higa T, Benardinelli G, Jefford CW (1989) Sulfur-containing polybromoindoles from the red alga *Laurencia brongniartii*. Tetrahedron 45:7301
799. El-Gamal AA, Wang WL, Duh CY (2005) Sulfur-containing polybromoindoles from the Formosan red alga *Laurencia brongniartii*. J Nat Prod 68:815
800. Tanaka J, Higa T, Bernardinelli G, Jefford CW (1988) Itomanindoles A and B, methylsulfinylindoles from *Laurencia brongniartii*. Tetrahedron Lett 29:6091
801. Fang HY, Chiou SF, Uvarani C, Wen ZH, Hsu CH, Wu YC, Wang WL, Liaw CC, Sheu JH (2014) Cytotoxic, anti-inflammatory and antibacterial sulfur-containing polybromoindoles from the Formosan red alga *Laurencia brongniartii*. Bull Chem Soc Jpn 87:1278
802. Liu Y, Gribble GW (2002) Syntheses of polybrominated indoles from the red alga *Laurencia brongniartii* and the brittle star *Ophiocoma erinaceus*. J Nat Prod 65:748
803. Kubota NK, Iwamoto H, Fukazawa Y, Uchio Y (2005) Five new sulfur-containing polybrominated bisindoles from the red alga *Laurencia brongniartii*. Heterocycles 65:2675
804. Wright AD, König GM, Angerhofer CK, Greenidge P, Linden A, Desqueyroux-Faúndez R (1996) Antimalarial activity: the search for marine-derived natural products with selective antimalarial activity. J Nat Prod 59:710
805. Mikami D, Kurihara H, Kim SM, Takahashi K (2013) Red algal bromophenols as glucose 6-phosphate dehydrogenase inhibitors. Mar Drugs 11:4050
806. Valdebenito H, Bittner M, Sammes PG, Silva M, Watson WH (1982) A compound with antimicrobial activity isolated from the red seaweed *Laurencia chilensis*. Phytochemistry 21:1456
807. Qin JC, Su H, Zhang YM, Gao JM, Zhu L, Wu XA, Pan HY, Li XA (2010) Highly brominated metabolites from marine red alga *Laurencia similis* inhibit protein tyrosine phosphatase 1B. Bioorg Med Chem Lett 20:7152
808. Zhong Y, Su J, Zeng L, Yan S, Luo B (1996) Studies on the chemical constituents of *Laurencia karlae* collected from the South China Sea. Chem J Chin Univ 17:249
809. Su JY, Xu XH, Zeng LM, Wang CJ (1997) A new iodolactone from *Laurencia majuscula*. Chem J Chin Univ 18:1333
810. Kavita K, Singh VK, Jha B (2014) 24-Branched $\Delta^5$ sterols from *Laurencia papillosa* red seaweed with antibacterial activity against human pathogenic bacteria. Microbiol Res 169:301
811. Al-Iihaibi SS, Ayyad SEN, Al-Wessaby E, Alarif WM (2010) 3β,7β-Dihydroxy-5-α-cholestane skeleton from *Laurencia papillosa*. Biochem Syst Ecol 38:861
812. Alarif WM, Al-Lihaibi SS, Abdel-Lateff A, Ayyad SEN (2011) New antifungal cholestane and aldehyde derivatives from the red alga *Laurencia papillosa*. Nat Prod Commun 6:1821
813. Xu XH, Su JY (1997) A new cytotoxic dihydroxysterol from *Laurencia majuscula*. Chin Chem Lett 8:235
814. Fukuzawa A, Kumagai Y, Masamune T, Furusaki A, Katayama C, Matsumoto T (1981) Acetylpinnasterol and pinnasterol, ecdysone-like metabolites from the marine red alga *Laurencia pinnata* Yamada. Tetrahedron Lett 22:4085
815. Fukuzawa A, Miyamoto M, Kumagai Y, Masamune T (1986) Ecdysone-like metabolites, 14α-hydroxypinnasterols from the red alga *Laurencia pinnata*. Phytochemistry 25:1305
816. Kobayashi M, Murata O (1992) Marine sterols. 23. 2a-Oxa-2-oxo-5$\alpha$-hydroxy-3,4-dinorcholestane from the Arabian Sea red alga *Laurencia obtusa*. Tetrahedron Lett 33:519
817. San-Martín A, Rovirosa J, Muñoz O, Chen MHM, Guneratne RD, Clardy J (1983) The isolation and structure determination of chilenone A, a putative dimer of 2-methyl-3(2*H*)-furanone from the marine alga *Laurencia chilensis*. Tetrahedron Lett 24:4063
818. San-Martín A, Rovirosa J, Xu C, Lu HSM, Clardy J (1987) Further structural studies on the 2-methyl-3(2*H*)-furanone derived metabolites of the marine alga *Laurencia chilensis*. Tetrahedron Lett 28:6013
819. Bittner M, González F, Valdebenito H, Silva M, Paul VJ, Fenical W, Chen MHM, Clardy J (1987) A novel tetracyclic polyketal from the marine red alga *Laurencia chilensis*. Tetrahedron Lett 28:4031

820. Uenishi J, Mimura Y, Yonemitsu O (1996) Synthetic studies on cyclic polyketides isolated from red alga. Symp Chem Nat Prod 38:667
821. Wiedenfeld H, Knoch F, Koch M (1985) The crystal structure of *spiro*-bis-pinnaketal, a new *spiro* compound from *Laurencia pinnatifida* Lamouroux. Arch Pharmacol 318:289
822. Bernart MW, Gerwick WH, Corcoran EE, Lee AY, Clardy J (1992) Laurencione, a heterocycle from the red alga *Laurencia spectabilis*. Phytochemistry 31:1273
823. De Kimpe N, Georgieva A, Boeykens M, Lazar L (1995) Synthesis of laurencione, a labile dihydro-3(2*H*)-furanone derivative from the red alga *Laurencia spectabilis*. J Org Chem 60:5262
824. De Kimpe N, Georgieva A, Boeykens M, Kozekov I, Aelterman W (1996) New formal synthesis of laurencione, a labile dihydrofuranone derivative from the red alga *Laurencia spectabilis*. Synthesis 28:1131
825. Aelterman W, De Kimpe N, Kalinin V (1997) One-step synthesis of laurencione. J Nat Prod 60:385
826. Astashko D, Habrus Y, Yurevich S, Tyvorskii V (2013) Formal synthesis of laurencione via $MgBr_2$-catalyzed rearrangement of α,β-epoxy ketones to 1,2-diketones. Chem Heterocycl Compd 49:676
827. Arroyo P, Valencia E, Valenzuela E, Zarraga M (1995) A new cyclic ether from *Laurencia chilensis*. Bol Soc Chil Quim 40:221
828. Li YX, Li Y, Qian ZJ, Kim MM, Kim SK (2009) In vitro antioxidant activity of 5-HMF isolated from marine red alga *Laurencia undulata* in free radical mediated oxidative systems. J Microbiol Biotechnol 19:1319
829. Zeng L, Zhong Y, Su J, Wu H, Ma K (1996) A novel secondary metabolite of Chinese red alga *Laurencia karlae*. Chin J Chem 14:370
830. Du B, Zhong XY, Liao XJ, Xu WJ, Zhou XL, Xu SH (2010) A new antitumor arabinopyranoside from *Laurencia majuscula* induces G2/M cell cycle arrest. Phytother Res 24:1447
831. Li Y, Li Y, Lee S, Qian Z, Kim S (2010) Inhibitors of oxidation and matrix metalloproteinases, floridoside, and D-isofloridoside from marine red alga *Laurencia undulata*. J Agric Food Chem 58:578
832. Aplin RT, Durham LJ, Kanazawa Y, Safe S (1967) 2-*O*-*α*-D-Galactopyranosylglycerol from *Laurencia pinnatifida*. J Chem Soc C:1346
833. Siddhanta AK, Mody KH, Ramavat BK, Chauhan VD, Sharma M, Garg HS (1995) Characterization of sulfonoglycolipid from the red alga *Laurencia pedicularioides*. Bot Mar 38:329
834. Barma DK, Lu B, Baati R, Mioskowski C, Falck JR (2008) Convenient preparation of (Z)-α-halo-α,β-unsaturated aldehydes: synthesis of a *Laurencia flexilis* toxin. Tetrahedron Lett 49:4359
835. Basavaiah D, Suguna Hyma R (1996) Synthetic applications of the Baylis-Hillman reaction: simple synthesis of (2*E*)-2-butyloct-2-enal and (2*E*)-2-tridecylheptadec-2-enal. Tetrahedron 52:1253
836. Cardellina JH II, Moore RE (1978) Sphingosine derivatives from red algae of the Ceramiales. Phytochemistry 17:554
837. Higgs MD, Mulheirn LJ (1981) Hybridalactone; an unusual fatty acid metabolite from red alga *Laurencia hybrida* (Rhodophyta; Rhodomelaceae). Tetrahedron 37:4259
838. Corey EJ, De B, Ponder JW, Berg JM (1984) The stereochemistry and biosynthesis of hybridalactone, an eicosanoid from *Laurencia hybrida*. Tetrahedron Lett 25:1015
839. Corey EJ, De B (1984) Total synthesis and stereochemistry of hybridalactone. J Am Chem Soc 106:2735
840. Ota K, Sugata N, Ohshiro Y, Kawashima E, Miyaoka H (2012) Total synthesis of marine eicosanoid (–)-hybridalactone. Chem Eur J 18:13531
841. Hickmann V, Kondoh A, Gabor B, Alcarazo M, Fürstner A (2011) Catalysis-based and protecting-group-free total syntheses of the marine oxylipins hybridalactone and the ecklonialactones A, B and C. J Am Chem Soc 133:13471

842. Maru N, Ohno O, Koyama T, Yamada K, Uemura D (2010) Papillamide, a novel fatty acid amide from the red alga *Laurencia papillosa*. Chem Lett 39:366
843. Higgs MD (1981) Antimicrobial components of the red alga *Laurencia hybrida* (Rhodophyta; Rhodomelaceae). Tetrahedron 37:4255
844. Bernart M, Gerwick WH (1988) Isolation of 12-(*S*)-HEPE from red alga *Murrayella periclados* and revision of structure of an acyclic icosanoid from *Laurencia hybrida*. Implications to the biosynthesis of the marine prostanoid hybridolactone. Tetrahedron Lett 29:2015
845. Feng MT, Yu XQ, Yang P, Yang H, Lin K, Mao SC (2015) Two new antifungal polyunsaturated fatty acid ethyl esters from the red alga *Laurencia okamurai*. Chem Nat Comp 51:418
846. Khotimchenko SV, Gusarova IS (2004) Red algae of Peter the Great Bay as a source of arachidonic and eicosapentaenoic acids. Russ J Mar Biol 30:183
847. Vlietinck AJ, Apers S (2001) Biological screening methods in the search for pharmacologically active natural products. In: Tringali C (ed) Bioactive compounds from natural sources. Taylor & Francis, London, p 1
848. Kittakoop P (2015) Anticancer drugs and potential anticancer leads inspired by natural products. In: Atta-ur-Rahman (ed) Studies in natural products chemistry, vol 44. Elsevier, Amserdam, p 251
849. Cragg GM, Grothaus PG, Newman DJ (2009) Impact of natural products on developing new anti-cancer agents. Chem Rev 109:3012
850. Bagya SK, Rajashree PV, Sam KG (2011) Preliminary anticancer screening and standardization of some indigenous medicinal plants using cell-biology and molecular biotechnology based models. Res J Med Plant 5:728
851. Fedorov SN, Shubina LK, Bode AM, Stonik VA, Dong Z (2007) Dactylone inhibits epidermal growth factor-induced transformation and phenotype expression of human cancer cells and induced $G_1$-S arrest and apoptosis. Cancer Res 67:5914
852. San-Martin BA, Rovirosa RJ, Darias JJ, Astudillo SYL (1996) Semisintesis y actividad biologicade derivados del sesquiterpeno pacifenol. Bol Soc Chil Quim 41:403
853. Liu J, Ma L, Wu N, Liu G, Zheng L, Lin X (2014) Aplysin sensitizes cancer cells to TRAIL by suppressing P38 MAPK/surviving pathway. Mar Drugs 12:5072
854. Kim MM, Mendis E, Kim SK (2008) *Laurencia okamurai* extract containing laurinterol induces apoptosis in melanoma cells. J Med Food 11:260
855. König GM, Wright AD, Franzblau SG (2000) Assessment of antimycobacterial activity of a series of mainly marine derived natural products. Planta Med 66:337
856. Lang KL, Silva IT, Zimmermann LA, Lhullier C, Mañalich Arana MV, Palermo JA, Falkenberg M, Simões CMO, Schenkel EP, Durán FJ (2012) Cytotoxic activity of semisynthetic derivatives of elatol and isoobtusol. Mar Drugs 10:2254
857. Campos A, Souza CB, Lhullier C, Falkenberg M, Schenkel EP, Ribeiro-do-Valle RM, Siqueira JM (2012) Anti-tumour effects of elatol, a marine derivative compound obtained from red algae *Laurencia microcladia*. J Pharm Pharmacol 64:1146
858. Su JH, Dai CF, Huang HH, Wu YC, Sung PJ, Hsu CH, Sheu JH (2007) Terpenoid-related metabolites from a Formosan soft coral *Nephthea chabrolii*. Chem Pharm Bull 55:594
859. Huang XC, Sun YL, Salim AA, Chen ZS, Capon RJ (2013) Parguerenes: marine red alga bromoditerpenes as inhibitors of P-glycoprotein (ABCB1) in multidrug resistant human cancer cells. Biochem Pharmacol 85:1257
860. Awad NE (2004) Bioactive brominated diterpenes from the marine red alga *Jania rubens* (L.) Lamx. Phytother Res 18:275
861. Kumar SC, Gadewal N, Mohammed SMM (2013) Identification of leads from marine seaweeds against human β-tubulin. Lett Drug Design Discov 10:67
862. Athinaios N, Kazantzis A, Putzker K, Lewis J, Pitsinos EN (2009) Synthesis of novel laurenditerpenol analogues and their evaluation as HIF-1 activation inhibitors. Lett Org Chem 6:269

863. Matsuzawa S, Suzuki T, Suzuki M, Matsuda A, Kawamura T, Mizuno Y, Kikuchi K (1994) Thyrsiferyl 23-acetate is a novel specific inhibitor of protein phosphatase PP2A. FEBS Lett 356:272
864. Matsuzawa S, Kawamura T, Mitsuhashi S, Suzuki T, Matsuo Y, Suzuki M, Mizuno Y, Kikuchi K (1999) Thyrsiferyl 23-acetate and its derivatives induce apoptosis in various T- and B-leukemia cells. Bioorg Med Chem 7:381
865. Morita H, Kishi E, Takeya K, Itokawa H, Iitaka Y (1993) Squalene derivatives from *Eurycoma longifolia*. Phytochemistry 34:765
866. Pec MK, Moser-Thier K, Fernández JJ, Souto ML, Kubista E (1999) Growth inhibition by dehydrothyrsiferol — a non-Pgp modulator, derived from a marine red alga — in human breast cancer cell lines. Int J Oncol 14:739
867. Pec MK, Hellan M, Moser-Thier K, Fernández JJ, Souto ML, Kubista E (1998) Inhibitory effects of a novel marine terpenoid on sensitive and multidrug resistant KB cell lines. Anticancer Res 18:3027
868. Pec MK, Aguirre A, Fernández JJ, Souto ML, Dorta JF, Villar J (2002) Dehydrothyrsiferol does not modulate multidrug resistance-associated protein 1 resistance: a functional screening system for MRP1 substrates. Int J Mol Med 10:605
869. Pec MK, Aguirre A, Moser-Thier K, Fernández JJ, Souto ML, Dorta J, Diáz-González F, Villar J (2003) Induction of apoptosis in estrogen dependent and independent breast cancer cells by the marine terpenoid dehydrothyrsiferol. Biochem Pharmacol 65:1451
870. Pec MK, Artwohl M, Fernández JJ, Souto ML, Alvarez de la Rosa D, Giraldez T, Valenzuela-Fernández A, Díaz-González F (2007) Chemical modulation of VLA integrin affinity in human breast cancer cells. Exp Cell Res 313:1121
871. Souto ML, Manríquez CP, Norte M, Leira F, Fernández JJ (2003) The inhibitory effects of squalene-derived triterpenes on protein phosphatse PP2A. Bioorg Med Chem 13:1261
872. Nishiguchi GA, Graham J, Bouraoui A, Jacobs RS, Little RD (2006) 7,11-*epi*-Thyrsiferol: completion of its synthesis, evaluation of its antimitotic properties, and the further development of an SAR model. J Org Chem 71:5936
873. Vera B, Rodríguez AD, La Clair JL (2011) Aplysqualenol A binds to the light chain of dynein type 1 (DYNLL1). Angew Chem Int Ed 50:8134
874. Brogi S, Kladi M, Vagias C, Papazafiri P, Roussis V, Tafi A (2009) Pharmacophore modeling for qualitative prediction of antiestrogenic activity. J Chem Inf Model 49:2489
875. The Lancet. Editorial (2009) Urgently needed: new antibiotics. Lancet 374:1868
876. Vairappan CS (2003) Potent antibacterial activity of halogenated metabolites from Malaysian red algae *Laurencia majuscula* (Rhodomelaceae, Ceramiales). Biomol Eng 20:255
877. Vairappan CS, Kawamoto T, Miwa H, Suzuki M (2004) Potent antibacterial activity of halogenated compounds against antibiotic-resistant bacteria. Planta Med 70:1087
878. Sims JJ, Donnell MS, Leary JV, Lacy GH (1975) Antimicrobial agents from marine algae. Antimicrob Agents Chemother 7:320
879. Kayser O, Kiderlen AF, Croft SL (2003) Natural products as antiparasitic drugs. Parasitol Res 90:S55
880. Veiga-Santos P, Pelizzaro-Rocha KJ, Santos AO, Ueda-Nakamura T, Dias Filho BP, Silva SO, Sudatti DB, Bianco EM, Perreira RC, Nakamura CV (2010) In vitro anti-trypanosomal activity of elatol isolated from red seaweed *Laurencia dendroidea*. Parasitology 137:1661
881. Desoti VC, Lazarin-Bidóia D, Sudatti DB, Pereira RC, Alonso A, Ueda-Nakamura T, Filho BPD, Nakamura CV, Silva SDO (2012) Trypanosomal action of (–)-elatol involves an oxidative stress triggered by mitochondria dysfunction. Mar Drugs 10:1631
882. Desoti VC, Lazarin-Bidóia D, Sudatti DB, Pereira RC, Ueda-Nakamura T, Nakamura CV, Silva SDO (2014) Additional evidence of the trypanocidal action of (–)-elatol on amastigote forms through the involvement of reactive oxygen species. Mar Drugs 12:4973
883. Bianco EM, Pires L, Santos GKN, Dutra KA, Reis TNV, Vasconcelos ERTPP, Cocentino ALM, Navarro DMAF (2013) Larvicidal activity of seaweeds from northeastern Brazil and of a halogenated sesquiterpene against the dengue mosquito (*Aedes aegypti*). Ind Crops Prod 43:270

884. Orhan I, Şener B, Kaiser M, Brun R, Tasdemir D (2010) Inhibitory activity of marine sponge-derived natural products against parasitic protozoa. Mar Drugs 8:47
885. Gautam R, Jachak SM (2009) Recent developments in anti-inflammatory natural products. Med Res Rev 29:767
886. Naik SR, Sheth UK (1976) Inflammatory process and screening methods for anti-inflammatory agents — a review. J Postgrad Med 22:5
887. Wijesinghe WAJP, Kim E-A, Kang MC, Lee WW, Lee HS, Vairappan CS, Jeon YJ (2014) Assessment of anti-inflammatory effect of 5$\beta$-hydroxypalisadin B isolated from red seaweed *Laurencia snackeyi* in zebrafish embryo in vivo model. Environ Toxicol Pharmacol 37:110
888. Mayer AMS, Paul VJ, Fenical W, Norris JN, de Carvalho MS, Jacobs RS (1993) Phospholipase $A_2$ inhibitors from marine algae. Hydrobiologia 260:521
889. Gil B, Ferrándiz ML, Sanz MJ, Terencio MC, Ubeda A, Rovirosa J, San-Martin A, Alcaraz MJ, Payá M (1995) Inhibition of inflammatory responses by epitaondiol and other marine natural products. Life Sci 57:25
890. Chatter R, Othman RB, Rabhi S, Kladi M, Tarhouni S, Vagias C, Roussis V, Lamia Guizane-Tabbane L, Kharrat R (2011) In vivo and in vitro anti-inflammatory activity of neorogioltriol, a new diterpene extracted from the red algae *Laurencia glandulifera*. Mar Drugs 9:1293
891. Kim M, Li YX, Dewapriya P, Ryu B, Kim SK (2013) Floridoside suppresses pro-inflammatory responses by blocking MAPK signaling in activated microglia. BMB Rep 46:398
892. Farrokhnia M, Nabipour I (2014) Marine natural products as acetylcholinesterase inhibitor: comparative quantum mechanics and molecular docking study. Curr Comput Aid Drug Des 10:83
893. Ge N, Liang H, Liu Y, Ma AG, Han L (2013) Protective effect of aplysin on hepatic injury in ethanol-treated rats. Food Chem Toxicol 62:361
894. Kaul PN, Kulkarni SK, Kurosawa E (1978) Novel substances of marine origin as drug metabolism inhibitors. J Pharm Pharmacol 30:589
895. Kaul PN, Kulkarni SK (1978) New drug metabolism inhibitor of marine origin. J Pharm Sci 67:1293
896. Kaul PN (1982) Biomedical potential of the sea. Pure Appl Chem 54:1963
897. Farrokhnia M, Nabipour I, Bargahi A (2012) A theoretical study of dactylyne stereoisomers: a marine natural product from *Aplysia dactylomela*. J Theor Comput Chem 11:833
898. Hay ME, Fenical W, Gustafson K (1987) Chemical defense against diverse coral-reef herbivores. Ecology 68:1581
899. Pereira RC, Da Gama BAP, Teixeira VL, Yoneshigue-Valentin Y (2003) Ecological roles of natural products of the Brazilian red seaweed *Laurencia obtusa*. Braz J Biol 63:665
900. Izac RR, Poet SE, Fenical W, Van Engen D, Clardy J (1982) The structure of pacifigorgiol, an ichthyotoxic sesquiterpenoid from the Pacific gorgonian coral *Pacifigorgia* cf. *adamsii*. Tetrahedron Lett 23:3743
901. Meyer BN, Ferrigni NR, Putnam JE, Jacobson LB, Nicols DE, McLaughlin JL (1982) Brine shrimp: a convenient bioassay for active plant constituents. Planta Med 45:31
902. Burgess JG, Boyd KG, Armstrong E, Jiang Z, Yan L, Berggren M, May U, Pisacane T, Granmo A, Adams DR (2003) The development of a marine natural product-based antifouling paint. Biofouling 19:197
903. Dobretsov S, Abed RMM, Teplitski M (2013) Inhibition of biofouling by marine microorganisms. Biofouling 29:423
904. De Nys R, Leya T, Maximilien R, Afsar A, Nair PSR, Steinberg PD (1996) The need for standardized broad scale bioassay testing: a case study using the red alga *Laurencia obtusa*. Biofouling 10:213
905. Novaczek I (2001) A guide to the common edible and medicinal sea plants of the Pacific Islands. The University of the South Pacific/Secretariat of the Pacific Community, Suva

**Maria Harizani** acquired a B.Sc. degree in Pharmacy from the National and Kapodistrian University of Athens, where she went on to obtain her M.Sc. degree in Pharmacognosy and the Chemistry of Natural Products. Her thesis focused on the isolation of secondary metabolites from species of red algae in the genus *Laurencia*. In further pursuing her research interests on bioactive natural products from marine organisms, she is currently a Ph.D. student at the Faculty of Pharmacy under the joint mentorship of Prof. Vassilios Roussis and Assistant Prof. Efstathia Ioannou.

**Efstathia Ioannou** received her B.Sc. degree from the Agricultural University of Athens in 2003 and her M.Sc. and Ph.D. degrees from the National and Kapodistrian University of Athens in 2005 and 2009, respectively, where she studied the isolation and structure elucidation of bioactive natural products from marine algae under the supervision of Prof. Vassilios Roussis. She carried out postdoctoral research on the chemistry of marine actinobacteria with Prof. William Fenical at Scripps Institution of Oceanography, before returning to Greece in 2010 to take up a lectureship at the Faculty of Pharmacy at the National and Kapodistrian University of Athens, where she is currently an Assistant Professor. Her research interests are on bioactive natural products from marine macro- and microorganisms.

**Vassilios Roussis** obtained his B.Sc. degree from the National and Kapodistrian University of Athens and his Ph.D. degree from the University of Iowa under the supervision of Prof. David F. Wiemer. Subsequently, he undertook postdoctoral research at Scripps Institution of Oceanography, La Jolla, California, with Prof. William Fenical and at the National Center for Scientific Research "Demokritos", Greece, with Dr. Basilis Mazomenos. In 1995, he joined the Faculty of Pharmacy at the National and Kapodistrian University of Athens, where in 2006 he was promoted to full professor. Since 2008, he has served as the Director of the Laboratory of Pharmacognosy at the Department of Pharmacognosy and Chemistry of Natural Products. His research interests focus on the chemical investigation of marine organisms from the eastern Mediterranean Sea.

MIX
Papier aus verantwortungsvollen Quellen
Paper from responsible sources
FSC® C105338

If you have any concerns about our products,
you can contact us on
**ProductSafety@springernature.com**

In case Publisher is established outside the EU,
the EU authorized representative is:
**Springer Nature Customer Service Center GmbH**
**Europaplatz 3, 69115 Heidelberg, Germany**

Printed by Libri Plureos GmbH
in Hamburg, Germany